HAROLD MAR

EPITAXIAL GROWTH
Part B

This is a volume in the
Materials Science and Technology series.
Editors: *A. S. Nowick and G. G. Libowitz*

A complete list of the books in the series appears at the end of the volume.

HAROLD MAR

EPITAXIAL GROWTH
Part B

Edited by

J. W. MATTHEWS

IBM Thomas J. Watson Research Center
Yorktown Heights, New York

Harold Mar

ACADEMIC PRESS

A Subsidiary of Harcourt Brace Jovanovich, Publishers

New York London Toronto Sydney San Francisco 1975

ACADEMIC PRESS, INC.
111 Fifth Avenue, New York, New York 10003

United Kingdom Edition published by
ACADEMIC PRESS, INC. (LONDON) LTD.
24/28 Oval Road, London NW1

Library of Congress Cataloging in Publication Data

Matthews, John Wauchope,
Epitaxial growth.

(Materials science and technology series)
Includes bibliographies and index.
1. Epitaxy. 2. Crystals–Growth. I. Title.
QD921.M37 548′.5 74-10191
ISBN 0–12–480902–2 (pt. B)

PRINTED IN THE UNITED STATES OF AMERICA

83 84 85 9 8 7 6 5 4 3 2

CONTENTS

LIST OF CONTRIBUTORS

Numbers in parentheses indicate the pages on which the authors' contributions begin.

C. A. B. Ball (493), *Department of Applied Mathematics, University of Port Elizabeth, Port Elizabeth, South Africa*

N. H. Fletcher (529), *Department of Physics, University of New England, Armidale, Australia*

Enrique Grünbaum (611), *Department of Physics and Astronomy, and School of Engineering, Tel-Aviv University, Ramat-Aviv, Tel-Aviv, Israel*

K. W. Lodge (529), *Department of Physics, University of New England, Armidale, Australia*

J. W. Matthews (559), *IBM Thomas J. Watson Research Center, Yorktown Heights, New York*

G. L. Price* (381), *School of Mathematical and Physical Sciences, University of Sussex, Brighton, England*

M. J. Stowell (437), *Tube Investments Research Laboratories, Hinxton Hall, Hinxton, Saffron Walden, Essex, England*

J. H. van der Merwe (493), *Department of Physics, University of Pretoria, Pretoria, South Africa*

J. A. Venables (381), *School of Mathematical and Physical Sciences, University of Sussex, Brighton, England*

* Present address: School of Physical Sciences, The Flinders University of South Australia, Bedford Park, South Australia.

PREFACE

This book is the result of a suggestion made to me several years ago by Professor A. S. Nowick. It is a collection of review articles that describe various aspects of the growth of single-crystal films on single-crystal substrates. The topics discussed are the historical development of the subject, the nucleation of thin films, the structure of the interface between film and substrate, and the generation of defects during film growth. The methods used to prepare and examine thin films are described and a list of the overgrowth–substrate combinations studied so far is given.

Invaluable help at all stages of the project was provided by Academic Press. Helpful discussions of a variety of topics that included the ways in which the subject could be subdivided, and who should be asked to write on each subdivision, were held with Dr. A. E. Blakeslee, Dr. S. Mader, Dr. M. J. Stowell, and Dr. J. A. Venables.

CONTENTS OF PART A

Chapter 3. **EXAMINATION OF THIN FILMS**

EPITAXIAL GROWTH
Part B

CHAPTER 4

NUCLEATION OF THIN FILMS

J. A. Venables and G. L. Price[†]

School of Mathematical and Physical Sciences
University of Sussex
Brighton, England

† Present address: School of Physical Sciences, The Flinders University of South Australia, Bedford Park, South Australia.

I. Introduction: The Mechanisms of Nucleation and Growth

Thin films are formed on a substrate by a process of nucleation and growth. This means that the initial stage is the formation of small clusters of the film material from individual atoms or molecules. As time progresses, more clusters are nucleated and these clusters grow, coalesce, and finally a continuous film is formed which then thickens. Because all these processes may be happening at once, the study of nucleation of thin films cannot truly be separated from a general study of the growth of crystals.

In many cases we can think of the study of nucleation as involving only the early stages of growth, extending, say, up to the point at which substantial rearrangement of the clusters occurs by coalescence. In other cases, nucleation events may be rate-limiting at much later stages of growth; for example, each atomic layer of the crystal may have to be nucleated independently, the rate of growth being thereby determined entirely by nucleation processes. In yet a third case, there may be essentially no nucleation step, if growth occurs at growth centers that already exist in the substrate.

A really general discussion of the "nucleation of thin films" should thus delve quite deeply into the different modes of crystal growth which occur. It should also be applicable to all the different methods of preparation of thin films described in Chapter 2. The nucleation processes occurring in these different methods of preparation may be qualitatively similar, but they are certainly very different in detail. In almost all of this chapter we shall limit the discussion to the nucleation of films prepared by vapor deposition, under conditions of fairly high supersaturation with respect to the bulk phase. We will, however, try to examine the different modes of growth that occur in this type of deposition process and to treat each of them quantitatively, where possible. We can only hope that those readers whose main interest is in films prepared by sputtering, chemical vapor deposition, electrodeposition, or liquid-phase epitaxy will find the approach useful by analogy.

The reasons research workers have for studying the nucleation of thin films are quite diverse. For some, it is the belief that the nucleation stage holds the key to the production of better-quality films for useful industrial applications; for others, better epitaxial films for other fields of scientific research is the goal. For more theoretically minded experimenters, the nucleation of thin films is an example of a nucleation and growth process in which the experimental variables can be fairly well controlled, and this means that such experiments can be used to test our ideas about nucleation and growth processes in general. Yet a fourth group of people is motivated mainly by the desire to express the theory of nucleation and growth mathematically as concisely and completely as possible. In this chapter we adopt a stance similar to these last two groups; this is, of course, largely as a result of personal bias, but also because we feel that the usefulness of the study of nucleation for the production of better (epitaxial) films is at the present time not really proven. In particular, the relationship between nucleation phenonomena and epitaxy has been somewhat elusive. These aspects of nucleation studies will be deferred until Section VII. The remainder of this chapter will be concerned with describing and understanding the nucleation phenomena per se.

The information obtainable in a nucleation experiment clearly depends on the technique used. However, high-resolution transmission electron microscopy is in many ways the most versatile technique available, and the work of Donohoe and Robins (1972) is an excellent illustration of the amount of information that is obtainable using this technique. They studied the nucleation of Ag and Au on several different alkali halide substrates. The substrates were freshly cleaved in UHV, and the metal was deposited at a known rate, R atoms $cm^{-2} sec^{-1}$, for a known time (t sec) onto the substrate held at temperature T. Immediately following deposition, the alkali halide was further cleaved to produce a fresh substrate surface, and the chip of alkali halide (with the metal on it) was caught and quickly cooled. After the run the chips were coated with carbon, and the alkali halide was dissolved, leaving the metal clusters on the amorphous carbon film. These films were then examined in the electron microscope.

A sequence of pictures obtained of Au deposited onto NaCl at $T = 250°C$, with an arrival rate $R = 1 \times 10^{13}$ atoms $cm^{-2} sec^{-1}$ is shown in Fig. 1. As can be seen, a great deal of information is contained in these pictures. First, the nucleation, growth, and coalescence stages are clearly seen as time proceeds. Second, one can measure the number of clusters $n_x(t)$ and the coverage of the substrate by clusters $Z(t)$; graphs of $n_x(t)$ are shown in Fig. 2. One can also see the (two-dimensional) shape of the clusters and determine their crystallographic orientation using electron diffraction. Given a reasonable assumption about the three-dimensional form of the clusters, the amount of metal condensed can be estimated,† and it is also possible to measure the size and spatial distributions of the clusters. When the spatial distribution is quite obviously nonrandom, as in the case of the various straight lines of clusters in Fig. 1, it is clear that we are dealing with nucleation of clusters at (or migration of clusters to) cleavage steps. However, nucleation of clusters at isolated point defects (or emergent dislocations) may well be impossible to detect in the absence of a technique for knowing where these substrate defects are. Since all the experimental measurements listed can be repeated at different values of R and T, and for different deposit–substrate combinations, it is clear that a large amount of data can be gathered in this type of experiment

† This can also be measured using a microbalance (Cinti and Chakraverty, 1972a; Fujiwara and Terajima, 1973).

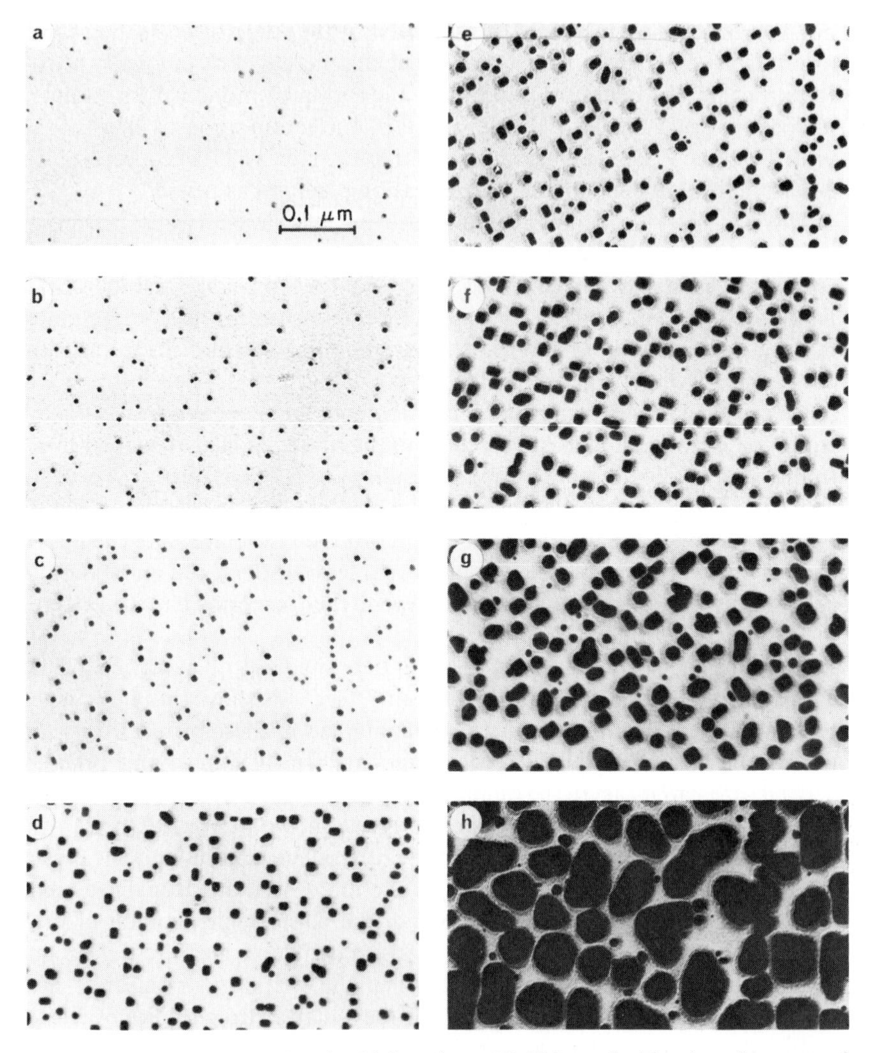

FIG. 1. Electron micrographs of gold deposits on NaCl formed with a deposition rate of 1×10^{13} atoms cm^{-2} sec^{-1}, substrate temperature of 250°C, and deposition times of (a) 0.5, (b) 1.5, (c) 4, (d) 8, (e) 10, (f) 15, (g) 30, and (h) 85 min (Donohoe and Robins, 1972).

(and, of course, in others also) and used to test our ideas about nucleation and growth processes.

Some of the processes that are thought to be most important are shown schematically in Fig. 3. Atoms arrive from the vapor at a rate R, and are accommodated; the accommodation coefficient can be measured by mass spectrometry experiments, and often has a value near unity. In what follows

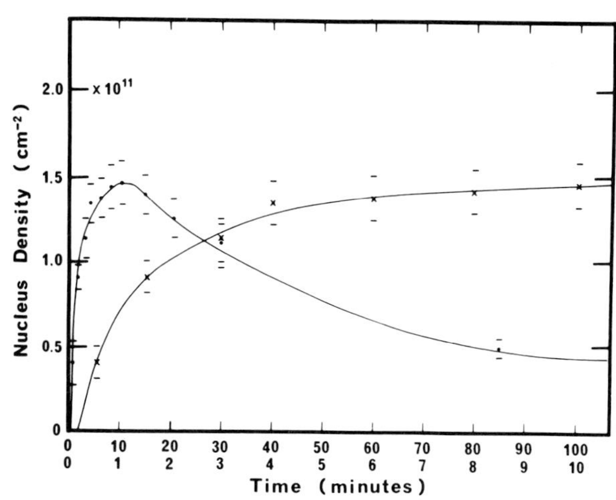

FIG. 2. Number density of nuclei versus deposition time $n_x(t)$ taken from the pictures of Fig. 1. The two different time scales enable one to see the short (×) and long (●) time behavior of $n_x(t)$ (Donohoe and Robins, 1972).

we shall neglect this factor. The population of single atoms n_1 (cm^{-2}) builds up and the atoms diffuse over the substrate at temperature T, with diffusion coefficient D ($\mathrm{cm}^2\ \mathrm{sec}^{-1}$). The diffusion coefficient may well be anisotropic and will be unless the surface plane (not the crystal) has three-, four-, or sixfold rotation symmetry. Reevaporation of single atoms may occur and is determined by their mean stay time τ_a.† The formation of large, stable clusters is a

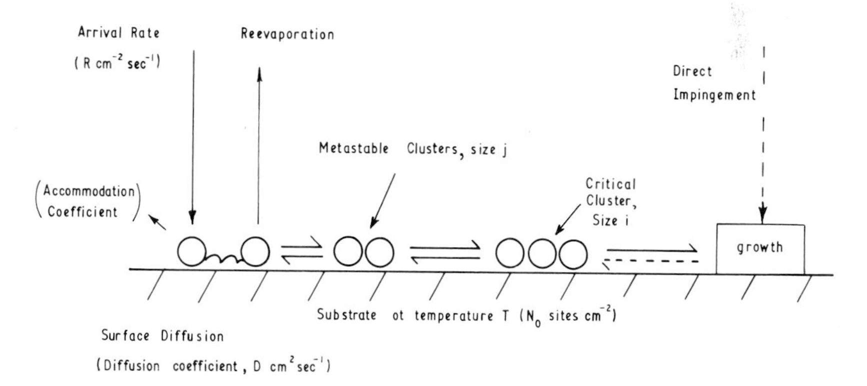

FIG. 3. Processes in the nucleation and growth of crystals on a substrate.

† Hamilton and Logel (1973) have recently given evidence that adatoms may also disappear by diffusing *into* the substrate. This process is clearly formally possible, and it may be important for open structures or when the condensate is highly soluble in the substrate. This process would have its own value of "τ_a."

chain reaction that starts with the formation of small clusters (n_j cm^{-2} of size j with binding energy E_j) which may be unstable. Above a certain "critical" size i growth becomes more probable than decay, and all larger clusters may be considered "stable," even if some atoms leave them during the growth process.

Several processes contribute to growth. Single atoms can diffuse across the substrate to join stable clusters, or they can impinge directly on the growing clusters and become incorporated into them. With these simple growth processes occurring at the same time as nucleation, the number of stable clusters $n_x(t)$ can only increase, or possibly saturate. Other processes, however, contribute to a fall in $n_x(t)$. The most obvious is that two clusters that grow into each other will coalesce into one; this process is clearly shown in the sequence of Fig. 1. If the clusters themselves are mobile, this process can happen at an earlier stage of growth. Moreover it is conceivable that stable clusters also evaporate directly, though experimental evidence for this process has not been reported.

Although the processes discussed above may be adequate for a discussion of nucleation on a perfect substrate containing N_0 (cm^{-2}) sites of equal adsorption energy E_a, it is well known that in many (if not most) important practical situations, the nucleation occurs at defect sites on the substrate surface. These defects can influence the nucleation process in several ways. Perhaps most important, they can speed up the nucleation process enormously by providing sites for single atoms that have adsorption energy greater than E_a, thus markedly increasing the population n_1 when reevaporation is important. Also, they can increase the binding energy E_i of critical clusters, and influence the effective diffusion constants of single atoms and clusters. These effects can be very marked, as shown by a large amount of work on the decoration of various types of defects: examples are the decoration of cleavage steps on alkali halide surfaces (Bethge, 1969; Bassett, 1958), vacancy etch pits in graphite (Hennig, 1966), or point defects produced by electron or ion irradiation of the substrate surface. Although it is possible to be sure in these cases that defect-induced nucleation is taking place, in general it is not possible to be sure that such effects are completely absent. Nonetheless, an understanding of the nucleation process must start from a study of homogeneous nucleation (i.e., nucleation on a perfect substrate).† A brief (and far from complete) examination of the role of defects is given in Section V.

We wish to emphasize that our quantitative understanding of the nucleation and growth processes is currently advancing rather rapidly, but is by no means complete. This advance is occurring because reliable experimental data, such

† The whole subject is often referred to as "heterogeneous nucleation." Following Frankl and Venables (1970) it seems preferable to refer to nucleation on substrates as "homogeneous" (two-dimensional) and leave the term "heterogeneous" for defect-induced nucleation.

as that shown in Figs. 1 and 2, is being obtained using UHV deposition systems, in conjunction with high-resolution electron microscopy, field ion microscopy, mass spectrometry, and other techniques. Some of this experimental work will be described in Section VI. At the same time, atomistic nucleation theory has been advanced by many authors, who have recognized that the older work ("classical" nucleation theory—see ,e.g., Hirth and Pound, 1963; Hirth and Moazed, 1967; Sigsbee and Pound, 1967; McDonald, 1962, 1963) is not well suited to discuss thin film nucleation kinetics. This is because the small size of the critical cluster (often only 1 or 2 atoms) makes it rather inappropriate to use classical thermodynamic variables (e.g., surface free energy, radius of critical cluster) to describe the nucleation process. Thus, in this chapter, classical nucleation theory will be ignored and a discussion of atomistic theory will be given in Sections III and IV. Before plunging into details, however, it is instructive to examine qualitatively the various regimes of condensation that can occur, and this is done in the next section. The insight obtained also enables one to see why an atomistic viewpoint is necessary in almost all cases.

II. Regimes of Condensation

In this section we wish to emphasize the different regimes that can exist in the nucleation and growth of crystals on a substrate. These regimes can be distinguished qualitatively using the parameters introduced in the last section. In addition, we shall need to express τ_a and D in terms of the adsorption energy E_a and the diffusion energy E_d

$$\tau_a^{-1} = \nu \exp(-\beta E_a) \tag{1}$$

and

$$D = (\alpha\nu/N_0) \exp(-\beta E_d) \tag{2}$$

where β is written for $(kT)^{-1}$, k is Boltzmann's constant, ν an atomic vibration frequency ($\sim 10^{11}$–10^{12} Hz), and α a constant $\simeq \frac{1}{4}$.

The binding energy E_j of a j-cluster is, in general, not simply related to the energy E_2 of a pair of atoms on the substrate. However, *for the purposes of this section only*, we shall consider binding energies of clusters to be calculable on the basis of summing nearest-neighbor bonds. With this highly simplified assumption both E_j and the adsorption energy of atoms on top of the condensate material are expressible in terms of E_2 only. This is illustrated schematically in Fig. 4 for the case of an epitaxial substrate–thin film combination in an orientation with threefold symmetry. As can be seen with reference to this figure, the adsorption energy of an isolated adatom on the substrate is E_a, whereas an atom bound to a kink site on the first monolayer has binding

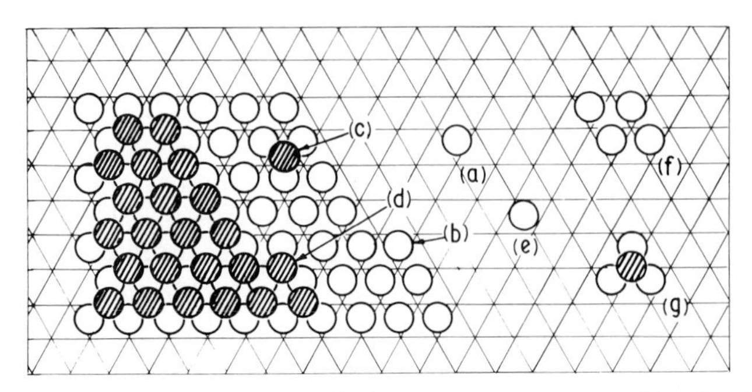

FIG. 4. Binding energies of atoms and clusters, on a nearest-neighbor bond model of nucleation and growth on a triangular lattice. Single atoms: (a) E_a; (b) E_a+3E_2; (c) $E_a' = 3E_2$; (d) $E_a'+3E_2 = 6E_2$; (e) E_a-E_d. Clusters: (f) $4E_a+5E_2$; (g) $3E_a+6E_2$.

energy E_a+3E_2. An isolated atom on the second or higher layers has adsorption energy E_a', and $E_a' = 3E_2$ in this simplified model; at a kink site an atom would be bound by $E_a'+3E_2$. The rate at which atoms arrive at such sites is determined by the diffusion energy E_d (E_d' on top of a monolayer). Thus the three most important activation energies are E_a, E_2, and E_d, and it is these energies, with the experimental variables R and T, that largely determine the regime observed. These regimes will be described with reference to Fig. 5, and experimental examples will be given to make the discussion more concrete. A very similar classification has been outlined by Bauer and Poppa (1972) following earlier work by Bauer (1958).

The regime appropriate to the lowest temperatures T and for highest arrival rates R is illustrated in Fig. 5a. This is describable by the condition $R/N_0^2D \gtrsim 1$; this means that atoms stay exactly where they arrive, and all other processes are thereby made irrelevant. An irregular bed of atoms is built up, and if atoms have directional bonds the layer will tend to be amorphous. Obvious examples are the covalent group IV elements C, Si, Ge, and many oxides (e.g., SiO, SiO_2, ZrO_2, etc.). Nucleation and growth of crystals occur, if at all, from within the layer; in the cases quoted they may occur both on annealing and by irradiation with fast particles (Parsons and Balluffi, 1964; Kelly and Naguib, 1970; Barna *et al.*, 1971). For atoms with nondirectional bonds, such as metals, this condition would lead to a very fine-grained polycrystalline film, since small-scale rearrangements cannot be prevented. Metals deposited on insulators at liquid helium temperatures may fall into this class. The recent observation (Rudee and Howie, 1972; Rudee, 1972) that even "amorphous" Ge and Si may be polycrystalline with a grain size of $\simeq 15$ Å, implies that this type of very fine-scale rearrangement probably can never be absolutely prevented. In fact, the condition $R/N_0^2D > 1$, which implies no movement after arrival, may be difficult to produce experimentally and is almost certainly too

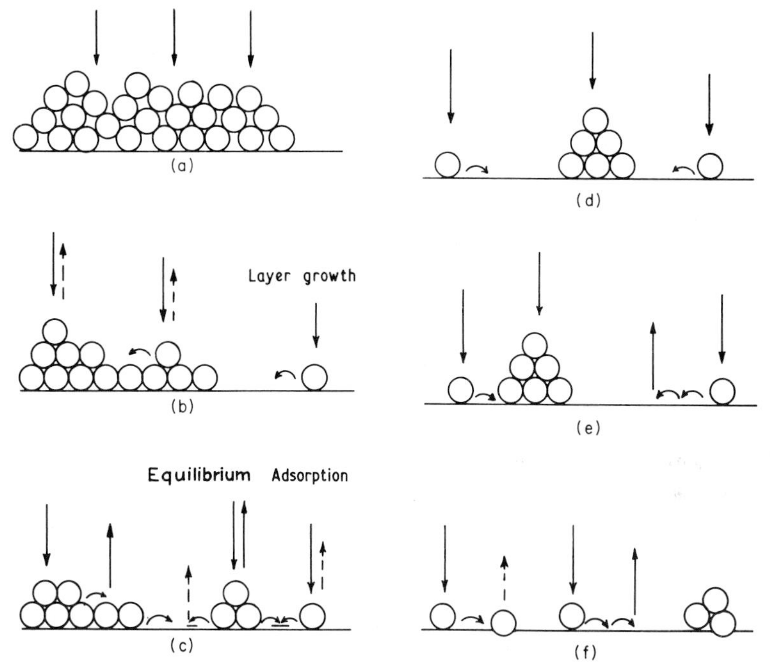

FIG. 5. Regimes of nucleation and growth. (See text for discussion.) (a) $R/N_0^2D \gtrsim 1$ (amorphous). (b) $E_a \gtrsim 3E_2$; $(R/N_0\nu)\exp(\beta 3E_2) \gtreqless 1$; $(R/N_0\nu)\exp(\beta 6E_2) > 1$. (c) $E_a \gtrsim 3E_2$; $(R/N_0\nu)\exp(\beta 6E_2) \lesssim 1$. (d) $3E_2 > E_a$; βE_a large; complete condensation. (e) $3E_2 > E_a$; βE_a small; incomplete condensation. (f) βE_2 or βE_a very small; defect nucleation only.

restrictive as a prescription for the production of amorphous or "quasi-amorphous" films. The difficulties of rearrangement, after small, relatively isolated, but noncrystalline nuclei have been formed, may be more important.

The second regime is illustrated in Fig. 5b. This is the case of layer-by-layer growth. It is characterized by $R/N_0^2D \ll 1$, and $E_a \gtrsim E_a'$, where E_a' is the adsorption energy of an atom on a monolayer of the condensate.

For the triangular geometry considered in Fig. 4, the condition for this regime is $E_a \gtrsim 3E_2$, though this condition does depend on the geometry assumed. For example, for the (100) simple cubic geometry (Kossel crystal) the condition is $E_a \gtrsim E_2$ (see, e.g., Kern, 1968). In addition the temperature must be low enough so that the two-dimensional crystal layers can grow [i.e., $(R/N_0\nu)\exp(\beta 6E_2) > 1$], even though single atoms may reevaporate [i.e., $(R/N_0\nu)\exp(\beta 3E_2) \gtreqless 1$].

In the case when the lattice of the deposited material does not exactly match that of the substrate, misfit dislocations become incorporated (see Chapters 5–8) as growth proceeds and special transition layers or surface alloys may be formed. This type of growth occurs in many cases of auto-epitaxy [e.g., Si on Si—see Section VI,C,3] and in cases of "almost auto-epitaxy" [e.g., Au on Ag, Ag on Cu (Bauer and Poppa, 1972)]. It also may occur in many cases where the thin film material is less strongly bound to itself than to the substrate (e.g., Cd on W—Section V,A—or rare gases on graphite—Section VI,C). However, the exact regime that is observed may well depend on the details of how E_a' decreases as subsequent layers are deposited. One would expect the layer-by-layer growth in its simplest form if E_a' decreases monotonically to the value appropriate for adsorption on a bulk crystal of the deposited material; other possibilities are discussed at the end of this section.

Although two-dimensional nuclei must first be formed on the substrate in this case, only field ion microscopy (FIM) and field emission microscopy (FEM) are at present able to see these nuclei directly, and then only under somewhat restrictive conditions, including having a minute area of substrate for examination. FIM is, however, an extremely powerful technique for studying the individual atoms in such small nuclei and for studying the detailed arrangements of atoms in the first few layers (see Section VI,B). The FEM technique is somewhat difficult to interpret (see, e.g., Wagner and Voorhoeve, 1971) but many studies have been done on this regime of condensation using the technique; this work has been reviewed by Gretz (1969) and Pollock (1972).

The third regime, illustrated in Fig. 5c, is the regime appropriate to equilibrium adsorption. The same restriction as in Fig. 5b, $E_a \gtrsim E_a' \simeq 3E_2$, applies, only the temperature T is higher or the arrival rate R lower. In this regime $(R/N_0 \nu)\exp\{\beta(E_a'+3E_2)\} \lesssim 1$, so that, at least from the highest layers, atoms reevaporate sufficiently fast so that the film growth only proceeds to a certain thickness and bulk crystals do not grow. In addition, the layers themselves become unstable and "evaporate" in two dimensions, parallel to the substrate. Thus, in this case, there is no stable microstructure of the film in the long term, and adsorption is usually described solely in thermodynamic (equilibrium) terms. When $E_a \gg E_2$, even liquid and gaslike adsorbed multilayers can be formed. However, the initial microstructure of a solid adsorbed film must bear a strong resemblance to the previous case of layer-by-layer growth (Fig. 5b), and the transition from adsorption to crystal growth is an interesting one that has not been investigated experimentally and is not well understood at present.

The regimes illustrated in Figs. 5d and e are probably more familiar to most growers of thin films. In these cases the film forms initially in the form of three-dimensional clusters on the bare substrate as was illustrated in the Au–NaCl case of Fig. 1. The reason for this is that the relative magnitudes of E_a and E_2 are reversed from the previous two regimes: If $3E_2 \gtrsim E_a$, then the equilibrium form of the growing cluster will be cap-shaped, as discussed in some detail by Lewis (1967), even though the initial nucleus may be a two-dimensional disk. For the stronger condition $E_2 > E_a$, the clusters will have a three-dimensional shape essentially from the start of growth. For example, the form of a four-atom cluster can be seen with reference to Fig. 4. On this nearest-neighbor bond argument the planar cluster has energy $4E_a + 5E_2$, whereas the tetrahedral arrangement is bound by $3E_a + 6E_2$; thus if $E_2 > E_a$, the three-dimensional form is preferred.

The condition $E_2 > E_a$ is often well satisfied in practice for metals condensing onto insulators. These systems include some of the most important for the production of expitaxial films. For example, the metals Ag, Au, Cu, Pt, etc. condensed onto alkali halides, mica, MoS_2, or MgO all fall into this class. The strong preference of the metals for binding to other metal atoms rather than to the substrate leads to the production of near-spherical clusters that initially are compact but not necessarily crystallographic. Clusters with five- and sevenfold symmetry and a great variety of multiply twinned configurations have been found experimentally (e.g., Allpress and Sanders, 1970; Gillet and Gillet, 1969, 1973; Ino, 1966; Ino and Ogawa, 1967; Komoda, 1968; Ogawa and Ino, 1972; Sato and Shinozaki, 1970; Stirland, 1967; Tillett and Heritage, 1971). In order to explain the occurrence of these clusters, pair bonding and stacking models, which are more sophisticated versions of the arguments used in this section, have been used. In most of these calculations, the energy E_a has been totally neglected with respect to E_2, and the cluster energies evaluated as if the clusters were not on a substrate at all (e.g., Allpress and Sanders, 1970; Dave and Abraham, 1971; Gillet and Gillet, 1972; Hoare and Pal, 1971a, b, 1972a, b; Ino, 1969). That reasonable results can be obtained from this type of approach is another indication that $E_2 \gg E_a$ for many of these metal–insulator systems. This condition also makes it possible to understand qualitatively that the clusters themselves can be mobile, since they are so weakly bound to the substrate (e.g., Kern *et al.*, 1971; Masson *et al.*, 1971a, b; Métois *et al.*, 1972a, b; Schwabe and Hayek, 1972).

For the regime illustrated in Fig. 5d, all atoms condense, even initially. The values of the parameters for which this occurs are discussed in detail in Section IV, but one can visualize that this happens when the rate of formation of new clusters, or loss of atoms to existing clusters, is fast compared to the rate of reevaporation. In Fig. 5e, the condensation is incomplete. Single adatoms reevaporate from the substrate but stick when they join a cluster, either by diffusion or direct impingement. Figures 5d and e therefore represent low- and high-temperature regimes with the same systems. For many of the metal–insulator systems mentioned, the transition between the two often occurs in the region 100–300°C.

The critical nucleus size i is influenced by all the material and experimental parameters but is primarily determined by the size of the binding energy E_i, or in the present simple picture by E_2; the larger the value of E_2, the smaller the critical cluster size, and vice versa. Since for these metal–insulator systems E_2 is the largest energy in the problem, the critical nucleus size is usually extremely small, often consisting of only one atom. As will be seen later in Section III, the nucleation rate of stable clusters when atoms reevaporate (Fig. 5e) is $\sim n_1^{i+1}$, and the single population n_1 is very small when E_a/kT is small. Thus, for small E_a/kT, the critical cluster size *has* to be small or the nucleation rate is immeasurably small. This implies that an atomistic theory is necessary to describe the processes that can actually occur. The opposite case, when E_a/kT is large but E_2/kT is small, may apparently allow the critical nucleus size to be large; but the arguments given in relation to Figs. 5b and c suggest that in this case we may be dealing with crystal growth at very low supersaturation or even with equilibrium physical adsorption. Thus, for these regimes, we should also be concerned with an atomistic theory. In any case, the nuclei will be in the form of monolayer disks that have to be described in terms of a slightly hypothetical "edge energy" using classical nucleation theory.

If either E_2/kT or E_a/kT (or both) is very small, or even if they are both only moderately small, then nucleation will essentially be impossible on a perfect substrate. In this regime, illustrated in Fig. 5f, the effect of defect sites on the substrate surface will be most dramatic. The main effects may be described by the increase of E_a and/or E_i at defective sites. Some experimental examples were quoted in the last section, and will be examined in a bit more detail in Sections V and VI, C, 2.

Regimes intermediate to those illustrated in Fig. 5 are, of course, also possible. Given that the simple relations between E_a, E_a', E_i, and E_2, assumed for purposes of illustration, are by no means exactly obeyed in practice, the number of such possibilities is, in fact, very large. A particularly important intermediate regime can occur if E_a' drops below the value appropriate for an infinite crystal after some finite number of layers have been deposited. In this case growth starts in the layer-by-layer mode (Fig. 5b) and three-dimensional crystals can nucleate on top of these layers (Figs. 5d or e). Bauer and Poppa (1972) and Mayer (1971) have given evidence for this type of behavior in the systems K, Na–W, Cs–Ni, In, and Al–Si among others. A particularly clear demonstration of this regime is shown by the recent work of Gueguen *et al.* (1973) on the Fe–Au system. First, fcc Fe grows on the Au substrate; then interface dislocations are introduced; and finally islands of bcc Fe are nucleated on top of the fcc Fe layer.

Bauer and Poppa (1972) term this regime the "Stranski–Krastanov growth mode"; the name derives from a calculation by Stranski and Krastanov (1938) in which it was shown that for a monovalent ionic crystal (M^+X^-) condensing onto a divalent ($M^{2+}X^{2-}$) substrate, the second layer of (M^+X^-) is less strongly bound than the surface of a bulk crystal of (M^+X^-), even though the first layer is more strongly bound. Such an oscillation of E_a' with increasing thickness deposited may well lead to this intermediate growth mode.

When a thin film is grown at high temperatures, surface alloying may take place or surface compounds may form, all of which are, of course, quite specific to the particular substrate–thin film combination studied. Clearly such effects complicate the description, and certainly one must beware of falsifying the real situation by placing too much reliance on these simplified models. Nevertheless, we feel that is useful to have this classification as a starting point and to keep in mind the relative values of E_a, E_d, and E_2 for the system under discussion.

The theory that will be described in some detail in the next two sections has been developed primarily to describe the cases illustrated by Figs. 5d and e; it can almost certainly be extended to describe defect-induced nucleation of stable clusters adequately (Fig. 5f and Section V), and with some modification it can hopefully be used in the nucleation and growth of the first layer only in a layer-by-layer growth regime (Figs. 5b and c). If lateral growth of these layers is extremely rapid, nucleation of subsequent layers will be describable in a similar manner, but there are many possible complications: in the Stranski–Krastanov mode, nucleation processes could be important at several stages and nucleation could be either homogeneous or defect induced. Some of these possibilities are discussed in Section IV,H. The theory of nucleation of crystals within amorphous layers (Fig. 5a) has, up to now, had more in common with three-dimensional classical nucleation of solids from liquids (e.g., Barna *et al.*, 1971; Hudson and Sandejas, 1969) and will not be discussed further here.

III. Quantitative Description of the Nucleation Process

A. Introduction

The incorporation of atoms into a growing cluster on a substrate is, in many ways, similar to a chemical (polymerization) reaction in which units of small size (usually single atoms or monomer units) are added to molecular entities that have a range of sizes. The quantitative description of these processes can therefore be formulated within the language of chemical kinetics. The most general way of doing this analytically is to formulate rate equations for the cluster densities n_j of different sized clusters ($1 \leqslant j < \infty$). These equations express the time rate of change of n_j in terms of the various processes which are thought to occur.

The distinguishing features of the description nucleation of thin films on a substrate are essentially twofold. First, the atoms have to arrive at the substrate from the vapor (or solution); then they have to nucleate stable clusters on the substrate surface. The fact that these two processes are in series, and that both have to occur at a reasonable rate for nucleation to take place, is the major reason for using an atomistic theory, as explained in the last section. The second feature is that when the supply of atoms is diffusion-limited, this corresponds (at least in the vapor-deposition case) to diffusion over the substrate; two-dimensional expressions for diffusion-limited capture should therefore be used. The two-dimensional diffusion fields, set up by the capture of atoms by stable clusters, have much longer range than in the corresponding three-dimensional nucleation problem. For this reason existing clusters are often in competition with each other for the available supply of atoms, and this can influence subsequent nucleation and growth rates dramatically, even at a very early stage of growth.

Lewis and Campbell (1967) were the first authors to realize the qualitative significance of this second point, and since then there has been a great surge of activity in an effort to obtain a quantitative description of the growth of thin films up to and including the stage when coalescence becomes important. Recent papers include those by Dettmann (1972), Frankl and Venables (1970), Halpern (1969), Joyce *et al.* (1967), Lewis (1970a, b), Logan (1969), Niedermayer (1971), Markov (1971), Robertson and Pound (1973), Routledge and Stowell (1970), Sigsbee (1971), Stowell (1970, 1972a, b, 1973), Stowell and Hutchinson (1971a, b), Vincent (1971), and Zinsmeister (1966, 1968, 1969, 1970, 1971). An attempt has been made recently by Venables (1973a) to incorporate the results of these authors into a single approach, and in this section and the next we follow this paper rather closely; in nearly all cases, the ideas were initiated by one of the other authors and this will be acknowledged in the text. However, following the argument of one paper does enable a single train of thought to be followed and a single notation to be used; a full examination of all the papers on thin-film nucleation kinetics would be a rather time-consuming task, fraught with notational and other difficulties.

Recently, several authors have independently started to study nucleation and growth on surfaces by Monte Carlo computer "experiments" (Johannesson and Persson, 1970a, b; Abraham and White, 1970a, b; Adams and Jackson, 1972; Binsbergen, 1972a, b; Gilmer and Bennema, 1972a, b; Michaels *et al.*, 1974). It is too early to review this work satisfactorily, but some features are already apparent. Because the shortest time in the problem, the diffusion jump time, is so much shorter than the other times involved (reevaporation, decay of subcritical clusters), only small areas of substrate and relatively high nucleation rates can be studied in a reasonable time, and only then on a very fast computer. Second, three-dimensional clusters, or clusters with a different lattice parameter from the substrate, greatly increase the computing time required.

We suspect that these limitations will mean that Monte Carlo methods will prove useful primarily for demonstration purposes and for attacking certain specific problems. For example, Michaels *et al.* (1974) have checked experimentally the Walton equation [Eq. (9)]; the work of Binsbergen (1972a, b) emphasizes that a single configuration for the critical nucleus is not to be expected when the critical nucleus size is large; and the "experiments" of Johannesson and Persson (1970a, b) may be useful in showing how the epitaxial orientation of an individual cluster is obtained.

In the remainder of this section, we set up rate equations that have been used to describe the nucleation and growth of thin films and simplify them sufficiently so that solutions can be given in Section IV. It should be borne in mind that, although the formulation appears general, it is primarily concerned with describing the situations illustrated in Figs. 5d and e as explained in the last section. Thus it has most direct relevance at present to the nucleation and growth of metals films on insulators.

B. Incorporation of Physical Processes into General Rate Equations

Rate equations that are more or less general have been written down many times (e.g., Zinsmeister, 1966; Frankl and Venables, 1970). When only single atoms are mobile on the substrate, and when coalescence is neglected, these can be written in the form:

$$dn_1/dt = R - (n_1/\tau_a) - 2U_1 - \sum_{j=2}^{\infty} U_j \tag{3}$$

$$dn_j/dt = U_{j-1} - U_j \qquad (j \geqslant 2) \tag{4}$$

The successive terms on the right-hand side of Eq. (3) represent the rate of arrival of single atoms at the substrate, their reevaporation,† the combination of two singles into a pair, and the net rate of capture of single atoms by j-clusters U_j. Equation (4) states that the density of all other size clusters j increases by capture of a single atom by a $j-1$ and decreases by a j-cluster capturing a single atom. The terms U_j are net rates, which should allow for capture and reemission of single atoms, and for direct impingement from the vapor (or solution). The capture may—or may not—be diffusion-controlled.

When coalescence and/or cluster mobility is included, the description clearly becomes more complicated. In this case we have to allow for the possibility that clusters of size j interact with clusters of all other sizes k. This can then be shown to give rise to other terms in Eq. (4), which Venables (1973a)

† A term of this type would also describe adequately the loss of adatoms into the substrate, provided they did not subsequently reappear on the substrate surface; this point is discussed by Venables (1973b).

has termed U_+ and U_-. Equation (4) then can be written formally as

$$dn_j/dt = U_{j-1} - U_j + U_+ - U_- \tag{5}$$

Both these new terms, however, have to be written as a sum of terms involving the densities n_k of clusters of all other sizes k. Thus these terms complicate a general description considerably, and so far progress toward including coalescence and cluster mobility has only been made using the simplified set of equations set out below.

The first step toward a simplification of Eqs. (3) and (4) was taken by Zinsmeister (1968) and the same approach has been taken by other authors since. This involves dividing the clusters into sizes $1 < j \leqslant i$ and $j > i$. The clusters of size less than the critical size i are "subcritical" and can decay; for $j > i$, the clusters are "stable" and do not decay. Then, by summing the densities of clusters for size greater than i and calling this the density of stable clusters n_x, Eq. (4) reduces to

$$dn_j/dt = U_{j-1} - U_j \qquad (1 < j \leqslant i) \tag{6}$$

and

$$dn_x/dt = U_i \tag{7}$$

In addition it is helpful to separate the term $\sum_{j=2}^{\infty} U_j$ in Eq. (3) into loss to unstable clusters $\sum_{j=2}^{i} U_j$ and loss to stable clusters, which is written as U_x. This separation of unstable and stable clusters is very convenient, because the density n_x can often be associated directly with the number density (per unit area) of clusters observed experimentally, for example, in Fig. 1. Also, it is often the case that Eq. (6) is approximately zero, because of the maintenance of equilibrium among the subcritical clusters of size j; this equilibrium is examined in Section III,C,1. In the limiting case, which often occurs in practice, when the critical cluster size $i = 1$, there are no equations in the set (6), so that this problem does not arise.

Although the inclusion of coalescence and cluster mobility into Eq. (5) introduces considerable complexity, it is possible to introduce these processes into Eq. (7) without too much difficulty. Coalescence will give a negative contribution, which one might anticipate will be proportional to n_x^2; for the moment we denote it by U_c. The incorporation of mobile clusters causes a bit more difficulty. In principle, mobile clusters of size j should be included in Eq. (6) in a variety of ways, such as loss to stable clusters, formation of j-sized clusters from $(j-k)$-clusters by diffusion of k-sized clusters, etc. However, provided the diffusion coefficients D_k of these clusters are substantially smaller than the single-atom diffusion coefficient D, these terms will be much smaller than U_{j-1} and U_j; hence Eq. (6) will be essentially unchanged by cluster mobility. The contribution to Eq. (7) has two main causes: the nucleation rate is increased by subcritical clusters diffusing to other subcritical clusters to form stable clusters; and the stable cluster density is decreased by stable clusters diffusing to each other and coalescing. The second effect is likely to be more important. In any case both can be included in Eq. (7); the first as part of U_i and the second as a mobile cluster term U_m. This means that Eq. (7) can be written in general as

$$dn_x/dt = U_i - U_c - U_m \tag{8}$$

The rate equations, (3), (6), and (8), can be solved to give a description of the nucleation process. Before this can be done in detail, however, some further simplifications are needed and explicit expressions are required for the rates U_1, U_j, U_i, U_c, and U_m, which appear in these equations. This is done in the following section.

C. Simplification of the Rate Equations

1. *The Walton Equilibrium Relation*

The density of small clusters that can easily decay, n_j for $1 < j \leqslant i$, is described by the set of equations (6). If these clusters decay much faster than their net rate of growth, local equilibrium will exist between the density of single atoms n_1 and the densities n_j. This means, first, that Eq. (7) can be set equal to zero, and second that n_j (and n_i) can be expressed explicitly in terms of n_1. This relationship between n_i and n_1 was first derived by Walton (1962) and is firmly based on ideas of equilibrium statistical mechanics, when $n_1 \gg n_i$ (see Frankl and Venables, 1970, and Venables, 1973a for a detailed discussion). In this case the density of critical clusters n_i is given by

$$(n_i/N_0) = C_i(n_1/N_0)^i \exp(\beta E_i) \tag{9}$$

where E_i is the binding energy of the critical cluster and C_i a statistical weighting factor. The values of C_i were estimated by Venables (1973a) and, for the triangular lattice illustrated in Fig. 3, would take the approximate values 1, 3, 2, 3, 6, 6, and 1 for $i = 1$ to 7, respectively, at low temperatures, when only the most compact type of cluster is important. At higher temperatures and for higher values of i, different forms of i-cluster will also be important and this leads to increased values of C_i.

Equation (9) has only been shown to be valid theoretically on a bare substrate with $n_1 \gg n_i$, when stable clusters are not present. In what follows we shall also need to use it when stable clusters are present. Although this has not been properly justified, computations by Frankl and Venables (1970) and Venables and Ball (1971) that retained Eqs. (6) showed that Eq. (9) holds approximately in this case also. Use of the Walton relation enables the nucleation rate to be expressed in terms of the density n_1, as explained below. The subject has been further discussed by Stoyanov (1973).

2. *Expressions for the Factors U_x, U_i, U_c, and U_m in the Rate Equations*

In order to formulate the rate equations explicitly, we need expressions for the various rates U_x, U_i, U_c, and U_m. The term U_x is the loss of single atoms to stable clusters. These clusters are of various sizes k with corresponding "radii" r_k; the area of the substrate covered by each cluster is a_k. The rate of removal of single atoms by direct impingement is therefore $Ra_k n_k$. The rate of removal of single atoms by diffusion may be written as $\sigma_k D_1 n_k$, where σ_k is referred to as a capture number, which will be discussed in Section III, D. Thus U_x is given by

$$U_x = \sum_k (\sigma_k D n_1 n_k + R a_k n_k) \tag{10a}$$

Defining σ_x and a_x as suitable averages of σ_k and a_k, respectively, allows one to write

$$U_x = \sigma_x D n_1 n_x + R a_x n_x \tag{10b}$$

The term U_i, the nucleation rate, can be expressed as

$$U_i = \sigma_i D n_1 n_i + \sum_k \sum_l \sigma_{kl} n_k n_l (D_k + D_l) \tag{11}$$

where, in the summations, k and l lie between 1 and i and $l+k > i$. The second term is expected to be small provided $D_k \ll D$. Again σ_i and σ_{kl} are capture numbers.

On a substrate devoid of stable clusters, the density of critical clusters n_i can be calculated by the Walton equilibrium relation as discussed in III, C, 1. As explained there, its use in general, when the coverage of the substrate by stable clusters is Z, is more problematical. As argued by Lewis (1970a) and Stowell and Hutchinson (1971a), the equilibrium relation should be used in a local sense, i.e., on the uncovered area of the substrate, and the nucleation rate multiplied by the factor $1-Z$. In the present notation, where n_1 is the density of single atoms over the entire substrate, this correction involves dividing the first term of (11) by $(1-Z)^i$. On inserting the equilibrium relation and neglecting the second term in (11), U_i can be written as

$$U_i = (1-Z)^{-i} \gamma_i D n_1^{i+1} \tag{12}$$

where $\gamma_i = \sigma_i C_i N_0^{1-i} \exp(\beta E_i)$. The correction for finite coverage Z is in fact rather uncertain, as is the Zeldovich factor (see Frankl and Venables, 1970, Section 5), which should be included in Eq. (12) to account for the departure from equilibrium of the chain of clusters $1 < j \leqslant i$, due to a nonzero rate of nucleation. This factor has been shown (Walton, 1962; Frankl and Venables, 1970) to lie between 1 and i^{-1}. There are thus four uncertain factors in the theoretical formulation of the nucleation rate: the correction for finite Z, the Zeldovich factor, and the exact values of C_i and σ_i. For the important case of

$i = 1$, the uncertainties are minimized, but they still make absolute calculations of nucleation rates rather difficult. This problem is, of course, common to all chemical kinetic studies.

The two terms that cause the stable cluster density to decrease, U_c and U_m, can be estimated as follows. The coalescence rate is estimated first, for circular clusters where $a_k = \pi r_k^2$. A circular k-sized cluster will coalesce with an l-cluster when the distance between their centers becomes equal to $r_k + r_l$. By considering all possible sets of coalescences, one can show, for low substrate coverages, that the rate of removal of randomly positioned stable clusters by coalescence is given by

$$U_c = \tfrac{1}{2} \sum_k \sum_l n_k n_l \, d\pi (r_k + r_l)^2 / dt \tag{13a}$$

If we make the oversimplification that $\pi(r_k + r_l)^2 = 4\pi r_x^2 = 4a_x$, then U_c can be expressed simply as

$$U_c = n_x^2 \, d2a_x/dt \tag{13b}$$

Vincent (1971), in a more sophisticated analysis of coalescence (but one that did not include nucleation), concluded that the n_x was reduced by a factor $(1 - e^{-4Z})/4Z$, from its original value at a coverage Z, where $Z = n_x a_x$. Vincent's expression reduces to

$$U_c = 2n_x \, dZ/dt \tag{13c}$$

for small Z; Eq. (13c) is the same as (13b) for $dn_x/dt = 0$. Since the uncertainties in these formulas are only of order of factors $1 \pm Z$ (Venables, 1973a), we will use the form (13c), which enables comparison with the work of Stowell and Hutchinson (1971a) and Stowell (1972b) to be made most easily.

The term U_m can be written formally as

$$U_m = \tfrac{1}{2} \sum_k \sum_l \sigma_{kl} n_k n_l (D_k + D_l) \tag{14a}$$

where the summations run over all values of k, $l > i$, since this describes the diffusive encounters of all stable clusters. Not much is known about diffusion coefficients of clusters as a function of size, but if all clusters have a similar diffusion coefficient D_x, this term can be written as

$$U_m = \sigma_{xx} D_x n_x^2 \tag{14b}$$

If, on the other hand, only small stable clusters, density n_m, are mobile with diffusion coefficient D_m and $n_m \ll n_x$, then

$$U_m = \sum_m \sigma_{mx} D_m n_m n_x \tag{14c}$$

Again σ_{kl}, σ_{xx}, and σ_{mx} are capture numbers, which can be estimated using the methods discussed in Section III, D. It is also possible to consider more specific models of cluster mobility, as has been done by Lewis (1970b), Kern *et al.* (1971), Robertson (1973), and Stowell (1974). As shown by Venables (1973a) these more detailed models may well fall into one of the extreme cases represented by Eqs. (14b) or (14c), as far as the effect of cluster mobility on nucleation kinetics is concerned.

3. *The Growth Rate of Stable Clusters*

The expressions given above enable the rate equations to be formulated explicitly in terms of the various σ's and the coverage $Z = n_x a_x$. In order to solve these equations we need an equation describing the change of Z (or a_x) as a function of time. In order to calculate how Z evolves, it has been noted (Stowell and Hutchinson, 1971a; Stowell, 1972b; Venables, 1973a) that if ω_x is the mean number of atoms in a stable cluster, then the total number of atoms in stable clusters ($n_x \omega_x$) can be evaluated. This equation is

$$d(n_x \omega_x)/dt = (i+1)\,U_i + U_x = (i+1)\,U_i + \sigma_x D n_1 n_x + R n_x a_x \qquad (15)$$

where Eq. (10b) has been used for U_x. The three terms on the right-hand side of (15) represent the rate of addition of atoms to stable clusters by the nucleation process itself, by diffusion across the substrate, and by direct impingement, respectively. The first term is unimportant except initially and can be neglected (Venables, 1973a) or included (Stowell, 1972b) without influencing the results.

The relationship between the quantities ω_x, a_x, and r_x used in the above equations clearly depends on the mode of growth of the stable clusters, and only simple (and constant) shapes of cluster can be treated analytically. The most important distinction is between growth such that the shape of the cluster is constant (i.e., $\omega_x \sim r_x^3$, $a_x \sim r_x^2$) and growth when the thickness is constant, when $\omega_x \sim a_x \sim r_x^2$. These cases produce different equations, which correspond to growth of three-dimensional clusters and two-dimensional monolayer disks, respectively. For three-dimensional growth as a hemisphere $\omega_x \simeq 2\pi r_x^3/(3\Omega)$, $a_x = \pi r_x^2$, where Ω is the atomic volume of the atoms in the cluster. For the monolayer disk case, $a_x = \pi r_x^2 = \omega_x \Omega^{2/3}$. Equation (15) can be cast in a more convenient form by using the coverage $Z = n_x a_x$ as a variable. Using Z, the left-hand side of (15) can be expressed as $(\frac{2}{3}\pi^{1/2}\Omega)\, d(Z^{3/2}/n_x^{1/2})/dt$ or $\Omega^{-2/3}\, dZ/dt$ for the three- and two-dimensional cases, respectively. Thus Eq. (15) can be considered as a rate equation for the coverage, when written in the form

$$\frac{dZ}{dt}\left(1 - \frac{1}{3}\frac{d(\ln n_x)}{d(\ln Z)}\right) = \Omega\left(\frac{\pi n_x}{Z}\right)^{1/2} [(i+1)\,U_i + \sigma_x D n_1 n_x + RZ] \qquad (16a)$$

suitable for three-dimensional growth, or

$$dZ/dt = \Omega^{2/3}[(i+1)U_i + \sigma_x Dn_1 n_x + RZ] \tag{16b}$$

in the two-dimensional case. It should be remembered that the exact shape of the clusters is often not measurable, a point emphasized by Terajima *et al.* (1973).

D. Capture Numbers and Diffusion-Limited Growth

In order to solve even the simplified coupled rate equations (3), (8), and (16), it is necessary to use specific values for the capture numbers σ_i and σ_x (and maybe σ_{xx} and σ_{mx}) that appear in the equations. This subject has been much discussed recently (Halpern, 1969; Logan, 1969; Frankl and Venables, 1970; Lewis, 1970a, b; Niedermayer, 1971; Sigsbee, 1971; Zinsmeister, 1971; Dettmann, 1972; Stowell, 1972a). This work has been summarized and extended by Venables (1973a) for the case where an atom experiences no difficulty in attaching itself to a growing stable cluster. In this case of diffusion-limited capture by clusters, he has shown that two types of approximate expressions can be derived for σ_x that represent upper and lower bounds to the true value; he has called these expressions the "lattice approximation" and the "uniform depletion approximation," respectively. More detailed calculations, which can be used for computations with arbitrary distributions of clusters, have been given by Lewis and Rees (1974).

The detailed mathematics involved will not be discussed here and the interested reader is referred to the original literature. For the general reader it is sufficient to realize that the solutions emerge as expressions involving Bessel functions (because of the two-dimensional geometry) and that σ_x and σ_i depend somewhat on the regime of condensation involved and the coverage. For the complete condensation regime (Fig. 5d) σ_x is a function of the coverage Z only for both approximations. A comparison of these approximations in the complete condensation regime is shown in Fig. 6, and it is seen that these two expressions determine σ_x to within a factor of about two. In the incomplete condensation regime at low values of Z there is much less uncertainty. In this case both σ_x and σ_i can be expressed in the form

$$\sigma = 2\pi X K_1(X)/K_0(X) \tag{17}$$

where K_1 and K_0 are modified Bessel functions and the argument X is $(r_k{}^2/D\tau_a)^{1/2}$, where $r_k \simeq r_x$ in the expression for σ_x and $r_k \simeq r_i$ for σ_i. If the stable clusters are also mobile and coalesce when they meet (but do not themselves evaporate from the substrate), the capture numbers involved are similar to σ_x in the complete condensation regime, with $\sigma_{xx} \gtrsim \sigma_{mx} \gtrsim \sigma_x$.

The diffusion-limited growth that gives rise to these expressions for σ_x also

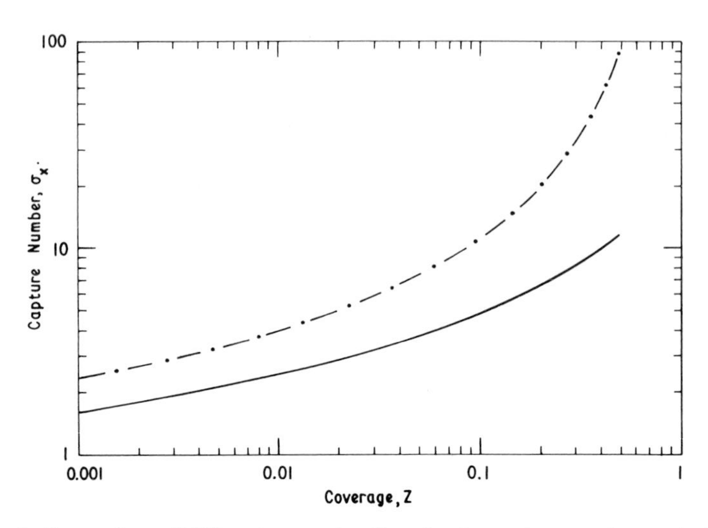

FIG. 6. Comparison of different approximations for the capture number σ_x, in the complete condensation regime (Venables, 1973a). Solid line: uniform depletion approximation (lower bound); dashed line: lattice approximation (upper bound).

gives rise to a depression of the local concentration of atoms around stable clusters and consequently to a reduction of the nucleation rate in such regions. This type of effect will lead to a nonrandom distribution of clusters even on a defect-free substrate, and is just beginning to be investigated both theoretically and experimentally (Stowell, 1970; Markov, 1971; Venables, 1973a; Venables and Ball, 1971; Schmeisser and Harsdorff, 1973; Schmeisser, 1974b). If diffusing atoms attach themselves to clusters only with difficulty, then diffusion zones are not formed around stable clusters and, superficially, it is easier to evaluate the capture numbers, which are then proportional to the radius of the cluster and an "attachment coefficient" μ. This approximation has often been made (Zinsmeister, 1966, 1971; Frankl and Venables, 1970), and is very common in other chemical kinetic treatments. The trouble is that this treatment is only true if $\mu \ll 1$, and in order to evaluate μ, one would need to know the details of the attachment mechanism. The condition $\mu \ll 1$ seems unreasonable for the highly stable clusters of metals grown on insulators (Figs. 5d and e), but it may be more appropriate for a discussion of the layer-by-layer growth cases (Figs. 5b and c) when growth is very slow (i.e., at the border line between layer growth and adsorption). This latter case will not be discussed further here.

Despite their relatively complex form, these capture numbers are just numbers, which do not depend strongly on the other parameters involved. As shown by Stowell (1972a) and Venables (1973a), σ_x is in the range 5–10, and σ_i in the range 1–3 in many situations encountered in practice.

IV. Solutions of the Rate Equations

A. Introduction

The simplified rate equations formulated in the last section can be solved to give expressions for experimentally observable quantities, such as those derivable from pictures such as Fig. 1. In particular, the equations can be used to give expressions for $n_x(t)$ and $Z(t)$, from which one can extract expressions for the maximum density of stable clusters and the time and coverage at this maximum. Expressions can also be obtained for the amount of material condensed as a function of time. Although it is not possible to derive exact expressions for the size distribution and spatial distribution of clusters using these simplified equations, nevertheless they enable a semiquantitative feel for the problems to be obtained.

Let us first review the rate equations that are obtained when the equilibrium assumption (Section III, C, 1) is made. They are, for three-dimensional growth,

$$dn_1/dt = R - (n_1/\tau_a) - d(n_x\omega_x)/dt \tag{18}$$

$$dn_x/dt = (1-Z)^{-i}\gamma_i Dn_1^{i+1} - 2n_x\,dZ/dt - U_m \tag{19}$$

$$\frac{dZ}{dt} = \frac{\Omega}{(1-m/3)}\left(\frac{\pi n_x}{Z}\right)^{1/2}\frac{d(n_x\omega_x)}{dt} \tag{20}$$

with

$$d(n_x\omega_x)/dt = (i+1)\,U_i + \sigma_x Dn_1 n_x + RZ \tag{21}$$

Equation (18) is the simplification of (3) that arises from the equilibrium assumption; Eq. (19) is (8) with (12) and (13c) inserted; and Eqs. (20) and (21) are (16a) and (15), respectively. The parameter m has been written for $d(\ln n_x)/d(\ln Z)$, i.e., $n_x \sim Z^m$.

The solutions to these equations can be divided into an initial "transient" and a later "steady state" nucleation stage; within the steady state stage the coalescence or cluster mobility terms eventually become important and cause a maximum (or a saturation) of the cluster density. Equations (18)–(21) enable us to follow this evolution quantitatively.

B. Transient Nucleation

The "transient" stage (Stowell and Hutchinson, 1971b; Venables, 1973a) occurs initially because at $t = 0$, $dn_1/dt = R$, and all other terms in Eq. (18) are zero. Thus $n_1 = Rt$. This stage continues until the other terms in (18) limit n_1. The end of this stage occurs approximately at a time τ, where τ is defined by

$$dn_1/dt = R(1-Z) - (n_1/\tau) \tag{22}$$

and

$$\tau^{-1} = \tau_a^{-1} + \sigma_x Dn_x + [(i+1)\,U_i/n_1] \tag{23}$$

These equations are obtained by substituting (21) in (18). The last term in (23) is numerically unimportant, and Z is always small at this stage. The nucleation that arises during this transient stage can be obtained from the first term of (19):

$$dn_x\, dt = \gamma_i D(Rt)^{i+1} \tag{24}$$

Thus at the end of the stage, $t \simeq \tau$, and

$$n_x \simeq \gamma_i DR^{i+1}\tau^{i+2}/(i+2) \tag{25}$$

Inserting Eq. (23)—neglecting the last term—for τ gives

$$n_x(1+\sigma_x D\tau_a n_x)^{i+2} \simeq \gamma_i D\tau_a (R\tau_a)^{i+1}/(i+2) \tag{26}$$

In Eq. (23) and (26) we see for the first time that it is the lumped parameter $\sigma_x D\tau_a n_x$ that determines whether condensation is "complete" or "incomplete," i.e., whether we are dealing with the regime illustrated in Fig. 5d or 5e. For $\sigma_x D\tau_a n_x \ll 1$, condensation is incomplete, but then $n_x(\tau)$ is negligibly small, and transient nucleation is totally unimportant. For $\sigma_x D\tau_a n_x \gtrsim 1$, condensation is complete and $n_x(\tau)$ may be significant. It is given by

$$n_x(\tau) \simeq \left\{\frac{\gamma_i}{i+2}\,\frac{(R/D)^{i+1}}{\sigma_x^{i+2}}\right\}^{1/(i+3)} \tag{27}$$

This formula was first derived by Stowell (1970), and was obtained in a similar manner to that given here by Stowell and Hutchinson (1971b) and Frankl and Venables (1970). In the complete condensation regime, the end of the transient stage corresponds to the stage at which the diffusion field around one cluster overlaps the neighboring clusters. Thus after this stage, the clusters are in competition for the diffusing adatoms. Initially it was thought that n_x then became constant (Stowell, 1970; Logan, 1969; Frankl and Venables, 1970, especially Fig. 7), but this depends on the form of σ_x (Venables, 1973a). For the Bessel function solutions described in Section III, D, saturation does not occur at this point, though there is a sharp decrease in the nucleation rate in the "steady state" nucleation stage that follows.

C. Steady State Nucleation

For times $t \gtrsim \tau$, Eq. (22) can be expressed simply as

$$n_1 = R\tau(1-Z) \tag{28}$$

with τ given by Eq. (23). This means that Eq. (18) is no longer needed, and the coupled equations (19) and (20) can be solved to obtain solutions for the experimental variables $n_x(t)$ and $Z(t)$. Before discussing the coalescence or cluster mobility-induced maxima in $n_x(t)$, the initial behavior of $n_x(t)$ and $Z(t)$ is worth noting. For $\tau \simeq \tau_a$, i.e., incomplete condensation, the nucleation rate dn_x/dt is, using (19) and (28), simply $(1-Z)\gamma_i D(R\tau_a)^{i+1}$. Using expression (12) for γ_i this can be written

$$dn_x/dt = (1-Z)\sigma_i R[C_i(R/N_0{}^2D)^i \exp(\beta E_i)](D\tau_a N_0)^{i+1} \tag{29a}$$

$$= \alpha\sigma_i C_i(1-Z)R(R/N_0\nu)^i \exp\beta\{E_i+(i+1)E_a - E_d\} \tag{29b}$$

where α and ν were introduced in Eqs. (1) and (2). Equation (29b) emphasizes that a measurement of the initial nucleation rate at high temperature as a function of the rate of arrival R enables the critical nucleus size i to be determined; the temperature dependence of the nucleation rate gives the activation energy $\{E_i+(i+1)E_a-E_d\}$.† These points have been realized for a long time (e.g., Walton, 1962; Walton *et al.*, 1963).

However, even in this high-temperature regime, the behavior of dZ/dt is not simple. Initially the first and second terms on the right-hand side of Eq. (21) must dominate: atoms arrive at clusters by diffusion. For very small values of $D\tau_a$, the third term then becomes more important, and most growth occurs by direct impingement. In the low-temperature regime (complete condensation) the equation for the nucleation rate is not particularly simple. The nucleation rate is proportional to τ^{i+1}, and τ is inversely proportional to both σ_x and n_x. Hence the nucleation rate is smaller for forms of σ_x that increase more rapidly with Z and decrease as n_x increases. However, in this low-temperature limit the right-hand side of Eq. (21) is just equal to the rate of arrival R.

Stowell and Hutchinson (1971a) and Stowell (1972b) have shown that n_x can be most usefully expressed in terms of the coverage Z. This involves combining Eqs. (19) and (20) by writing $dn_x/dZ = (dn_x/dt)/(dZ/dt)$. It is also extremely convenient to use the variables $\mathcal{N} = n_x/N_0$ and $\mathcal{T} = Rt/N_0$, and to use the lumped material parameters $A = D\tau_a N_0$ and $B_i = C_i e^{\beta E_i}(R/N_0{}^2D)^i$. For example, in Eq. (29a), one can see that the high-temperature nucleation rate can be expressed in this form. The parameters A and B_i have a relatively simple physical meaning. The number of jumps an atom makes before desorption is $\simeq A$; $2A^{1/2}$ is the root mean square number of jumps it is away from its arrival point when it desorbs. The parameter B_i indicates the stability of the critical nucleus i. For a given value of τ (i.e., for a given n_x, Z, and A) the critical nucleus size i is the size that makes B_i a minimum. For the important case of $i=1$, $B_1 = R/N_0{}^2D$; this parameter is the inverse of the number of jumps an atom makes before the next one arrives (from the vapor) at the same spot.

† For accurate evaluation of these energies, any temperature dependence of σ_i, C_i, and ν should strictly be considered as well.

Using these variables, equations for $\mathcal{N}(Z)$ and $\mathcal{T}(Z)$ can be derived from Eqs. (19) and (20). These are

$$\frac{d\mathcal{N}}{dZ} = \left\{(1-Z)\,\sigma_i\, B_i\left(\frac{A}{1+\sigma_x A\mathcal{N}}\right)^{i+1} - \frac{U_m}{R}\right\}\frac{d\mathcal{T}}{dZ} - 2\mathcal{N} \tag{30}$$

and

$$\frac{d\mathcal{T}}{dZ} = \frac{(1-m/3)}{(\pi\Omega^2 N_0{}^3)^{1/2}}\left(\frac{Z}{\mathcal{N}}\right)^{1/2}\left\{\frac{1+\sigma_x A\mathcal{N}}{Z+\sigma_x A\mathcal{N}}\right\} \tag{31}$$

An expression for the total amount of material condensed $\mathcal{T}_c$ expressed in the same (monolayer) units as $\mathcal{T}$, can be simply obtained by noting that the rate of removal of atoms from the substrate is just n_1/τ_a. After some simple manipulation one obtains

$$d\mathcal{T}_c/dZ = d\mathcal{T}/dZ\{1-(1-Z)/(1+\sigma_x A\mathcal{N})\} \tag{32}$$

Other "average quantities," such as the mean radius of the clusters $Z^{1/2}/\pi\mathcal{N}$, can be easily derived by solving the coupled equations (30) and (31). Equations of this type were first derived by Stowell and Hutchinson (1971a) and Stowell (1972b), and were given in the present form by Venables (1973a). They enable the experimental observables $\mathcal{N}$, $\mathcal{T}$, and $\mathcal{T}_c$ to be computed as a function of the coverage Z for any values of the parameters A and B_i, using any approximation for σ_i and σ_x, such as the expressions discussed in Section III, D and in more detail by Venables (1973a).

D. High- and Low-Temperature Regimes of Condensation

For a given value of i and B_i the behavior of Eqs. (30) and (31) depends only on the one parameter $\sigma_x A\mathcal{N}$ $(=\sigma_x D\tau_a n_x)$. For $\sigma_x A\mathcal{N} \gg 1$ we have complete condensation; for $\sigma_x A\mathcal{N} \ll Z$ we have the regime Stowell and Hutchinson (1971a) have termed "extreme incomplete condensation," when growth occurs by direct impingement of atoms on to the cluster only. For $1 > \sigma_x A\mathcal{N} > Z$ we have an intermediate regime that arises initially under virtually all conditions. In this regime condensation is incomplete, but cluster growth occurs mainly by diffusion of atoms over the substrate.

For the extreme incomplete condensation regime Eqs. (30) and (31) simplify, when $U_m = 0$, to

$$\frac{d\mathcal{N}}{dZ} = \frac{(1-Z)(1-m/3)}{(\pi\Omega^2 N_0{}^3)^{1/2}}\frac{\sigma_i B_i}{(Z\mathcal{N})^{1/2}}A^{i+1} - 2\mathcal{N} \tag{33}$$

and

$$\frac{d\mathcal{T}}{dZ} = \frac{(1-m/3)}{(\pi\Omega^2 N_0{}^3)^{1/2}}\left(\frac{1}{Z\mathcal{N}}\right)^{1/2} \tag{34}$$

Following Stowell and Hutchinson (1971a), one can write

$$\begin{aligned}\mathcal{N}(Z) &= \eta(Z)\{\sigma_i B_i A^{i+1}/(\Omega N_0^{3/2})\}^{2/3} \\ &= \eta(Z)\left(\frac{\sigma_i \alpha C_i}{\Omega N_0^{3/2}}\right)^{2/3}\left(\frac{R}{N_0 \nu}\right)^{2i/3} \exp\frac{2\beta}{3}\{E_i + (i+1)E_{\mathrm{a}} - E_{\mathrm{d}}\} \qquad (35)\end{aligned}$$

and

$$\mathcal{T}(Z) = \kappa(Z)\{\sigma_i B_i A^{i+1}\}^{-1/3} \qquad (36)$$

where $\eta(Z)$ and $\kappa(Z)$ are pure numbers that are solutions of the differential equations

$$d\eta/dZ = (1-m/3)(1-Z)(\pi Z\eta)^{-1/2} - 2\eta \qquad (37)$$

and

$$d\kappa/dZ = (1-m/3)(\pi Z\eta)^{-1/2} \qquad (38)$$

The σ_i can be incorporated into Eqs. (35) and (36) because the σ_i are not functions of Z in this regime. These equations only have simple solutions when $Z \ll 1$ and the coalescence term 2η in Eq. (37) is unimportant. Under these conditions m $(=d\ln\eta/d\ln Z) = \frac{1}{3}$ and $\eta = \kappa = (\frac{8}{3}\pi^{1/2})^{2/3}Z^{1/3}$. Under these same conditions the amount of material condensed T_{c} [from Eq. (32) with $\sigma_x A\mathcal{N} \ll Z$] equals $Z\mathcal{T}(Z)/4$. At a later stage the coalescence term 2η in Eq. (37) produces a maximum in $\eta(Z)$. The curve $\eta(Z)$ is shown in Fig. 7a. It has a maximum value of 0.59 at $Z = Z_0 = 0.23$ (Stowell and Hutchinson, 1971a; Venables, 1973a).

In the complete condensation regime, $\sigma_x A\mathcal{N} \gg 1$. In an exactly similar way one can derive from Eqs. (30) and (31) that

$$\begin{aligned}\mathcal{N}(Z) &= \eta(Z)\{\sigma_i B_i/(\Omega N_0^{3/2})\}^{2/(2i+5)} \\ &= \eta(Z)\left(\frac{\sigma_i C_i}{\Omega N_0^{3/2}}\right)^{2/(2i+5)}\left(\frac{R}{N_0 \nu}\right)^{2i/(2i+5)} \exp\frac{2\beta}{(2i+5)}\{iE_{\mathrm{d}} + E_i\} \qquad (39)\end{aligned}$$

and

$$\mathcal{T}(Z) = \kappa(Z)\{\sigma_i B_i\}^{-1/(2i+5)} \qquad (40)$$

if we take σ_i to be independent of $\mathcal{N}$ and Z, which is approximately true (Venables, 1973a). Again $\eta(Z)$ and $\kappa(Z)$ are pure numbers that can be found by integrating the coupled equations

$$d\eta/dZ = (1-Z)(1-m/3)(Z/\pi\eta)^{1/2}(\sigma_x\eta)^{-(i+1)} - 2\eta \qquad (41)$$

and

$$d\kappa/dZ = (1-m/3)(Z/\pi\eta)^{1/2} \qquad (42)$$

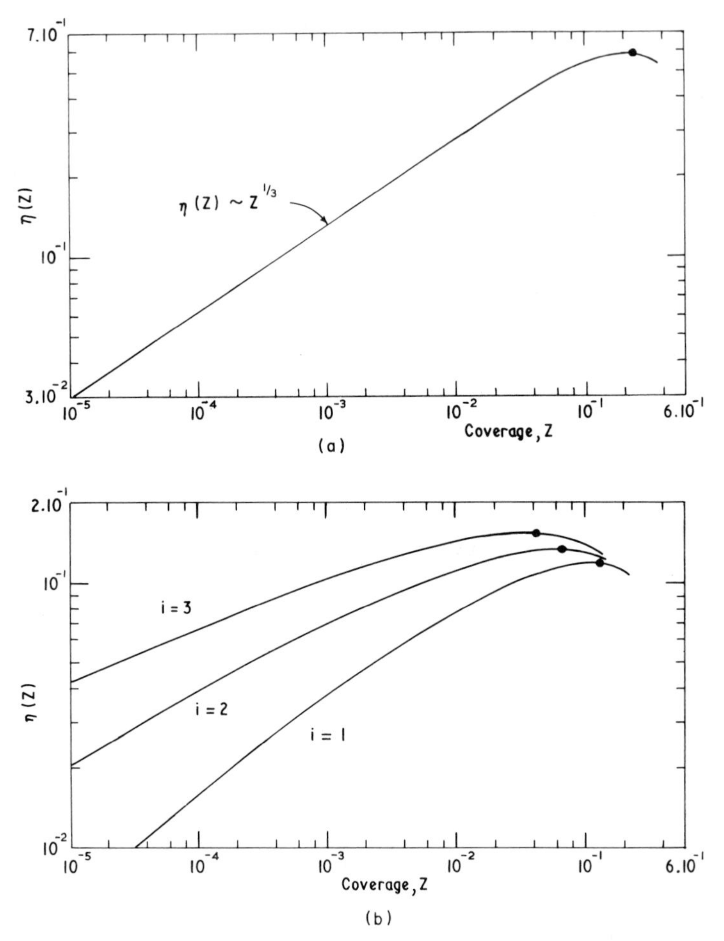

FIG. 7. Number of clusters versus coverage in the high- and low-temperature regimes (Stowell and Hutchinson, 1971a; Venables, 1973a). (a) $\eta(Z)$ in the extreme incomplete condensation regime according to Eq. (37). (b) $\eta(Z)$ in the complete condensation regime, using the lattice approximation for σ_x in Eq. (41).

Even without the coalescence correction -2η there is no simple approximation to these equations, as σ_x is a complicated function of Z. The form of $\eta(Z)$ using the lattice approximations for σ_x is shown in Fig. 7b for $i = 1$ to 3 (Stowell and Hutchinson, 1971a; Venables, 1973a). Note that increasing the value of i decreases the nucleation rate and also shifts the coalescence-induced maximum in $\eta(Z)$ to smaller values of Z; values of the coverage Z_0 at the maximum range from $Z_0 \simeq 0.12$ for $i = 1$, to $Z_0 = 0.04$ for $i = 3$, using this approximation for σ_x. Further details are given by Stowell and Hutchinson (1971a) and Venables (1973a).

If one wants to follow the evolution of $\mathcal{N}$ and Z with time in general, there is no alternative but to integrate the coupled equations (30) and (31) directly, starting with the solutions at the end of the "transient" nucleation stage. These integrations can easily be performed using a small computer, and examples have been given by Stowell (1972b). Venables (1973a) has emphasized that there are uncertainties in some of the "constants" appearing in these equations. In particular the value of $(\Omega N_0^{3/2})$ appropriate to the system studied should be used. There is also a spread of possible approximations for σ_x and in particular σ_i. Curves of $\mathcal{N}(T)$ and $\mathcal{T}(Z)$ are shown in Figs. 8a and 8b. Venables (1973a) showed that if the only uncertainty were in σ_x, then the curves for the uniform depletion and lattice approximations (Section II, D) would bracket the correct solution. However, one can see from the curves that the uncertainty in σ_i is probably more important, and that the two approximations, when used self-consistently for both σ_x and σ_i give values that are very similar.

E. General Equations for the Maximum Cluster Density

The maximum in the cluster density as a function of coverage or time is one of the most easily measurable quantities. It can be measured, for example, even in the complete condensation case, when the initial nucleation rate is much too large to be measured. A general expression for $\mathcal{N}$ at $Z = Z_0$ can be written simply by equating $d\mathcal{N}/dZ$ to zero in Eq. (30). At the maximum, $d(\ln \mathcal{N})/d(\ln Z) = m$ is also zero, and thus, for immobile clusters ($U_m = 0$),

$$\frac{(1-Z_0)Z_0^{1/2}}{2} \cdot \frac{\sigma_i B_i A^{i+1}}{(\pi\Omega^2 N_0{}^3)^{1/2}} = \mathcal{N}^{3/2}(1+\sigma_x A\mathcal{N})^i (Z_0+\sigma_x A\mathcal{N}) \tag{43}$$

An equation of this form was first derived by Stowell (1972b), and as Venables (1973a) has emphasized, the only approximation is the use of the steady-state approximation [Eq. (28)] for n_1, which is well satisfied.

In order to use Eq. (43), a value of Z_0 has to be supplied and a particular approximation for σ_i and σ_x used. The value of N is, however, rather independent of the assumed value of Z_0, and any reasonable approximation for σ_i and σ_x will give satisfactory results. Examples of the use of this equation to analyze experimental data have been given by Stowell (1972b), Venables (1973a), and Stowell (1974). Some examples are also given here in Section VI.

It has also been shown (Venables, 1973a) that an equation similar to (43) is obtained even when the clusters are allowed to be mobile ($U_m \neq 0$), and the corresponding term in Eq. (30) is retained. This equation has the form

$$A_1(Z_1) \cdot (1 - U_{\mathrm{m}}/U_i) = A_2(Z_1) \tag{44}$$

where $A_1(Z_1)$ and $A_2(Z_1)$ are the left- and right-hand sides of Eq. (43) evaluated for the reduced coverage Z_1 at which the maximum in the cluster

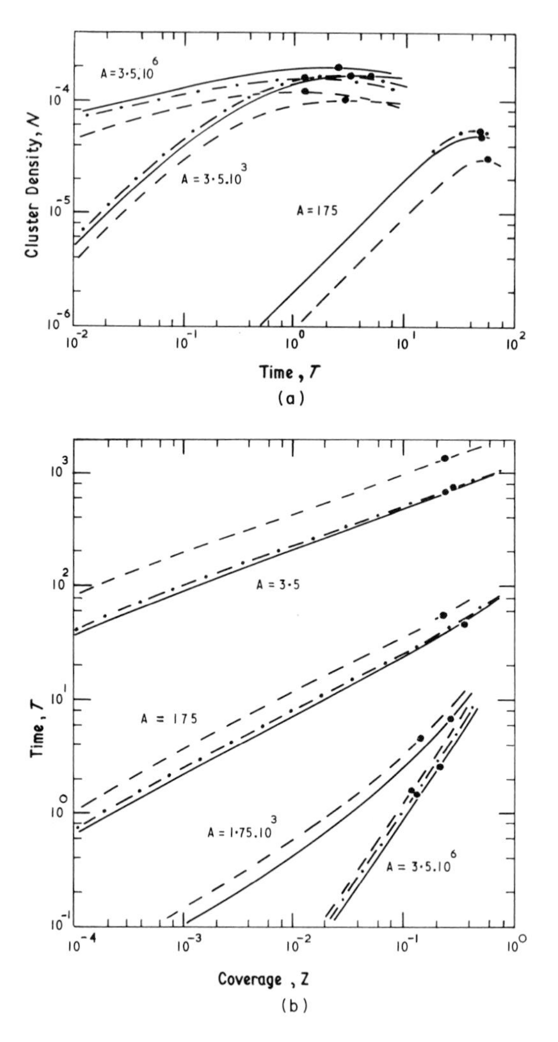

FIG. 8. (a) Cluster density $\mathcal{N}$ vs $\mathcal{T}$, and (b) time $\mathcal{T}$ vs Z, for $i = 1$, $R/N_0{}^2D = 3\times10^{-11}$, $\Omega N_0^{3/2} = 1$, and the values of A indicated (Venables, 1973a). The black circles indicate the points corresponding to the maxima in the $\mathcal{N}(\mathcal{T})$ curve. Solid lines indicate the uniform depletion approximation for σ_i and σ_x; dot–dash lines the lattice approximation for σ_i and σ_x; and dashed lines the lattice approximation for σ_x and $\sigma_i = 1$ (Stowell, 1972b). [For $A = 1.75\times10^3$, the solid and dot–dash lines effectively coincide.]

density occurs. By comparison with direct integrations of Eqs. (30) and (31) and by an approximate analytical argument he was able to show that, to a good approximation,

$$Z_1/Z_0 = (1-U_m/U_i)/(1-CU_m/U_i) \tag{45}$$

where the constant C is in the range $\frac{1}{3}$–$\frac{1}{2}$. With this approximation, Eq. (44) can conveniently be used to analyze maximum cluster density data even with mobile cluster effects present.

F. Mobile Cluster Effects

Several authors have shown experimentally that stable clusters can move over the substrate during deposition. Perhaps the most complete set of quantitative experiments have been those of Kern and collaborators (Masson *et al.*, 1971a, b; Métois *et al.*, 1972a, b). The incorporation of such effects into a nucleation theory has been attempted by Lewis (1970b) and Venables (1973a). In the latter paper it was shown that the two extreme types of expression, given by Eqs. (14b) or (14c), result in quite different types of $n_x(t)$ curves. A curve with a definite maximum in $n_x(t)$ is predicted by (14b) because $U_m \sim n_x^2$. On the other hand, Eq. (14c) predicts a saturation in n_x because $U_m \sim n_x$. The calculations of Lewis (1970b) in which a specific model of cluster mobility as a function of size was assumed, also gave a saturation rather than a maximum. A measurement of the full $n_x(t)$ curve when mobile clusters are important would help to decide which expression to use.

Cluster mobility will only be important when the clusters are sufficiently mobile to cause the maximum (or saturation) of $n_x(t)$ before coalescence takes place. The coverage Z_1 at which this maximum occurs can be very small, and Venables (1973a) has given examples from the work of Robinson and Robins (1970) where $Z_1 \lesssim 0.01$; this example is also given in Section VI. Further work on cluster mobility is undoubtedly needed: Stowell (1974) has recently explored a model where dimer mobility is included.

G. Cluster Size Distributions

In many ways, the ultimate test of a nucleation theory is whether it describes the observed cluster size distribution accurately. In the theory described here, information about the size distribution is implicitly contained in Eqs. (4) and (5) but this detail has been thrown away in making the transition to the simplified equation (7). This simplification is necessary, unless we are prepared to solve a very large (in theory infinite) set of rate equations. Hirth and Pound (1963) refer to Courteney's (1961) solution of the first 100 rate equations for three-dimensional nucleation; Abraham (1969) has studied 111 rate equations for the three-dimensional case; Robertson and Pound (1973) have classified clusters into 40 size classes and have solved this number of coupled rate equations for two-dimensional nucleation on a substrate. In any case, a full numerical calculation is a major undertaking.

Simplified treatments of the size distributions rest on various approximations. Zinsmeister (1969) has treated the case for $i = 1$ where the capture number σ_k of a stable k-cluster is assumed to be constant, and direct impingement, coalescence, and cluster mobility are neglected. In this case the distribution of the radii r_k of three-dimensional clusters $\rho(r_k, t)[= dn_k(t)/dr_k]$ is $\sim r_k^2$. This result can also be obtained from a simple steady-state argument; a general form of σ_k can be used and direct impingement can be included (Venables, 1973a). This makes $\rho(r_k, t) \sim r_k^n$, where n lies between 1.66 and 1.32 at low temperatures; at high temperatures (where direct impingement is most important) all radii become equally likely; this would also be true if diffusion into the substrate were dominant (Hamilton and Logel, 1973).

If coalescence and/or cluster mobility are included, the problem is much more complex, and closed treatments have not yet been given. Venables (1973a) has argued that it may be fairly easy to see qualitatively what will happen. In the coalescence case the clusters which coalesce will tend to be somewhat above average in size. Thus sizes above average will become depleted, and a peak at roughly twice this size will grow. The distribution will therefore probably be bimodal, since the small size peak will be fed by new nucleation. This type of effect has been seen experimentally by Sacédon (1972), Sacédon and Martin (1972), and Donohoe and Robins (1972). Vincent (1971) has shown theoretically when coalescence alone is considered that a solution may be obtained; but since his calculation did not include new nucleation, his distribution is unimodal. If mobile cluster effects predominate, the distribution will almost certainly be unimodal. This will be especially so if small clusters move faster than large ones, for then new nucleation is completely suppressed. Donohoe and Robins (1972) and Schmeisser (1974a) have observed distributions of this type. Unfortunately, it is difficult to be much more precise using the present formulation without detailed numerical computation. This is clearly an area in which much more work remains to be done.

H. The Theory of Layer Growth

The theory presented in some detail in the previous sections was developed largely for the description of nucleation and growth of cluster on a bare substrate, where the stable clusters have a compact three-dimensional form. As indicated in Section II the same theory could be used to discuss adequately the nucleation of the first monolayer in a layer-by-layer growth situation, as only the relative magnitudes of E_a and $3E_2$ are reversed (see Figs. 5b and c). The growth-rate equation appropriate to two-dimensional growth [Eq. (16b)] would have to be used, and the form of the solutions so obtained have been given by Stowell and Hutchinson (1971a). There is no doubt, however, that this description is not as satisfying or nearly as complete as in the three-dimensional case presented here; some of the difficulties have been discussed by Voorhoeve (1971). The reasons for this seem to be essentially threefold.

First, a general microscopic technique is not presently available for studying the spatial distribution of nuclei within the first of these monolayers over a sufficiently large area. Second, the first layers may well be in the form of a surface alloy or compound that may well require a highly chemical-specific theoretical description. Third, the nucleation and growth stages are not well separated in the layer-by-layer growth regime. While the first layer is well past the coalescence stage, the second layer will be nucleating on top of the first; in general, nucleation on the higher layers will always be in competition with growth at existing kink and ledge sites. In many practical cases of growth from the vapor (e.g., Ag on Cu and Au, Si on Si), unless the deposition is carried out at very high temperatures, the condensation will be complete and no practical effects of the nucleation stage will be noticeable. In particular, the thin film will grow uniformly, incorporating defects and/or misfit dislocations in order to match with the substrate (see Chapters 5 and 8), but nucleation itself will be unimportant.

At higher temperatures, where condensation is incomplete, it is possible for nucleation of new layers to become rate-limiting. In the simplified language of Fig. 5b, this might occur if $(R/N_0 \nu)\exp(\beta 3E_2) < 1$ but $(R/N_0 \nu)\exp(\beta 6E_2) > 1$. If the critical clusters were large, lateral growth of the layers would become rapid compared to nucleation; this would mean that most of the time was spent waiting for the nucleation of a new layer to occur, and that this nucleation *could* be described by the two-dimensional version of the theory presented here. It seems likely that some of the condensed gas layers studied by the present authors or chemically vapor-deposited layer compounds may grow in this fashion. Kaischew and Budevski (1967) and Budevski (1972) have given highly convincing demonstrations of this effect occurring during electrochemical deposition onto dislocation-free single crystals. At the very highest temperatures, i.e., lowest supersaturations, the surface itself should become rough due to configurational entropy effects (e.g. Jackson, 1967; Gilmer and Bennema, 1972a, b); whether crystal growth from the vapor can actually occur at a measurable rate at such low supersaturations is rather doubtful.

There are, however, several mechanisms that may prevent this type of nucleation from being observed. In the first place, this is the regime in which growth at emergent screw dislocations, first suggested by Frank (1949) and Burton *et al.* (1951), is most favorable. Second, as emphasized recently by Wagner and Voorhoeve (1971), if the slowest growing face of the growing crystal is not parallel to the substrate, then there will be corners where this face meets the substrate, and the crystal can grow from these points without a nucleation barrier. They gave evidence for this type of effect, for the case of cadmium growing on tungsten, by combining molecular beam experiments and scanning electron microscopy of the surface (see Section VI, A).

The complexities of this layer-by-layer growth regime are intriguing. This regime may not be the most common for people working on thin films, but it seems to be the one most similar to other types of crystal growth, such as growth from the melt (e.g., Jackson, 1967) and electrodeposition from solution (e.g., Bertocci, 1969, 1972). In the latter paper a computer-generated picture shows very nicely the combination of step motion and new nucleation that can occur in this regime: this figure is reproduced in Fig. 9. [Other examples are given by van Leeuwen *et al.* (1974)]. However, there is no technique available at the moment for observing this structure directly.

In the case of intermediate regimes, such as the Stranski–Krastanov growth mode, which begins as in Fig. 5b and converts to Fig. 5d or e, there are many unknown features and a quantitative theory has not been given. In particular it is unknown whether nucleation processes within or on top of the first few layers determine the observed structures or whether they are homogeneous or defect-induced.

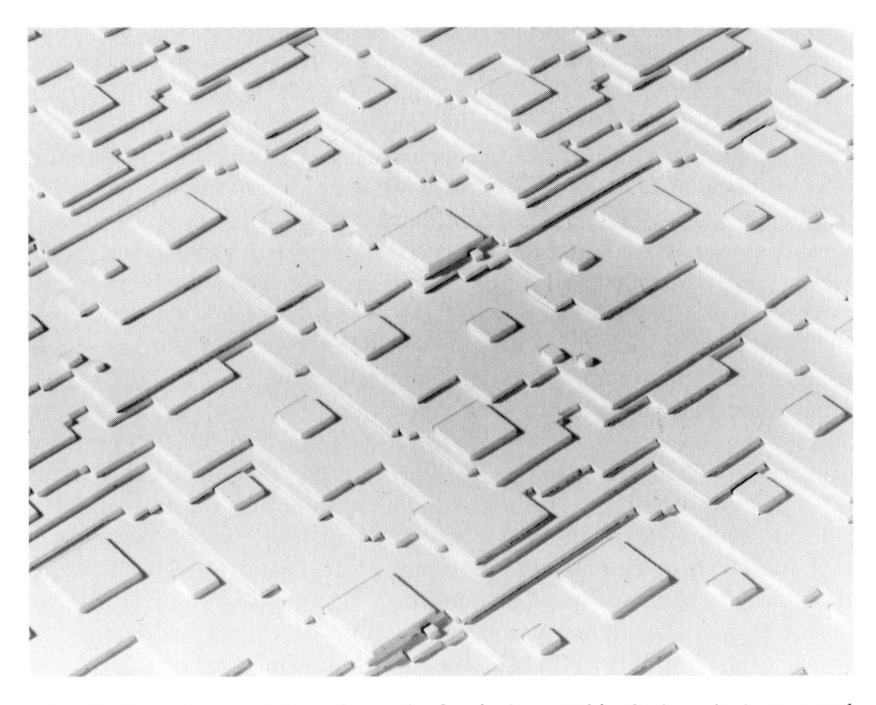

FIG. 9. Computer simulation of growth of a single crystal in the layer-by-layer growth regime (Bertocci, 1972). The picture shows random two-dimensional nucleation and step growth occurring simultaneously.

In Section VI a description of some experiments where the nucleation behavior can be followed will be given. Before that is done a brief description of the role of substrate defects on the nucleation process is given in the next section.

V. Effects of Imperfections

It is well known that imperfections can play a dominant role in the nucleation and growth of crystals. In the thin-film case, the most striking effects are the numerous examples of the decoration of cleavage steps, etch pit boundaries, and dislocation spirals by small stable clusters that have great difficulty in nucleating elsewhere on the substrate (e.g., Bassett, 1958; Bethge, 1969). The decoration of isolated point defects on the surface is also important. These point defects can be in the form of impurities or surface vacancies (e.g., Hennig, 1966) and may be created by quenching or by electron or ion bombardment (Palmberg *et al.*, 1967, 1968; Stirland, 1966, 1967, 1969; Stroud, 1972a, b). In all these examples, we are dealing with metals condensing onto insulators, and the effects are most marked in the incomplete condensation regime, illustrated in Figs. 5e and 5f, where the adsorption energy E_a is low, but E_2 is high.

Defects also have a strong influence on the layer growth case, and at low supersaturation, growth can often only occur when screw dislocations are present, as mentioned in the last section. It is commonly observed that dislocations and stacking faults present in the substrate are incorporated into crystals that grow from the vapor in a layer-by-layer fashion. However, it is doubtful that these defects are necessary for growth to occur, as in contrast to the case of growth from the liquid, such films are usually grown under conditions of high supersaturation with respect to the vapor. Impurities are also important in the layer growth regime, though they may primarily affect the subsequent growth rather than the nucleation stage. This is discussed briefly in Section VII and is not pursued here.

Little has been done toward a quantitative theory of nucleation at defect sites. As Frankl and Venables (1970) pointed out, the variety of possible influences is so vast that a comprehensive treatment would be virtually useless. They give, however, a discussion of some of the effects of point and line defects, which will not be repeated here. More recently, Stowell and Hutchinson (1971b) have attempted to treat the effect of N_d (cm^{-2}) isolated point defects on the substrate, which bind j-sized clusters by an extra energy ΔE_j. The equilibrium density of critical clusters of size i at defect sites n_{id} is related to the number n_{in} when no defects are present by a relation given by Rhodin and Walton (1963):

$$n_{id}/N_d = (n_{in}/N_0) \exp(\beta \Delta E_i) \qquad (46)$$

and is valid for $n_{id} \ll N_d \ll N_0$. The nucleation rate can therefore be written, analogously to Eq. (11), as

$$U_i = U_{in} + U_{id} = \sigma_{in} D n_{1n} n_{in} + \sigma_{id} D n_{1n} n_{id} \tag{47}$$

where σ_{in} and σ_{id} are capture numbers appropriate to normal and defect sites, and cluster mobility effects have been ignored. Inserting relation (46) gives

$$U_i = U_{in}(1+\alpha_{id}) \tag{48}$$

where

$$\alpha_{id} = (\sigma_{id}/\sigma_{in})(N_d/N_0)\exp(\beta\,\Delta E_i) \tag{49}$$

Thus the parameter α_{id} (Stowell and Hutchinson's α) determines the importance of defect sites.

The value of n_{in} is determined by an equation similar to (3) or its simplification (18). As pointed out by Stowell and Hutchinson (1971b), if the defects only trap the moving atoms weakly, then in steady-state conditions there are no extra terms in (18). Under these conditions the initial nucleation rate is increased by the factor $1+\alpha_{id}$, and in expressions for the maximum cluster density and time to the maximum such as (35), (36), (39), and (40), all factors in $\exp(\beta E_i)$ should be multiplied by $1+\alpha_{id}$.

However, this defect nucleation problem has not yet been solved rigorously, and one should beware of applying these equations in an uncritical fashion. For example, defect sites are only important if they influence the nucleation rate strongly, i.e., if $\alpha_{id} > 1$. This is then not the weak trapping limit. In order to approach the problem rigorously, extra loss terms should be incorporated into the equation for n_{1n}, and a rate equation set up for the density n_{1d} of single atoms trapped at defect sites. This extra equation could include the continuous generation of defect sites if necessary and may, in special cases, yield the result that the densities n_{1n} and n_{1d} are in equilibrium with each other.

If n_{1n} and n_{1d} are in local equilibrium and n_1 is the total single density, then

$$n_{1n} = n_1/(1+\alpha_{1d}) \qquad \text{and} \qquad n_{1d} = \alpha_{1d} n_1/(1+\alpha_{1d}) \tag{50}$$

where α_{1d} is defined analogously to α_{id} in Eq. (49). Thus, in the initial, transient nucleation stage, where $n_1 = Rt$, n_{1n} is reduced by the factor $1+\alpha_{1d}$, and this can be described equivalently by retaining n_1 and using an effective diffusion constant $D/(1+\alpha_{1d})$. This approach has been adopted by Frankl and Venables (1970) and Brown *et al.* (1969). In the strong trapping limit many of the clusters will be nucleated in the transient nucleation stage under conditions of complete condensation, but not when incomplete condensation conditions obtain. This means that none of the above treatments give an adequate general account of nucleation at defect sites, and that more work needs to be done on this problem. However, it is clear from the form of the equations for n_x given here

(Stowell and Hutchinson, 1971b; Frankl and Venables, 1970) that it will not always be easy to distinguish between homogeneous and defect-induced nucleation. Effective activation energies replace the pure substrate values and the absolute magnitudes of the expressions can be considerably different; but in the absence of reliable and independently known values to insert for these material parameters, the form of the expressions, in particular their dependence on R and T, are similar and cannot be easily distinguished.

The extreme case of strong trapping is represented by $i = 0$. In that case single atoms, once trapped, cannot escape from the defect site before another atom arrives, and all atoms that arrive subsequently stick to the growing cluster. Under these conditions, the *initial* nucleation rate is given by $U_{0d} = \sigma_{0d} D n_1 N_d$, and under incomplete condensation (steady-state) conditions this is

$$U_{0d} = \sigma_{0d} D N_d R \tau_a = (\sigma_{0d} \alpha N_d / N_0) R \exp \beta (E_a - E_d) \tag{51}$$

Thus a nucleation rate at high temperatures proportional to R is indicative of defect-induced nucleation. The temperature dependence will give $E_a - E_d$, or will be steeper than this if N_d is itself temperature-dependent, because a spectrum of defect sites with different binding energies exists on the substrate. If the clusters, once nucleated, are immobile, then the number of clusters saturates at $n_x = N_d$. In the complete condensation case this number will be reached in a very short time, but for incomplete condensation, the time scale is set by Eq. (51). If the clusters are mobile, there is no a priori reason why the maximum density of stable cannot be larger than N_d. It is clear that more work remains to be done on this problem, but a start has been made by Markov and Kaschiev (1973).

In the next section some recent experimental examples are described. These include examples of both island and layer growth and of homogeneous and defect-induced nucleation.

VI. Experimental Examples

A selective review of nucleation experiments to 1969 was given by Frankl and Venables (1970). In this section we shall concentrate on recent developments and particularly on those experiments that give quantitative information about the nucleation process. As mentioned in Section I, there are several techniques that can give quantitative information. The main techniques in use at present consist of:

(1) Measurement of desorbed atom fluxes by mass spectrometry and molecular beam techniques. Experiments of this type will be described in Section VI, A.

(2) Observation of single-atom mobility and of interactions between atoms by field ion microscopy. Work that has been done in this area is reviewed briefly in Section VI, B.

(3) Observation of cluster densities and growth morphologies by high-resolution electron microscopy. Several examples of this type of experiment will be described in Section VI, C.

In all cases we will attempt to see how these experimental examples relate to the conceptual and theoretical framework established in the previous sections.

A. Mass Spectrometry Studies

A particularly good example of the use of mass spectrometry (and replica electron microscopy) to study nucleation and growth of thin films is the work of Wagner and Voorhoeve (1971) on the growth of cadmium on clean tungsten. A source of Cd vapor and a mass spectrometer were arranged symmetrically with respect to a polycrystalline tungsten ribbon, in a UHV apparatus with a base pressure of 2×10^{-10} Torr. The mass spectrometer is used to measure the flux of Cd evaporating from the tungsten, and the type of information obtainable is shown in Fig. 10.

When the tungsten is at high temperature, the Cd vapor is accommodated but then reevaporates; measurement of the evaporation rate calibrates the rate of arrival, termed V_i in Fig. 10. At lower temperatures, the Cd crystals grow and the desorbing flux has a lower value V_{ei}. By means of a shutter the desorbing flux with the beam off V_0 can also be measured. The integrated difference $\int_0^t (V_i - V_{ei})\, dt$ enables one to calculate the amount of material condensed as a function of time t and this is referred to as the coverage σ in Fig. 10. [This is not the coverage in the same sense as the Z of Section IV.]

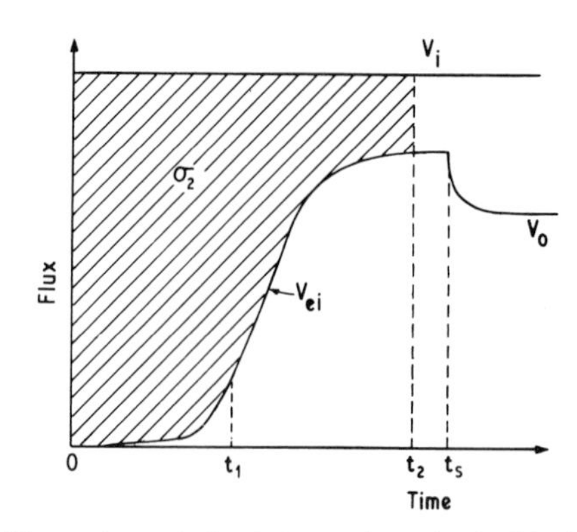

FIG. 10. Schematic illustration of the information obtainable in mass spectrometry experiments using a molecular beam (Wagner and Voorhoeve, 1971). (See text for discussion.)

By making detailed measurements as a function of T, R (i.e., V_i in their notation), and time, one is clearly finding out a great deal about the adatom population n_1 and lifetimes τ_a of the adsorbed Cd atoms, both on the bare substrate and on top of multilayers of solid Cd.

Wagner and Voorhoeve concluded that the Cd–W system is a case of layer-by-layer growth where the adsorption energy decreases by a factor of two within the first 3–4 monolayers to the bulk Cd value. Within the first monolayer there are at least five different adsorption energies, which are probably correlated with different tungsten crystal faces present in the polycrystalline substrate. There is no evidence for a nucleation barrier, in that the growth rate does not increase with time in the sudden fashion, which it would do if nucleation of each subsequent layer were rate-limiting. They ascribe this lack of a nucleation stage, however, to the strong influence of the substrate orientation on the subsequent growth. When the initial monolayers are forced, by the substrate, into an orientation that differs from that of the slowest growing face ((0001) in this case), this face will grow out of the initial deposit at an angle to the substrate and an island and/or ridge structure will be produced. They observed this type of structure, which depended on the orientation of the underlying tungsten grain, by replica and scanning electron microscopy. The Cd–W system thus seems to be a good example of Fig. 5b where the face from which atoms are reevaporating $[(R/N_0 \nu)\exp(\beta E_a') < 1]$ may be tilted with respect to the substrate. Wagner and Voorhoeve indicated that there might well have been a nucleation barrier if a single-crystal tungsten substrate had been used and (0001) cadmium grew parallel to the substrate surface: this would then be exactly the case illustrated in Fig. 5b.

Voorhoeve and Wagner (1971) also studied the growth of Cd on tungsten covered with monolayer quantities of oxygen and hydrogen. Although hydrogen had no effect, the effect of oxygen was dramatic. Less than a monolayer of oxygen strongly reduced the growth rate, and evidence of a nucleation barrier (a maximum in V_{ei} with time) was obtained. Scanning electron microscopy revealed three-dimensional clusters, and the adsorption energy of Cd was observed to drop to less than half the value on clean tungsten. Thus it seems that a strongly adsorbing impurity (O_2) is inducing a transition from the case illustrated in Fig. 5b to that of Fig. 5e by reducing the value of E_a. A further example of this type of effect seems to be the case of rare gases on "clean" and "dirty" graphite, described in Section VI, C, (d).

Voorhoeve *et al.* (1972a, b) have also studied Cd–Ge by the same methods; Anh *et al.* (1972) have studied In–Si; Lo and Hudson (1972) Ag–W. It is clear from these papers that mass spectrometry is a powerful way of measuring the total number of atoms condensed as a function of time, adsorption stay times, and adsorption energies. Coupled with a suitable microscopic technique, this is a most valuable way of studying nucleation and growth processes.

B. Field Ion Microscopy Observations

One of the most impressive and intriguing techniques for studying the initial stages of thin film growth is field ion microscopy (FIM). Individual atoms can be seen, and can be observed to move and to form clusters, which themselves can be observed to move, rearrange, and dissociate. There is clearly a great deal of quantitative information that can be obtained from such studies, which are only in their infancy. The main drawback to the technique is that the area under examination is minute (rarely as much as 100 Å square), and the materials that can be studied have been, until recently, limited to the high-melting-point refractory metals. Progress in instrumentation and image intensification techniques is, however, quite rapid, and FIM is becoming applicable to lower-melting-point metals such as gold.

Recently, the migration and interactions of Ta, Mo, W, Re, Ir, and Pt atoms on tungsten filed ion tips have been studied by two groups of authors (Bassett and Parsley, 1969, 1970; Bassett, 1970, 1973; Tsong, 1972, 1973). The technique involves evaporation of atoms onto a clean tip and taking pictures at low temperature, interspersed with heating the tip to a given constant temperature for a known length of time. As an example of this type of work, the diffusion of single Re atoms along a surface channel in a tungsten (123) plane is shown in Fig. 11 (Tsong, 1972). In this case a one-dimensional diffusion analysis gives $E_d = 0.88$ eV, a result that was also obtained by Bassett and Parsley (1969). In this type of observation one sees very directly that the diffusion coefficient (and E_d) is not isotropic, even on a cubic crystal, if the face itself does not have three- or fourfold symmetry.

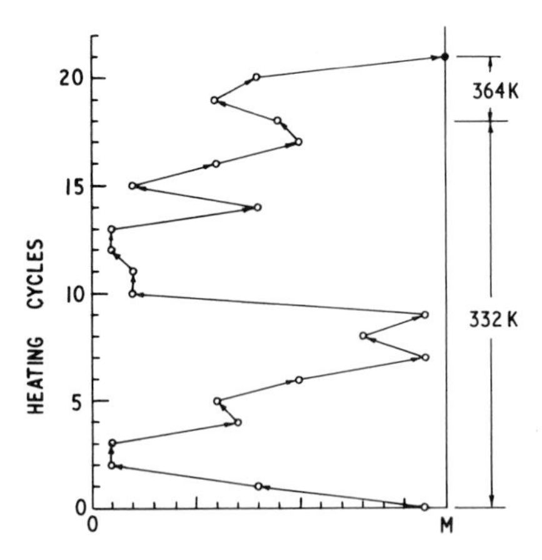

FIG. 11. One-dimensional discrete random walks of a Re atom on a tungsten (123) plane; the diffusion is along a surface channel. Each step is 2.74 Å and the channel contains a total of 17 steps (Tsong, 1972).

When this type of measurement is performed with more than one atom on a particular crystallographic face at a time, binding energies of clusters can be estimated by observing what proportion of the time atoms spend bound to each other. Surface traps can also be identified in this way, though the technique is not so direct if the trap itself is invisible. Binding energies† of dimers turn out to be surprisingly small [$\sim$0.1 eV for Ir_2, $\sim$0.14 eV for Pt_2 and 0.29 eV for W_2 on a W (110) plane (Bassett and Parsley, 1969; Tsong, 1973)] and Re_2 dimers are unstable though Re_3 clusters are stable. The effects observed depend critically on which face of the tungsten tip is investigated. The complexities exhibited by even the limited numbers of systems so far studied indicate the need for careful measurements and for considerable scepticism about reliance on simplified models (such as the ones presented here in Sections II and III). In particular, they underline the fact that a simple pair-binding model for evaluating binding energies of small clusters is quite inadequate for these refractory metal systems. Often small clusters consist of linear arrays of atoms, as, for example, Pt and Ir on (110) W (Tice and Bassett, 1974). A detailed model of the electronic interaction of the adsorbed atom with the substrate would be required to rationalize the diffusion energies and the binding energies of these clusters, and this is not currently available. Detailed reviews of this work have been given by Bassett (1970, 1973).

Another elegant example of the use of the FIM to study the initial stages of growth is the work of Nishikawa and Utsumi (1973a, b). The detailed positions of Ga atoms in the first layer of Ga condensed onto W and Mo can be seen clearly on several different crystal faces of the substrate. In the case of Ga–Mo a surface compound was observed whereas in Ga–W it was not; again this is a highly chemical-specific result. Epitaxy of the first layers of Pd, Pt, Rd, and Ir on Ir and Rh has also been observed over a wide temperature range by Graham *et al.* (1972).

The relationship of these FIM studies to nucleation of thin films is not clear at present. Since tungsten is so strongly bound to itself, most other metals that are condensed onto it will be less strongly bound; an extreme case is the Cd–W case described in Section VI, A. Most of these metals should therefore eventually grow in a layer-by-layer growth mode or possibly in the Stranski–Krastanov mode. It may be that, as in the Cd–W case, the orientation of the first layer determines, in a complex way, how the crystals eventually grow. Whether the orientation of this layer and the subsequent growth mode is eventually traceable to the form of the initial nucleus of a few atoms in size is, however, not known at present.

† In Tsong's (1972) paper, incorrect pair binding energies E_2 appear to have been deduced by associating E_2 with the energy needed for dissociation, which is actually (E_2+E_d). This correction (Tsong, 1973) brings his values of E_2 down to the same order of magnitude as those of Bassett and Parsley (1969).

C. Transmission Electron Microscopy Studies

High-resolution electron microscopy is, as indicated in Section I, probably the most useful technique presently available for the study of nucleation processes. Especially in the cases where clusters grow as compact three-dimensional islands, much quantitative information can be obtained about the density of stable clusters and their sizes, as illustrated in Fig. 1. In many ways the information is complementary to that obtainable from mass spectrometry (Section VI, A) since information about the single adatom density n_1 and corresponding adsorption times τ_a is the one piece of information *not* obtainable directly by electron microscopy. In the following subsections, recent work using transmission electron microscopy, divided according to the materials studied, is reviewed.

1. *Noble Metals on Alkali Halides*

A large amount of detailed work has been performed recently on the deposition of the noble metals, gold and silver, onto alkali halide surfaces, cleaved and annealed in UHV conditions. Robinson and Robins (1970) studied the Au–KCl system. More recently, Donohoe (1972), Donohoe and Robins (1972), Robins and Donohoe (1972), and Robinson and Robins (1974) have studied Au–NaCl, Ag–KCl, Ag–KBr, and Ag–NaCl. In addition Masson *et al.* (1971a) and Métois *et al.* (1972b) have studied the mobility of small clusters in these systems, and the data obtained by Robinson and Robins (1970) have been analyzed by Stowell (1972b) and Venables (1973a). Thus much information is available on these systems, and despite the known instabilities of alkali halide surfaces (e.g., Gallon *et al.*, 1970; Vermaak and Henning, 1970a, b) these data are probably the best available at the time of writing.

All the authors measured the nucleation rate as a function of the arrival rate R and substrate temperature T. In all cases where the nucleation rate was small enough to be measurable, they found it to be proportional to R^2. This is unequivocal evidence that the critical nucleus size $i = 1$, up to the highest temperatures. Robinson and Robins (1970) also measured the maximum cluster density as a function of T, for a fixed value of R and as a function of R for a few values of T. Donohoe and Robins (1972) have also determined the full $n_x(t)$ curve in several cases (as shown in Fig. 1), and have thereby extracted the time to the maximum as a function of temperature. These authors and Schmeisser and Harsdorff (1973) and Schmeisser (1974a,b) have also started to study size distributions and spatial correlations of clusters.

With all this information available, a start can be made to check in some detail the nucleation theory described in Sections III and IV. There are, of course, various ways to attempt this. Stowell (1972b) has chosen to analyze the

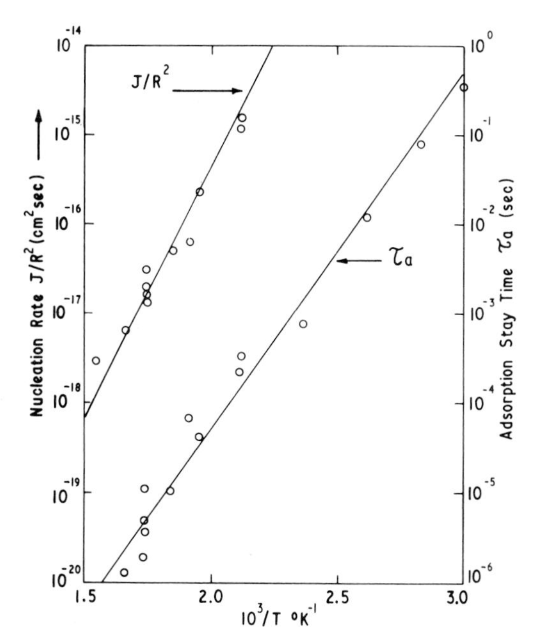

FIG. 12. Experimental nucleation rate data of Robinson and Robins (1970), plotted as J/R^2, and the values of τ_a derived from the maximum cluster density measurements (Stowell, 1972b).

Au–KCl data by Robinson and Robins (1970) by using the expression for the initial nucleation rate [Eq. (29)] and the maximum cluster density [Eq. (43)]. With σ_x and σ_i put equal to suitable constants, he then determined a value of τ_a at each temperature. The resulting values of the nucleation rate J divided by R^2 and the derived value of τ_a are shown in Fig. 12. The points give an idea of the experimental scatter and the straight lines correspond to the parameter values $E_a = 0.80$ eV, $E_d = 0.45$ eV, and $\nu = 2 \times 10^{12}$ Hz for the constant $\Omega N_0^{3/2}$ taken equal to one. Venables (1973a) analyzed the same data, but chose to use two nucleation rate points and two low-temperature points on the maximum cluster density versus $1/T$ curve to deduce the values of the four material parameters E_a, E_d, ν, and α [in the expression for D, Eq. (2)]. He used self-consistent values for σ_i and σ_x, calculated according to the methods outlined in Section III, D. On the assumption that gold atoms jump between two equivalent sites on the (100) KCl face ($\Omega N_0^{3/2} = 0.193$), this yielded parameter values $E_a = 0.69$ eV, $E_d = 0.28$ eV, $\nu = 1.2\ 10^{11}$ Hz, and $\alpha = 0.15$. If he assumed four such sites ($\Omega N_0^{3/2} = 0.545$), E_a, E_d, and ν were unchanged, but $\alpha = 0.30$. Use of different approximations for σ_i, σ_x, and different assumed values of the coverage at the maximum density Z_0 gave values of E_a and E_d that varied by less than 0.02 eV, and values of ν and α that varied by a factor of two. These curves for Au–KCl are shown in Fig. 13. Use of maximum and minimum slopes permitted by the data gave a wider spread with $0.65 < E_a < 0.74$ eV, $0.23 < E_d < 0.42$ eV, $0.6.10^{11} < \nu < 10^{12}$ Hz, and $0.1 < \alpha < 2$ for the Au–KCl data.

The values deduced by Stowell (1972b) lie at the upper end of the allowed range and, when plotted as in Fig. 13, give a rather steep slope at low temperature, and somewhat too large an activation energy for the nucleation rate. Venables (1973a) has argued that the lack of agreement at high temperatures in the Au–KCl case is due to the effect of cluster mobility, and that better agreement can be obtained using Eqs. (44) and (45). Strong evidence that mobile clusters are, in fact, involved comes from the observation that the coverage at which the maximum in the cluster density occurs is much lower than that needed to cause substantial coalescence. Indeed, Donohoe (1972) has even found that the time required to reach the maximum in the cluster density *decreased* at high temperatures as shown in Fig. 14 for Au–NaCl. If coalescence were the mechanism that determined the maximum cluster density then this time should *increase* at high temperatures as given by Eq. (36). Stowell (1974) has argued that the mobility of dimers is largely

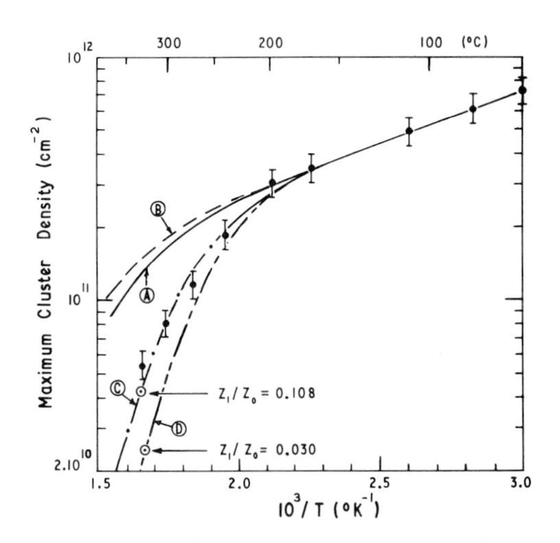

FIG. 13. Experimental maximum cluster density data of Robinson and Robins (1970), as analyzed by Venables (1973a) for $R = 3 \times 10^{12}$ cm^{-2} sec^{-1}. The curves have all been fitted to the nucleation rate, and to the maximum cluster density at low temperatures. Curve (A): uniform depletion approximation for σ_i and σ_x with $E_a = 0.69$ eV, $E_d = 0.28$ eV, $\alpha = 0.15$, $\nu = 1.2 \times 10^{11}$ Hz, $Z_0 = 0.21$. Curve (B): lattice approximation for σ_i and σ_x *or* $\sigma_i = 1.9$, $\sigma_x = 6.8$. Curves (C) and (D): all clusters mobile at high temperatures with cluster diffusion energy = 1.0 eV, curve (C) or 1.2 eV, curve (D). Cluster mobility results in the maximum density occurring at a reduced coverage Z_1 [see Eq. (45)].

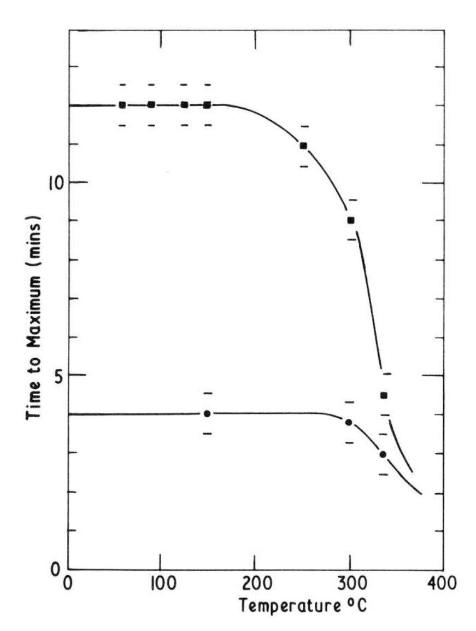

FIG. 14. The time required to reach the maximum in the $n_x(t)$ curve as a function of temperature for two different arrival rates, 3×10^{13} cm^{-2} sec^{-1} (●) and 1×10^{13} cm^{-2} sec^{-1} (■) (Donohoe, 1972).

responsible, but this alone is not sufficient to account for the reduced value of the coverage at the maximum cluster density.

More information can be obtained if one also measures the size distribution of clusters and/or the cluster separations (e.g., Schmeisser and Harsdorff, 1973, and Schmeisser, 1974a, b). As an example, two size distributions of Donohoe and Robins (1972) are shown in Fig. 15. As explored briefly by these authors and by Venables (1973a) [see Section IV, 6] these two distributions are probably typical of nucleation with coalescence (Fig. 15a) and of cluster mobility, which effectively suppresses new nucleation (Fig. 15b). The spatial distribution of clusters seems to give especially clear indication of mobile cluster effects (Schmeisser, 1974b), as does the evolution of the size distributions of Métois *et al.* (1974).

The binding energy of a pair of gold atoms in free space is about 2.23 eV (Zinsmeister, 1966). Since the value of E_a is much smaller than this for the alkali halide substrates ($\simeq 0.7$ eV), it is highly likely that the pair-binding energy on the substrate E_2 is of the same order as in free space. Hence it is understandable that the critical nucleus size is only one atom throughout the whole temperature range studied. These noble metal–alkali halide systems therefore definitely fall into the regimes illustrated in Figs. 5d and 5e; since $E_2 > E_a$, the clusters must be three-dimensional from the beginning.

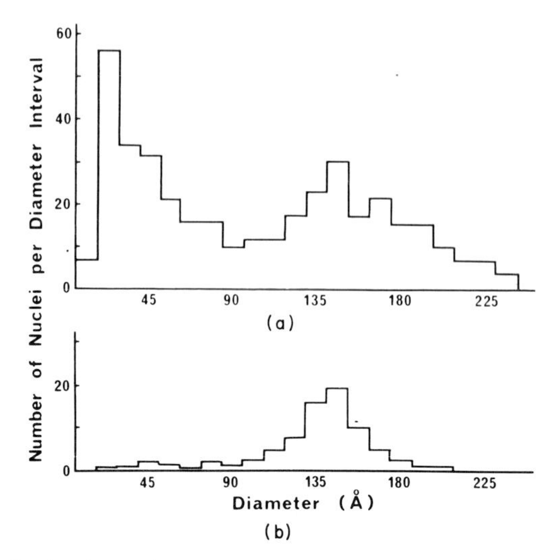

FIG. 15. Size distributions of clusters at two different temperatures; data on Au–NaC (Donohoe and Robins, 1972). (a) Low-temperature deposit, 150°C; 20 min at $R = 1 \times 10^{13}$ atoms cm^{-2} sec^{-1}. (b) High-temperature deposit, 300°C; 30 min at $R = 1 \times 10^{13}$ atoms cm^{-2} sec^{-1}.

2. *Noble Metals on Other Insulators*

Nucleation studies have also been done on the Ag–MoS_2 system (Corbett and Boswell, 1969), Au–graphite (Hennig, 1966), and Au–mica (Poppa *et al.*, 1971, 1972; Elliott, 1973a, b; Morris and Hines, 1970). In all these cases there is considerable evidence that E_a is even smaller than in the alkali halide cases discussed above. All the substrates are cleaved to expose close-packed faces with threefold symmetry, which is the orientation in which an adatom is bound to the substrate by the minimum number of "bonds." In Corbett and Boswell's work, the condensation was incomplete for temperatures as low as 300°K, and the nucleation rate was proportional to R. As discussed by Frankl and Venables (1970) and in Section V, this almost certainly corresponds to defect-induced nucleation with $i = 0$, and the number of defects and sites $N_d \simeq 10^{11}$ cm^{-2}. Analysis of the temperature dependence of the nucleation rate on the basis of Eq. (51) gave $E_a - E_d$ to be only 0.03 eV, which is, of course, an extremely small value. Since it is most unlikely that $E_d > 2E_a/3$, this also implies that $E_a \lesssim 0.1$ eV. This value is almost certainly too low (due to experimental uncertainty) as even the residual van der Waal's force between an adsorbed Ag atom and the MoS_2 surface would be expected to be in the range 0.1–0.2 eV. Nonetheless the result does indicate that defect-induced nucleation (Fig. 5f) is really the only possibility, since the density of single atoms on normal sites on the surface would be so low with this value of E_a. The coverage at the maximum density was also observed to be low, and this almost certainly is the result of cluster mobility. If the clusters are mobile, the defect sites will be regenerated, and this can help to explain that the maximum cluster density can be greater than the number of defect sites present. Mobility of clusters in this system has, in fact, been observed directly by Bassett (1964).

The more qualitative studies of the Au–graphite system by Hennig (1966) demonstrate conclusively that nucleation at defect sites predominates on clean graphite surfaces. Surface vacancies, surface steps, and even boron impurities have been decorated; the adsorption energy is almost certainly due to van der Waal's forces only with E_a in the order of 0.2 eV. The Au–MgO system studied by Robins and Rhodin (1964) is another example that can be analyzed in terms of $i = 0$, as is the recent work by Cinti and Chakravery (1972a, b) on Ag–silica. In this latter case, it seems that the "effective" number of defect sites is itself temperature-dependent, due to finite defect binding energies. The work of Cinti and Chakraverty and also that of Fujiwara and Terajima (1973) demonstrates that a sensitive microbalance is a very useful supplementary tool in nucleation studies, enabling one to measure the total amount of material condensed with considerable precision.

The recent studies of Poppa *et al.* (1971, 1972) and Elliott (1974) on the Au–mica system seem to show both defect and random nucleation occurring on the same samples. The defect sites quickly get used up, and then random nucleation proceeds at a slower rate. This random nucleation is proportional to R^p, where $p < 2$. This, therefore, seems to be a case of competition between nucleation with $i = 1$ and previous defect nucleation with $i = 0$. The number density of defect sites was temperature-dependent, which indicated a range of defect binding energies. The defects were found to be inhibited by cleaving in air (instead of UHV) and could be regenerated by preheating the substrate in oxygen. Although it seems that this system is in almost all respects intermediate between the Au–MoS_2 and Au–alkali halide systems described above, some further analysis is required to describe the case when defect-induced and random nucleation occur together, before full use can be made of the extensive amount of data available. The work of Morris and Hines (1970) also indicates regions with $i = 0$ and 1, although the appearance of $i = 0$ at high temperatures giving rise to a nucleation rate higher than that in the $i = 1$ case seems to indicate an increase in the number of defect sites with increasing temperature, in a way they were unable to explain.

A survey of recent work on the nucleation of Ag and Au on a whole range of substrates, including the ones discussed here, has been made by Bauer and Poppa (1972). This paper indicates very clearly that a large amount of new experimental information is becoming available. The results presented span the range from island growth to layer growth and included examples of intermediate cases as well. The complexities of the intermediate cases are well illustrated; as in the present chapter, it is really only the nucleation data in the island growth case that can be analyzed quantitatively at present.

3. *Growth of Semiconductor Layers*

The nucleation of Si on Si has been studied by Joyce and co-workers (Joyce and Bradley, 1966; Joyce *et al.*, 1967, 1969a, b; Booker and Joyce, 1966; Watts *et al.*, 1968). They studied the deposition of Si from a molecular beam of silane. Initially (Joyce *et al.*, 1967, 1969a) they observed islands of Si and the results had the right R and T dependencies for the current nucleation theory (Joyce *et al.*, 1967) for $i = 2$ on a (111) surface and $i = 3$ on (100). Despite this apparent agreement, the absolute magnitudes of the nucleation rate were about six orders of magnitude too low. In the last paper (Joyce *et al.*, 1969b), they reported that removal of a carbon impurity led to the growth proceeding in a smooth layer-by-layer fashion with no islands present at all. This result has recently been confirmed by Henderson and co-workers (Henderson and Helm, 1972; Henderson *et al.*, 1971). At this stage it is not entirely clear whether the carbon acts primarily as a nucleating center or as an inhibitor of regular layer growth, but the effect is certainly striking. As the surface was purified by heating in the range 1150–1250°C, island growth gave way to layer growth. It is worth mentioning that in cases of autoepitaxy such as these, the qualitative arguments of Section II suggest that layer growth (Fig. 4b) should be observed, and that this seems to occur once the experimental conditions are sufficiently clean. The cleanliness required is, however, rather daunting. Charig and Skinner (1969) and Henderson and Helm (1972) claim that surface carbon must be removed to well below the sensitivity limit ($\simeq 10^{13}$ atoms cm^{-2}) of present Auger electron spectroscopy techniques. A more detailed review of this work has been given recently by Joyce (1974).

4. *Rare Gas Crystals*

The condensation of rare gas crystals is currently being studied by the present authors (Price and Venables, 1972, 1974; Kramer and Venables, 1972), following previous work by Venables and Ball (1971). In the previous work, island growth was observed on both amorphous carbon and graphite substrates, and the density of stable clusters was observed to be about the same in both cases, despite a fair amount of scatter. The maximum density results were fitted to a nucleation theory, but, unlike other experiments reported, the parameter values E_a, E_d, and E_2 were obtained independently from semi-empirical calculations. Reasonable agreement was obtained between theory and experiment on the assumption of monolayer disk nuclei with $i = 2$ and 6. Despite this agreement, the problem remained that, with the high relative values of E_a and E_2 ($E_a \simeq 10E_2$), layer growth ought to have been observed. An ingenious, but not necessarily correct, model was put forward to explain why three-dimensional growth was observed. There was also some experimental uncertainty as to whether the graphite surface was really clean.

In repeating the work, Price and Venables (1974) found that cleaning the graphite by heating to 1500°C for a few minutes resulted in epitaxial layer growth of Xe and Ar at all temperatures studied, even below 10°K. On amorphous carbon that was also heated they found, however, island growth as before.

As in the Si–Si case described above, cleaning the surface of the graphite substrate results in the regime being observed that one would expect on the simple qualitative arguments given in Section II. In this sense the problem of understanding what is happening does seem to be getting simpler as time proceeds. Nonetheless, the drastic differences that are produced in all the experimental examples described by very small amounts of surface defects and impurities makes one cautious about being sure that a "final" understanding of nucleation and growth is just around the corner. This understanding is particularly tentative when one discusses the relationship of nucleation phenomena to epitaxy. A few possibilities are discussed in the next section.

VII. Nucleation and Epitaxy

The relevance of nucleation mechanisms to epitaxy and the production of good epitaxial films has long been a subject of debate. While much of what has been discussed in this chapter has little bearing on the problem, some recent experiments and ideas seem to shed some light, even though important anomalies and areas in which understanding is lacking remain. The island growth (Figs. 5d–f) and layer growth cases (Fig. 5b) will be discussed separately.

In the island growth case, there has been a strong move in recent years to the belief that epitaxy is a postnucleation phenomenon. Indeed, since critical nucleus sizes of 1 (or even 0 !) seem to be the most common (Section VI, C) this is almost a matter of definition, since one cannot sensibly speak of an "epitaxial" pair of atoms. However, this is more than mere semantics, and it is connected with the demonstration that clusters that from the nucleation point of view are quite "stable," can reorient themselves and move over the substrate. This type of movement has been seen by many authors and has been made into a quantitative study by Masson, Métois, Kern, and co-workers. They clearly believe that it is largely the migration (and rotation and coalescence) of small stable crystallites that eventually produces a film in epitaxial orientation. A review of this work has been given by Masson *et al.* (1971b).

Even more recently they have shown (Métois *et al.*, 1972) that in the case of Au–KCl (100) at low temperature (20°C) the (100) gold orientation results, on annealing at $T \gtrsim 200$°C from the coalescence of (111) oriented islands that migrate into each other and coalesce. There is thus no doubt that cluster rotation and migration processes are important in establishing epitaxial orientation. All the experiments performed by Masson *et al.* are in the form of a low-temperature deposition followed by annealing. If deposition were carried out at higher temperatures, as would normally be the case in the production of epitaxial films, these processes must be going on during deposition. However some doubt must be expressed about whether the migration–coalescence processes are really necessary to establish epitaxy in a high-temperature deposition. Clusters are often seen to be in epitaxial orientation at all (visible) stages of the deposition, and cluster rotation and some migration without significant coalescence may be all that is required.

Bombardment of the surface to produce surface defects has been shown to improve epitaxy and it is tempting therefore to devise specific models of the "epitaxial" nuclei (e.g., Palmberg *et al.*, 1968; Lewis and Jordan, 1970; Lewis, 1971; Lord and Prutton, 1974). However this may be misleading as bombarded surfaces, particularly of alkali halides, may exhibit a great range of chemical compositions and of microstructure (Gallon *et al.*, 1970; Chambers *et al.*, 1970; Elliott and Townsend, 1972). The one method of producing good epitaxial films from island clusters that seems to be soundly based is the "pulse–anneal" method. In this method the crystals are allowed to find their epitaxial orientation while they are still small, either by annealing a high density of clusters at a temperature higher than the subsequent growth temperature (Chadderton and Anderson, 1968) or by interrupting the stream of the condensing material (Harsdorff, 1968). It could be that bombardment of the surface functions in a similar way, by causing a higher density of nucleating sites, and hence relatively slower growth rate of each cluster, and also by increasing cluster mobility, but it is clear that this remains a controversial subject.

In the layer-growth regime, the indications are, as outlined in Sections VI, C, 3 and VI, C, 4 that epitaxy results in almost all circumstances provided the substrate surface is clean enough. In this regime the substrate has a very strong influence on the form of the thin film produced, and the growing film has little option but to choose the best (i.e., necessarily epitaxial) orientation in which to grow. A recent example of this is a LEED examination of growth of xenon crystals on iridium by Ignatiev *et al.* (1972). The xenon crystals grow in (111) orientation on both the 1×1 and the 1×5 Ir (100) surface structures but in quite different azimuthal orientations in the two cases. The slowest growing face of xenon will also be (111), which is parallel to the substrate (see Figs. 4b and c). The case of Cd–W discussed in Section VI, A seems to be similar, except that the slowest growing face (0001) is not necessarily parallel to the tungsten surface. Thus these "epitaxial" films will not be smooth, as is the case in the other examples (Xe–graphite, Xe–Ir, Si–Si) quoted.

The reasons why impurities (maybe in less than 0.01 monolayer concentration) can destroy smooth layer-by-layer growth is at present not at all clear. The impurities could affect the nucleation kinetics, or the subsequent growth by reducing the binding energy at kink sites, or conceivably by favoring twin or stacking fault formation. In the Si–Si case, Joyce *et al.* (1969b) and Henderson and Helm (1972), while acknowledging that the evidence is slight, favor an explanation in terms of the poisoning of growth due to adsorption of impurities at kink sites at the edges of the growing layers. It is worth noting that, though the crystals grow in the form of islands in this case, they are still in epitaxial orientation. In order to destroy the epitaxial growth itself, more defects and/or impurities must be present than in the case discussed above.

Despite these general trends, there are some experimental examples that are difficult to fit into this framework at present. Perhaps the best known example is the work of Distler (1970) and Henning (1970) who have observed epitaxial growth of metals on top of amorphous carbon layers (40–100 Å thick) on crystalline substrates. An explanation in terms of long-range electrical interactions has been offered by Distler. The observations themselves are however controversial and Dümler and Marrapodi (1972) and others have shown that 10 Å of carbon is sufficient to destroy epitaxy in the case of Au on Ag, and explain the persistence of epitaxy in terms of holes in the carbon films.

Another example is the observation of epitaxy of gold islands on UHV-cleaved ZnO at 20°K by Wassermann and Polacek (1970, 1971). One possible explanation is simply that the surface diffusion energy E_d is so low—say $\lesssim 0.03$ eV—that there is no problem. This might be so, since other similar systems such as Ag–MoS_2 and possibly Au–MgO, described in Section VI, B, seem to have very low values of E_a and E_d. On the other hand, cleavage of an insulator at low temperature may well produce a surface with locally trapped electric charges; the resulting long-range fields may be substantial and may take a very long time to decay. In such a case the forces between charged atoms and islands would be large, quite enough to cause island formation without thermal activation. Some estimates of the magnitude of these effects have been made by Marcus and Joyce (1971, 1972).

The relation between nucleation processes and the occurrence of epitaxy is, therefore, not yet completely clear. However, in the island growth case, the evidence strongly suggests that epitaxy is a postnucleation phenomenon involving rotation, migration, and rearrangement of "stable" clusters. In the layer growth case, it seems probable that one can "destroy" epitaxy by influencing the growth process at a later stage with impurities or defects incorporated into the growing film. The initial layers are, however, more or less forced to be related epitaxially to the substrate, provided that surface diffusion is sufficiently rapid that we are not dealing with the growth of "amorphous" layers.

Finally, the influence of long-range electric fields at the substrate surface, which have not been discussed in this chapter at all, may have to be considered in the future in order to account for some of the current "anomalies."

ACKNOWLEDGMENTS

The authors are most grateful to D. W. Bassett, U. Bertocci, A. G. Elliott, B. A. Joyce, R. Kern, B. Mutaftschiev, J. L. Robins, M. J. Stowell, T. T. Tsong, and R. J. H. Voorhoeve for permission to use figures from their papers and for useful correspondence and conversations. Thanks are due to Mrs. L. Lammiman and Mme S. Hanania for typing the manuscript.

References

Abraham, F. F. (1969). *J. Chem. Phys.* **51**, 1632.
Abraham, F. F., and White, G. M. (1970a). *J. Appl. Phys.* **41**, 1841.
Abraham, F. F., and White, G. M. (1970b). *J. Appl. Phys.* **41**, 5348.
Adams, A. C., and Jackson, K. A. (1972). *J. Crystal Growth* **13/14**, 144.
Allpress, J. G., and Sanders, J. V. (1970). *Aust. J. Phys.* **23**, 23.
Anh, N. G. T., Cinti, R., and Chakraverty, B. K. (1972). *J. Crystal Growth* **13/14**, 174.
Barna, A., Barna, P. B., and Pocza, J. F. (1971). *Proc. Int. Conf. Amorphous Liquid Semicond. 4th*, Ann Arbor.
Bassett, D. W. (1970). *Surface Sci.* **23**, 240.
Bassett, D. W. (1973). *In* "Surface and Defect properties of Solids" (M. W. Roberts and J. M. Thomas, eds.), Vol. 2. Chem. Soc., London.
Bassett, D. W., and Parsley, M. J. (1969). *Nature (London)* **221**, 1046.
Bassett, D. W., and Parsley, M. J. (1970). *J. Phys.* D **3**, 707.
Bassett, G. A. (1958). *Phil. Mag.* **3**, 1042.
Bassett, G. A. (1964). *In* "Condensation and Evaporation of Solids" (E. Rutner, P. Goldfinger, J. P. Hirth, eds.), p. 599. Gordon and Breach, New York.
Bauer, E. (1958). *Z. Kristallogr.* **110**, 372.
Bauer, E., and Poppa, H. (1972). *Thin Solid Films* **12**, 167.
Bethge, H. (1969). *In* "Molecular Processes on Solid Surfaces" (E. Drauglis, R. D. Gretz, and R. I. Jaffee, eds.), p. 569. McGraw-Hill, New York.
Bertocci, U. (1969). *Surface Sci.* **15**, 286.
Bertocci, U. (1972). *J. Electrochem. Soc.* **119**, 822.
Binsbergen, F. L. (1972a). *J. Cryst. Growth* **13/14**, 44.
Binsbergen, F. L. (1972b). *J. Crystal Growth* **16**, 249.
Booker, G. R., and Joyce, B. A. (1966). *Phil. Mag.* **14**, 301.
Brown, L. M., Kelly, A., and Mayer, R. M. (1969). *Phil. Mag.* **19**, 721.
Budevski, E. (1972). *J. Crystal Growth* **13/14**, 93.
Burton, W. K., Cabrera, N., and Frank, F. C. (1951). *Phil. Trans. Roy. Soc.* **A243**, 299.
Chadderton, L. T., and Anderson, M. (1968). *Thin Solid Films* **1**, 229.
Chambers, A., Lord, D. G., and Prutton, M. (1970). *Thin Solid Films* **6**, R1.
Charig, J. M., and Skinner, D. K. (1969). *Surface Sci.* **15**, 277.
Cinti, R. C., and Chakraverty, B. K. (1972a). *Surface Sci.* **30**, 109.
Cinti, R. C., and Chakraverty, B. K. (1972b). *Surface Sci.* **30**, 125.

Corbett, J. M., and Boswell, F. W. (1969). *J. Appl. Phys.* **40**, 2663.
Courtney, W. G. (1961). Texas Inst. Inc. Rep. TM–1250 (unpublished).
Dave, J. V., and Abraham, F. F. (1971). *Surface Sci.* **26**, 557.
Dettman, K. (1972). Private communication.
Distler, G. I. (1970). *Krist. Tech.* **5**, 73.
Donohoe, A. J. (1972). PhD. Thesis, Univ. of Western Australia.
Donohoe, A. J., and Robins, J. L. (1972). *J. Crystal Growth* **17**, 70.
Dümler, I., and Marrapodi, M. R. (1972). *Thin Solid Films* **12**, 279.
Elliott, A. G. (1974). *Surface Sci.* **44**, 337.
Elliott, D. J., and Townsend, P. D. (1971a). *Phil. Mag.* **23**, 249.
Elliott, D. J., and Townsend, P. D. (1971b). *Phil. Mag.* **23**, 261.
Frank, F. C. (1949). *Discuss. Faraday Soc.* **5**, 48.
Frankl, D. R., and Venables, J. A. (1970). *Advan. Phys.* **19**, 409.
Fujiwara, S., and Terajima, H. (1973). *Phil. Mag.* **27**, 853.
Gallon, T. E., Higginbotham, I. G., Prutton, M., and Tokutaka, H. (1970). *Surface Sci.* **21**, 224.
Gillet, E., and Gillet, M. (1969). *Thin Solid Films* **4**, 171.
Gillet, E., and Gillet, M. (1972). *J. Crystal Growth* **13**/**14**, 212.
Gillet, E., and Gillet, M. (1973). *Thin Solid Films* **15**, 249.
Gilmer, G. H., and Bennema, P. (1972a). *J. Crystal Growth* **13**/**14**, 148.
Gilmer, G. H., and Bennema, P. (1972b). *J. Appl. Phys.* **43**, 1347.
Graham, W. R., Reed, D. A., and Hutchinson, F. (1972). *J. Appl. Phys.* **43**, 2951.
Gretz, R. D. (1969). *In* "Molecular Processes on Solid Surfaces" (E. Drauglis, R. D. Gretz, R. I. Jaffee, eds.), p. 425. McGraw-Hill, New York.
Gueghuen, P., Cahoreau, M., and Gillet, M. (1973). *Thin Solid Films* **16**, 27.
Halpern, V. (1969). *J. Appl. Phys.* **40**, 4627.
Hamilton, J. F., and Logel, P. C. (1973). *Thin Solid Films* **16**, 49.
Harsdorff, M. (1968). *Z. Naturforsch.* **23**, 1253.
Henderson, R. C., and Helm, R. F. (1972). *Surface Sci.* **30**, 310.
Henderson, R. C., Marcus, R. B., and Politto, W. J. (1971). *J. Appl. Phys.* **42**, 1208
Hennig, G. R. (1966). *Chem. Phys. Carbon* **6**, 1.
Henning, C. A. O. (1970). *Nature* (*London*) **227**, 1129.
Hirth, J. P., and Moazed, K. L. (1967). *Phys. Thin Films* **4**, 97.
Hirth, J. P., and Pound, G. M. (1963). "Condensation and Evaporation—Nucleation and Growth Kinetics." Pergamon, Oxford.
Hoare, M. R., and Pal, P. (1971a). *Nature* (*London*) **230**, 5.
Hoare, M. R., and Pal, P. (1971b). *Advan. Phys.* **20**, 161.
Hoare, M. R., and Pal, P. (1972a). *Nature* (*London*) **236**, 35.
Hoare, M. R., and Pal, P. (1972b). *J. Crystal Growth* **17**, 77.
Hudson, J. B., and Sandejas, J. S. (1969). *Surface Sci.* **15**, 257.
Ignatjevs, A., Jones, A. V., and Rhodin, T. N. (1972). *Surface Sci.* **30**, 573.
Ino, S. (1966). *J. Phys. Soc. Japan* **21**, 346.
Ino, S. (1969). *J. Phys. Soc. Japan* **27**, 941.
Ino, S., and Ogawa, S. (1967). *J. Phys. Soc. Japan* **22**, 1365.
Jackson, K. A. (1967). *Progr. Solid State Chem.* **4**, 53.
Johannesson, T., and Persson, B. (1970a). *Phys. Scripta* **2**, 309.
Johannesson, T., and Persson, B. (1970b). *Phys. Status Solidi* K 251.
Joyce, B. A., and Bradley, R. R. (1966). *Phil. Mag.* **14**, 289.
Joyce, B. A., Bradley, R. R., and Booker, G. R. (1967). *Phil. Mag.* **15**, 1167.
Joyce, B. A., Bradley, R. R., Watts, B. E., and Booker, G. R. (1969a). *Phil. Mag.* **19**, 403.

Joyce, B. A., Neave, J. H., and Watts, B. E. (1969b). *Surface Sci.* **15**, 1.
Joyce, B. A. (1974). *Rep. Prog. Phys.* **37**, 363.
Kaischew, R., and Budevski, E. (1967). *Contemp. Phys.* **8**, 489.
Kelly, R., and Naguib, H. M. (1970). *J. Nucl. Mater.* **36**, 293.
Kern, R. (1968). *Bull. Soc. Fr. Mineral. Cryst.* **91**, 247.
Kern, R., Masson, A., and Métois, J. J. (1971). *Surface Sci.* **27**, 483.
Kramer, H. M., and Venables, J. A. (1972). *J. Crystal Growth* **17**, 329.
Komoda, T. (1968). *Jap. J. Appl. Phys.* **7**, 27.
Lewis, B. (1967). *Thin Solid Films* **1**, 85.
Lewis, B. (1970a). *Surface Sci.* **21**, 273.
Lewis, B. (1970b). *Surface Sci.* **21**, 289.
Lewis, B. (1971). *Thin Solid Films* **7**, 179.
Lewis, B., and Campbell, D. S. (1967). *J. Vac. Sci. Tch.* **4**, 209.
Lewis, B., and Jordan, M. R. (1970). *Thin Solid Films* **6**, 1.
Lewis, B., and Rees, G. J. (1974). *Phil. Mag.* **29**, 1253.
Lo, C. M., and Hudson, J. B. (1972). *Thin Solid Films* **12**, 261.
Logan, R. M. (1969). *Thin Solid Films* **3**, 59.
Lord, D. G., and Prutton, M. (1974). *Thin Solid Films* **21**, 341.
Marcus, R. B., and Joyce, W. B. (1971). *Thin Solid Films* **7**, R3.
Marcus, R. B., and Joyce, W. B. (1972). *Thin Solid Films* **10**, 1.
Markov, I. (1971). *Thin Solid Films* **8**, 281.
Markov, I., and Kaschiev, D., (1973). *Thin Solid Films* **15**, 181.
Masson, A., Métois, J. J., and Kern, R. (1971a). *Surface Sci.* **27**, 463.
Masson, A., Métois, J. J., and Kern, R. (1971b). *In* "Advances in Epitaxy and Endotaxy" (V. Ruth and H. G. Schneider, eds.), Chapter 2.2. V.E.B. Deutscher Verlag für Grundstoffindustrie, Leipzig.
Mayer, H. (1971). *In* "Advances in Epitaxy and Endotaxy" (V. Ruth and H. G. Schneider, eds.), Ch. 1.2. V.E.B. Deutscher Verlag für Grundstoffindustrie, Leipzig.
McDonald, J. E. (1962). *Amer. J. Phys.* **30**, 870.
McDonald, J. E. (1963). *Amer. J. Phys.* **31**, 31.
Métois, J. J., Gauch, M., Masson, A., and Kern, R. (1972a). *Surface Sci.* **30**, 43.
Métois, J. J., Gauch, M., Masson, A., and Kern, R. (1972b). *Thin Solid Films* **11**, 205.
Métois, J. J., Zanghi, J. C., Erre, R., and Kern, R. (1974). *Thin Solid Films* **22**, 331.
Michaels, A. I., Pound, G. M., and Abraham, F. F. (1974). To be published.
Morris, W. L., and Hines, R. L. (1970). *J. Appl. Phys.* **41**, 2231.
Niedermayer, R. (1971). *In* "Advances in Epitaxy and Endotaxy" (V. Ruth and H. G. Schneider, eds.), Chapter 1.1. V.E.B. Deutscher Verlag für Grundstoffindustrie, Leipzig.
Nishikawa, O., and Utsumi, T. (1973a). *J. Appl. Phys.* **44**, 945.
Nishikawa, O., and Utsumi, T. (1973b). *J. Appl. Phys.* **44**, 955.
Ogawa, S., and Ino, S. (1972). *J. Crystal Growth* **13/14**, 48.
Palmberg, P. W., Todd, C. J., and Rhodin, T. N. (1967). *Appl. Phys. Lett.* **10**, 122.
Palmberg, P. W., Todd, C. J., and Rhodin, T. N. (1968). *J. Appl. Phys.* **39**, 4650.
Parsons, J. R., and Balluffi, R. W. (1964). *J. Phys. Chem. Solids* **25**, 263.
Pollock, H. M. M. (1972). *Thin Solid Films* **14**, 193.
Poppa, H., Heinemann, K., and Elliott, A. G. (1971). *J. Vac. Sci. Technol.* **8**, 471.
Poppa, H., Moorhead, R. D., and Heinemann, K. (1972). *Nucl. Instrum. Methods* **102**, 521.
Price, G. L., and Venables, J. A. (1972). *Proc. Eur. Reg. Conf. Electron. Microsc., 5th, Manchester* p. 338. Inst. of Phys., London.
Price, G. L., and Venables, J. A. (1974). *Surface Sci.* (in press).

Rhodin, T. N., and Walton, D. (1963). *In* "Metal Surfaces" (W. D. Robertson and M. A. Gjostein, eds.), p. 259. Amer. Soc. for Testing Metals, Philadelphia, Pennsylvania.
Robertson, D. (1973). *J. Appl. Phys.* **44**, 3924.
Robertson, D., and Pound, G. M. (1973). *J. Crystal Growth* **19**, 269.
Robins, J. L., and Donohoe, A. J. (1972). *Thin Solid Films* **12**, 255.
Robins, J. L., and Rhodin, T. N. (1964). *Surface Sci.* **2**, 346.
Robinson, V. N. E., and Robins, J. L. (1970). *Thin Solid Films* **5**, 313.
Robinson, V. N. E., and Robins, J. L. (1974). *Thin Solid Films* **20**, 155.
Routledge, K. J., and Stowell, M. J. (1970). *Thin Solid Films* **6**, 407.
Rudee, M. L. (1972). *Thin Solid Films* **12**, 207.
Rudee, M. L., and Howie, A. (1972). *Phil. Mag.* **25**, 1001.
Sacédon, J. L. (1972). *Thin Solid Films* **12**, 267.
Sacédon, J. L., and Martin, C. S. (1972). *Thin Solid Films* **10**, 99.
Sato, H., and Shinozaki, S. (1970). *J. Appl. Phys.* **41**, 3165.
Schmeisser, H. (1974a). *Thin Solid Films* **22**, 83.
Schmeisser, H. (1974b). *Thin Solid Films* **22**, 99.
Schmeisser, H., and Harsdorff, M. (1970). *Z. Naturforsch.* **257**, 1896.
Schmeisser, H., and Harsdorff, M. (1973). *Phil. Mag.* **27**, 739.
Schwabe, U. and Hayek, K. (1972). *Thin Solid Films* **12**, 403.
Sigsbee, R. A. (1971). *J. Appl. Phys.* **42**, 3904.
Sigsbee, R. A., and Pound, G. M. (1967). *Advan. Colloid Interface Sci.* **1**, 335.
Stirland, D. J. (1966). *Appl. Phys. Lett.* **8**, 326.
Stirland, D. J. (1967). *Thin Solid Films* **1**, 447.
Stirland, D. J. (1969). *Appl. Phys. Lett.* **15**, 86.
Stowell, M. J. (1970). *Phil. Mag.* **21**, 125.
Stowell, M. J. (1972a). *Phil. Mag.* **26**, 349.
Stowell, M. J. (1972b). *Phil. Mag.* **26**, 375.
Stowell, M. J. (1974). *Thin Solid Films* **21**, 91.
Stowell, M. J., and Hutchinson, T. E. (1971a). *Thin Solid Films* **8**, 41.
Stowell, M. J., and Hutchinson, T. E. (1971b). *Thin Solid Films* **8**, 411.
Stoyanov, S. (1973). *Thin Solid Films* **18**, 91.
Stranski, I. N., and Krastanov, L. (1938). *Acad. wiss Math.-Nat. KlIIb, Deut.* **146**, 797.
Stroud, P. T. (1972a). *Thin Solid Films* **9**, 273.
Stroud, P. T. (1972b). *Thin Solid Films* **11**, 1.
Terajima, H., Ozawa, S., and Fujiwara, S. (1973). *Thin Solid Films* **18**, S7.
Tice, D. R., and Bassett, D. W. (1974). *Thin Solid Films* **20**, 537.
Tillett, P. I., and Heritage, M. B. (1971). *J. Phys. D* **4**, 773.
Tsong, T. T. (1972). *Phys. Rev. B* **6**, 417.
Tsong, T. T. (1973). *Phys. Rev. B* **7**, 4018.
van Leeuwen, C., van Rosmalen, R., and Bennema, P. (1974). *Surface Sci.* **44**, 213.
Venables, J. A. (1973a). *Phil. Mag.* **27**, 697.
Venables, J. A. (1973b). *Thin Solid Films* **18**, S11.
Venables, J. A., and Ball, D. J. (1971). *Proc. Roy. Soc. Ser. A* **322**, 331.
Vermaak, J. S., and Henning, C. A. O. (1970a). *Phil. Mag.* **22**, 269.
Vermaak, J. S., and Henning, C. A. O. (1970b) *Phil. Mag.* **22**, 281.
Vincent, R. (1971). *Proc. Roy. Soc. Ser. A* **321**, 53.
Voorhoeve, R. J. H. (1971). *Surface Sci.* **28**, 145.
Voorhoeve, R. J. H., Carides, J. N., and Wagner, R. S. (1972a). *J. Appl. Phys.* **43**, 4876.
Voorhoeve, R. J. H., Carides, J. N., and Wagner, R. S. (1972b). *J. Appl. Phys.* **43**, 4886.
Voorhoeve, R. J. H., and Wagner, R. S. (1971). *Met. Trans.* **2**, 3421.

Wagner, R. S., and Voorhoeve, R. J. H. (1971). *J. Appl. Phys.* **42**, 3948.
Walton, D. (1962). *J. Chem. Phys.* **37**, 2182.
Walton, D., Rhodin, T. N., and Rollins, R. W. (1963). *J. Chem. Phys.* **38**, 2698.
Wassermann, E. F., and Polacek, K. (1970). *Appl. Phys. Lett.* **17**, 259.
Wassermann, E. F., and Polacek, K. (1970). *Surface Sci.* **28**, 77.
Watts, B. E., Bradley, R. R., Joyce, B. A., and Booker, G. R. (1968). *Phil. Mag.* **17**, 1163.
Zinsmeister, G. (1966). *Vacuum* **16**, 529.
Zinsmeister, G. (1968). *Thin Solid Films* **2**, 497.
Zinsmeister, G. (1969). *Thin Solid Films* **4**, 363.
Zinsmeister, G. (1970). *Krist. Tech.* **5**, 207.
Zinsmeister, G. (1971). *Thin Solid Films* **7**, 51.

CHAPTER 5

DEFECTS IN EPITAXIAL DEPOSITS

M. J. Stowell

Tube Investments Research Laboratories
Hinxton Hall, Hinxton, Saffron Walden
Essex, England

I. Introduction

Epitaxial deposits usually contain much higher densities of crystalline defects than do crystals grown by other methods. Also, the substrate imposes constraints on the deposit that are not met in other modes of crystal growth, and this sometimes leads to configurations of defects that are peculiar to epitaxial films. These factors lead to problems that are of both academic and technological interest. In this chapter, the most commonly encountered defects will be characterized, and emphasis will be placed on discussing by what means they are introduced during growth of epitaxial films. The defects considered here are the usual crystalline imperfections such as dislocations, stacking faults, and twins; in ordered alloys and in compound deposits, superdislocations and antiphase domain boundaries are also found. All these imperfections have been well studied in bulk crystals, so there is no need to detail their characteristics. Brief descriptions of their basic geometrical properties are given and the reader is referred to standard texts for further details.

Because the mode of growth of an epitaxial deposit is an important factor that influences the types of defects found and their densities, it is relevant to review briefly some of the prominent features of epitaxial growth mechanisms before considering the crystalline defects that are the main subject of this chapter.

II. Nucleation and Growth of Epitaxial Films

The mode of nucleation and initial growth of epitaxial films is governed strongly by the bonding between deposit and substrate. It is convenient to employ surface and interfacial free energies in the simplified discussion of the initial phase of film formation. If the deposit is in the form of a hemispherical cap, then, at equilibrium,

$$\sigma_{sv} = \sigma_{xs} + \sigma_{xv} \cos\theta$$

where σ_{sv}, σ_{xs}, and σ_{xv} are the free energies of the substrate–vapor, deposit–substrate, and deposit–vapor interfaces and θ is the contact angle. Three categories of initial film formation have been discussed in the literature (e.g., Bauer, 1958), and these are schematically illustrated in Fig. 1.

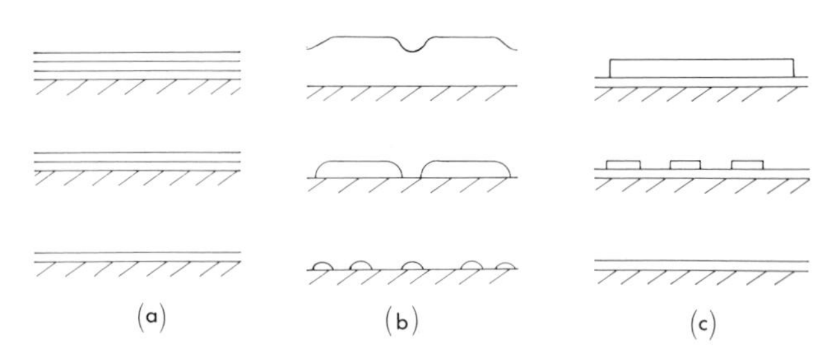

FIG. 1. Three forms of epitaxial growth: (a) layer-by-layer growth; (b) formation and growth from discrete nuclei; (c) initial formation of a tightly bound layer (or layers) followed by nucleation and growth.

The deposit "wets" the substrate when $\theta = 0$, i.e., $\sigma_{sv} = \sigma_{xs} + \sigma_{xv}$, and in this case it is to be expected that growth will occur in a layer-by-layer manner (Fig. 1a), as discussed by Frank and van der Merwe (1949). Ideal "autoepitaxy," for which $\sigma_{xs} = 0$, is a special case of this condition. When $\sigma_{xs} + \sigma_{xv} > \sigma_{sv}$, θ is finite; the deposit forms by the nucleation of discrete particles that grow three dimensionally (Fig. 1b) and eventually agglomerate into a continuous film. As θ increases, the bonding between substrate and deposit becomes progressively weaker than the cohesive binding in the deposit particles. The third category corresponds to $\sigma_{xs} + \sigma_{xv} < \sigma_{sv}$, and here the deposit–substrate binding is stronger than that within the deposit. Initially, the deposit grows as strongly bound layers on top of which discrete nuclei may form at a later stage, as depicted in Fig. 1c.

These arguments are, of course, over simplified, especially as we have neglected any anisotropy of the interfacial free energies. As Matthews (1967) has noted, these criteria often break down if one employs surface free energies measured on bulk specimens, but this general discussion serves to indicate at least qualitatively the dependence of the initial stages of thin-film growth on binding forces.

In all three of these cases, it is possible for the initial deposit to be strained elastically to match the substrate; this is referred to as "pseudomorphic" growth, a topic that is dealt with in detail elsewhere in this book (see Chapters 6–8). The dominant condition for the occurrence of this mode of growth is that the strain necessary to ensure coherency at the substrate–deposit interface is small ($\gtrsim 12\%$). Initially the deposit is elastically strained, but as it thickens the strain energy exceeds that at which it is energetically favorable to introduce misfit dislocations at the interface, so losing coherency. The defect structure and crystallography of the substrate play an important role in determining the types of misfit dislocation that are generated, and this growth mechanism also influences the defect structure in thick, incoherent layers (see Chapter 8).

The most closely studied mode of thin-film formation occurs in the moderate-to-weak binding situation ($\theta > 0$). Here the deposited atoms (or molecules) are initially equilibrated thermally with the substrate, on which they diffuse and interact to form polyatomic clusters (nuclei). This initial clustering process is discussed in detail in Chapter 4. The nuclei continue to grow, and, because of their nonuniform spatial distribution on the substrate, they grow together and agglomerate. For a few deposit–substrate systems, the agglomeration process has been studied in detail with the aid of in situ transmission electron microscopy (Pashley *et al.*, 1964; Bassett, 1964; Pocza, 1967; Coopersmith *et al.*, 1966; Sato *et al.*, 1969; Yagi *et al.*, 1969; Stowell, 1972). Agglomerating deposit islands can change shape in a manner that has been referred to as "liquidlike" (Bassett, 1960, Pashley and Stowell, 1962), a description that was coined because of the similarity of this process with that of the running

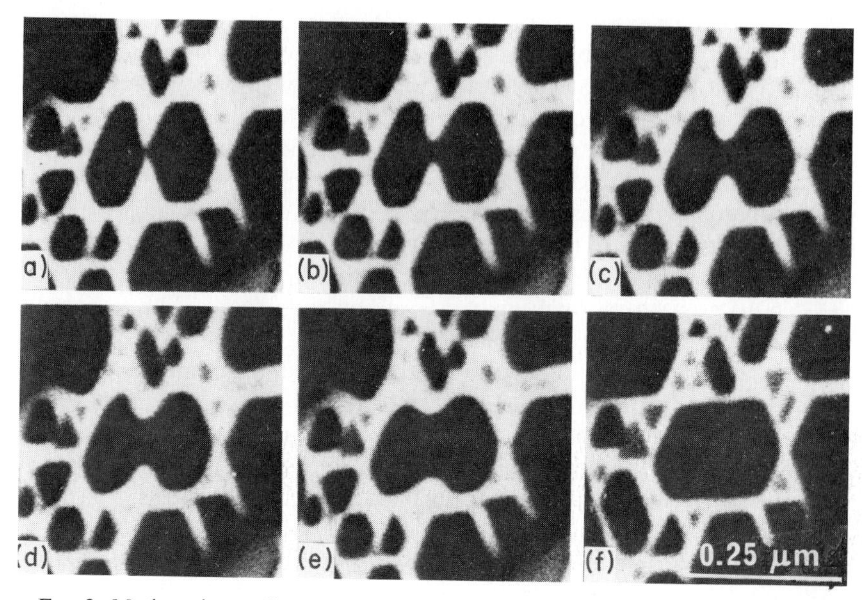

FIG. 2. Motion picture film sequence of two coalescing Au islands grown on MoS_2 at 400°C. Time scale: (a) zero; (b) 1/16 sec; (c) $\frac{1}{4}$ sec; (d) $\frac{3}{4}$ sec; (e) 2 sec; (f) 53.5 sec (Stowell, 1969, reproduced by permission of McGraw-Hill Book Co., Inc.).

together of liquid droplets. An example of coalescing islands of Au on a molybdenite substrate at 400°C is seen in Fig. 2. Analysis of the rates at which the necks between the coalescing islands are removed has shown this to be due to mass transfer by surface self-diffusion of the deposit atoms (Pashley *et al.*, 1964; Pashley and Stowell, 1966; Stowell, 1969). It will become evident later that agglomeration plays an important part in determining the defect structure of epitaxial films. As the size of the islands increases, the marked shape changes illustrated in Fig. 2 become less pronounced, and the deposit takes on a lacelike appearance with long, narrow channels of deposit-free substrate. At this stage the film is geometrically continuous but does not completely cover the substrate. The channels and holes later fill in to give a complete film that is not parallel-sided; the deposit–vapor interface is irregular, with grooves where the channels were in the incomplete film. A sequence of pictures illustrating discontinuous epitaxial growth is shown in Fig. 3. In some cases, facets that were formed on the islands persist in the continuous film and may become more pronounced as growth proceeds. As the substrate temperature is decreased, the influence of the diffusion-controlled coalescence processes diminishes, but the general description above of discontinuous growth still obtains.

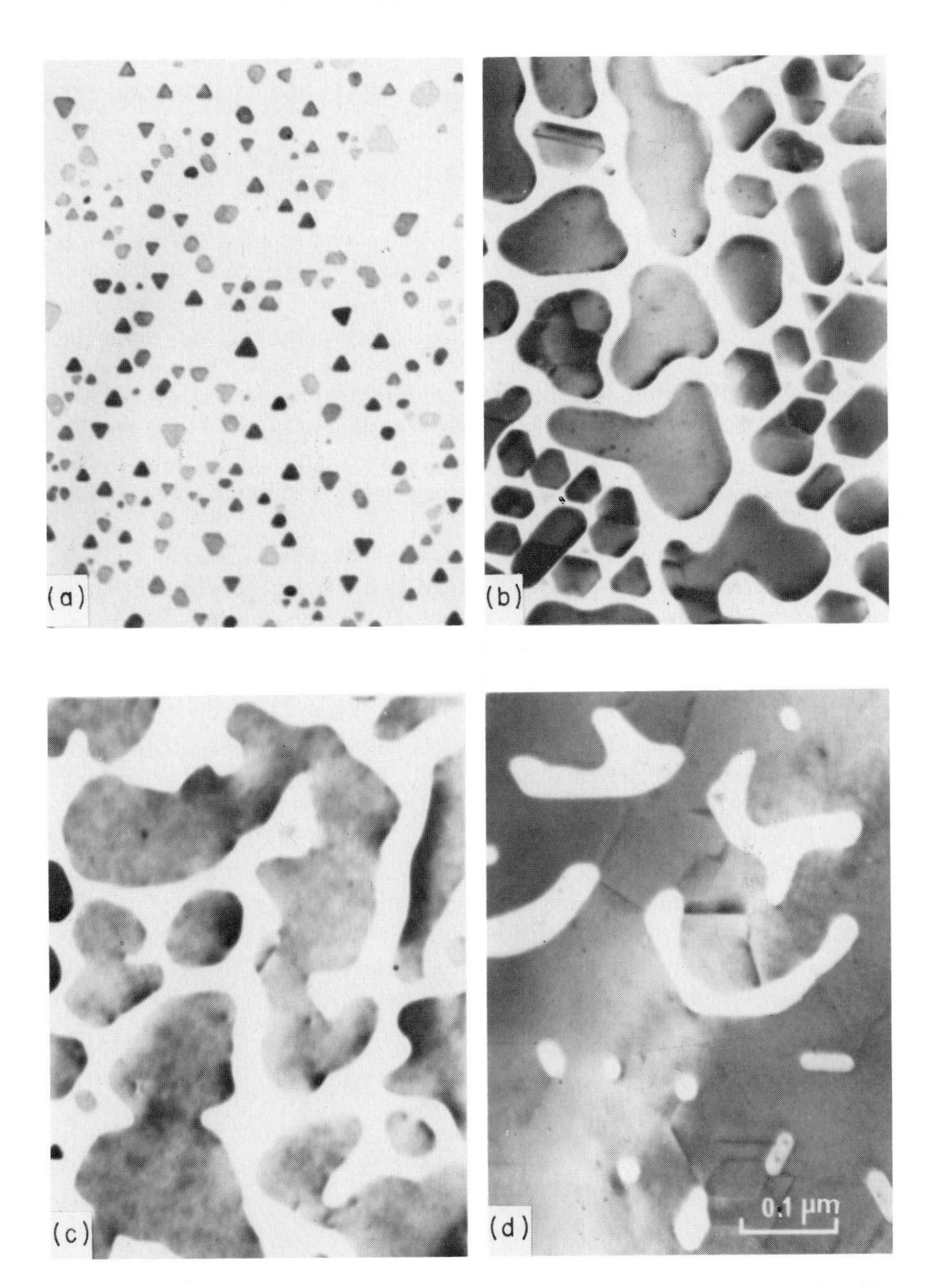

FIG. 3. Sequence of growth of Au on MoS_2 at 400°C (Pashley *et al.*, 1964, reproduced by permission of Taylor and Francis Ltd.).

At various points in the following sections, the modes of epitaxial growth that have just been briefly described will be seen to play important roles in determining defect structures of epitaxial deposits.

III. Dislocations

A dislocation is a *line* defect in a crystal and is characterized by a vector **l**, which defines (locally) the direction of the dislocation line, and by a vector **b**, the Burgers vector, which defines the atomic displacement needed to generate the dislocation from a *perfect* lattice (Fig. 4). The vector **l** may vary along the dislocation line, but **b** is invariant. *Perfect* dislocations have Burgers vectors that are translation vectors of the lattice. *Imperfect* or *partial* dislocations can exist, having Burgers vectors that are not translation vectors of the lattice; these will be ignored here since they are always associated with other imperfections such as stacking faults (Section IV) or antiphase boundaries (Section VI).

An important feature of a dislocation is its elastic strain field, and because of this dislocations are observable in an electron microscope. Atomic planes are significantly distorted in the vicinity of a dislocation, causing electrons to be diffracted more or less strongly than from the surrounding, more perfect crystal. Detailed accounts of diffraction contrast can be found in the standard texts on electron microscopy (Thomas, 1962; Hirsch *et al.*, 1965; Amelinckx, 1964; Amelinckx *et al.*, 1970; Heidenreich, 1964). Often one is concerned with dislocations in very thin crystals, in which case the method of imaging dislocations by the moiré fringe technique is very useful (Bassett *et al.*, 1958). This is illustrated for the case of two overlapping crystals via the reciprocal lattice section shown in Fig. 5. If a diffracted beam $\mathbf{g}_1$ (OP) is excited in the first crystal, and this excites a $\mathbf{g}_2$ reflection (OQ') in the second crystal, a doubly diffracted beam $\mathbf{g}_1 - \mathbf{g}_2$ (OQ'') is produced. Interference between the undiffracted and the doubly diffracted beams results in fringes in the electron image that are normal to $\mathbf{g}_1 - \mathbf{g}_2$ with a spacing

$$D = d_1 d_2/(d_1^{\,2} + d_2^{\,2} - 2d_1 d_2 \cos\theta)^{1/2}$$

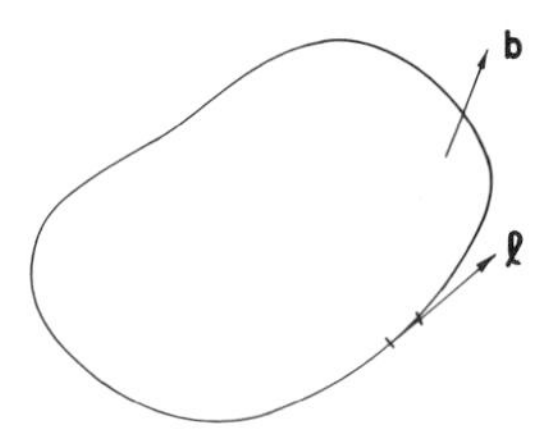

FIG. 4. Dislocation loop showing line vector **l** and Burgers vector **b**.

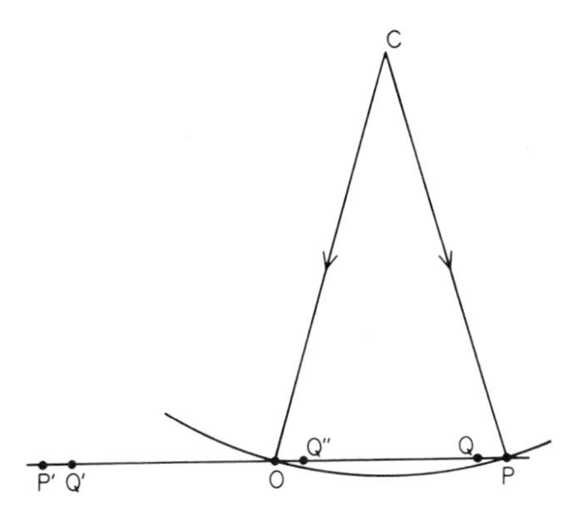

FIG. 5. Reciprocal lattice construction for the formation of moiré patterns (Stowell, 1966).

where θ is the angle between the beams diffracted by planes of spacing d_1 and d_2. If a dislocation is present in one of the crystals, it is seen in the moiré image, as illustrated in the optical analog of Fig. 6 (where $\theta = 0$). The number of extra fringes N is given by $\mathbf{g} \cdot \mathbf{b}$. Note that if $\mathbf{g} \cdot \mathbf{b} = 0$, the dislocation is not revealed in the moiré image; this also is the condition for dislocation invisibility in normal diffraction contrast (although in some cases this rule breaks down).

In this section, only elementary geometrical properties of dislocations will be considered. Detailed accounts of their elastic properties, their interactions, and mobility may be found in the standard texts (Read, 1953; Cottrell, 1953; Friedel, 1956; Nabarro, 1967; Hirth and Lothe, 1968).

Early electron-microscope observations of the substructure of epitaxial films revealed that they contain very high dislocation densities (10^9–10^{11} lines/cm^2) that are comparable to the values found in heavily worked metals. Most of these observations were of the moderate-to-weak binding systems such as Ag, Au, and Cu grown on NaCl or mica. In subsequent research, these types of systems have been extensively studied and, consequently, will

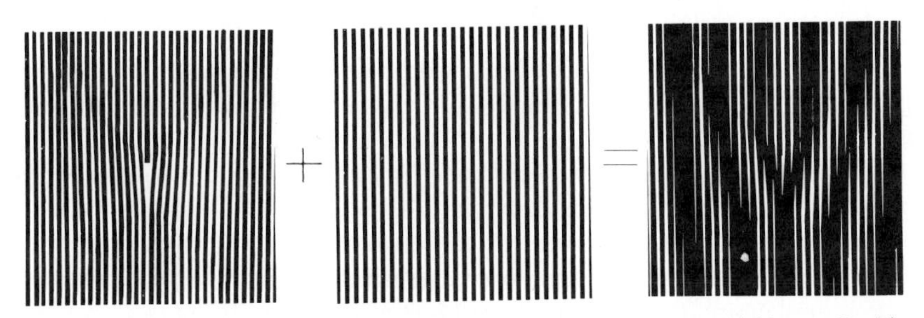

FIG. 6. Optical analog showing the appearance of a dislocation in a moiré image (Pashley, 1964, reproduced by permission from "Thin Films," p. 67, American Society for Metals).

be of major concern here. Dislocations in deposits that initially grow pseudomorphically are produced mainly as a result of coherency breakdown at the deposit–substrate interface and by the interaction of these misfit dislocations with those copied from the substrate; this topic is dealt with in detail elsewhere (Chapters 6–8).

Pashley (1964) has discussed the possibilities for dislocation introduction during epitaxial growth in the light of work that was largely carried out by Matthews (1959), Phillips (1960), Bassett *et al.* (1959), Bassett and Pashley (1959), Pashley (1959), and Bassett (1960). These are:

(i) the extension of substrate dislocations,
(ii) the accommodation of translational and rotational displacements between agglomerating islands that are close to epitaxial orientation,
(iii) the formation of dislocation loops by the aggregation of point defects,
(iv) plastic deformation of the film, both during growth and subsequent cooling and removal from the substrate.

The first mechanism must undoubtedly operate to some extent but, because most substrates in common use contain much lower densities ($\sim 10^5$–10^7 cm^{-2}) than are observed in overgrowths, it cannot always be the dominant mechanism. Pashley (1964) has argued that this mechanism is of smaller significance than might intuitively be imagined, especially in cases where bonding of atoms in the deposit is much stronger than bonding across the interface. In this case, the deposit atoms may prefer to adopt their regular lattice structure rather than distort and form an island with higher internal energy containing a dislocation line. If the dislocation in the substrate has a Burgers vector with a component normal to the interface, a step will occur in the substrate, and the deposit will be elastically strained. When the deposit–substrate bonding becomes comparable with that of the deposit and (or) the lattice misfit is small, the tendency to copy substrate dislocations will evidently increase. Dislocation copying is indeed observed in systems that initially grow pseudomorphically, and in some cases these dislocations play the dominant role in determining the conditions for the onset of coherency breakdown (Matthews, 1966a, 1967; Jesser and Matthews, 1967, 1968).

The second possibility, that of dislocation generation due to the accommodation of (small) translational and rotational displacements between agglomerating islands, has been shown experimentally to be a major factor. The most striking evidence was obtained by direct observations of epitaxial growth inside transmission electron microscopes.

It is fortunate that, in several of the epitaxial metal–inorganic deposit–substrate combinations that have been extensively studied by in situ electron microscopy, moiré fringes in the range 15–50Å can be produced (e.g., Au, Ag, and Pb on MoS_2 and Ag–MgO). In a detailed study of defects in Au films grown on MoS_2, Jacobs *et al.* (1966) observed that the initial nuclei were defect-free but, when coalescences occurred, dislocations were often found in the composite islands. Such an event is seen in Fig. 7, which shows a dislocation formed as a result of the coalescence of three islands (Fig. 7c). An important additional feature is that, at the temperatures used in this work (~400°C), these dislocations were observed to migrate out of the island (Fig. 7d). (By

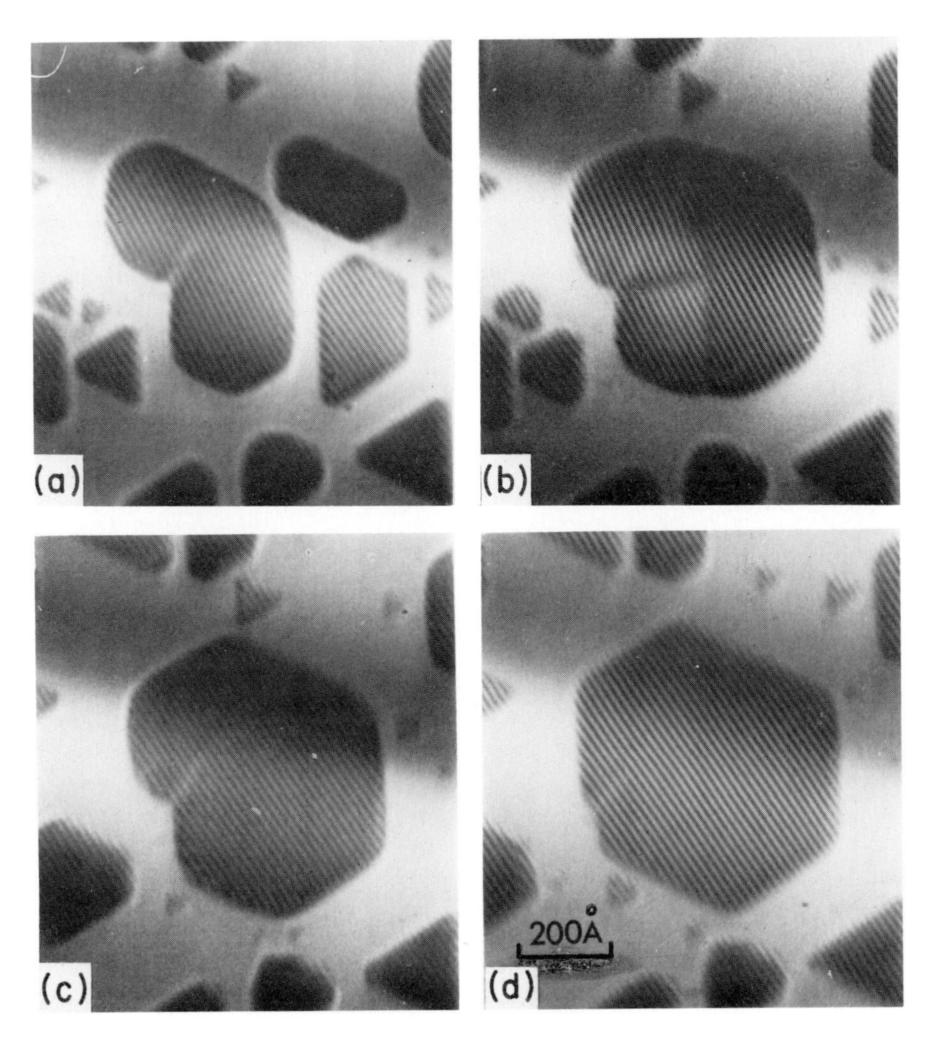

FIG. 7. Generation of a dislocation by coalescence of three islands and its subsequent removal (Jacobs *et al.*, 1966, reproduced by permission of Taylor and Francis Ltd.).

observing Fig. 7a obliquely, it is seen from the moiré fringes that the initial islands are rotationally misaligned.) As the island size increased, the dislocations were less able to migrate out of the deposit, and the dislocation density gradually increased. At the stage where only holes were left in the otherwise continuous film, incipient dislocations were observed, as in Fig. 8;

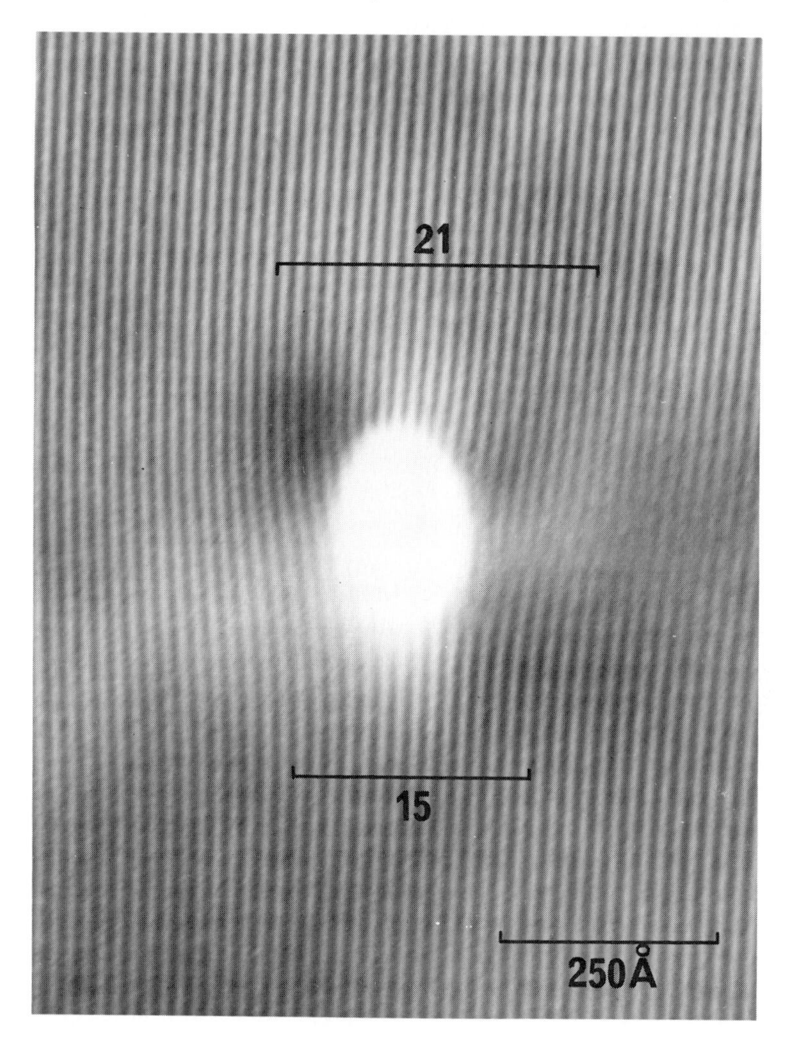

FIG. 8. Incipient dislocations in a hole; Au film grown on MoS_2 at 300°C (Pashley, 1959, reproduced by permission of Taylor and Francis Ltd.).

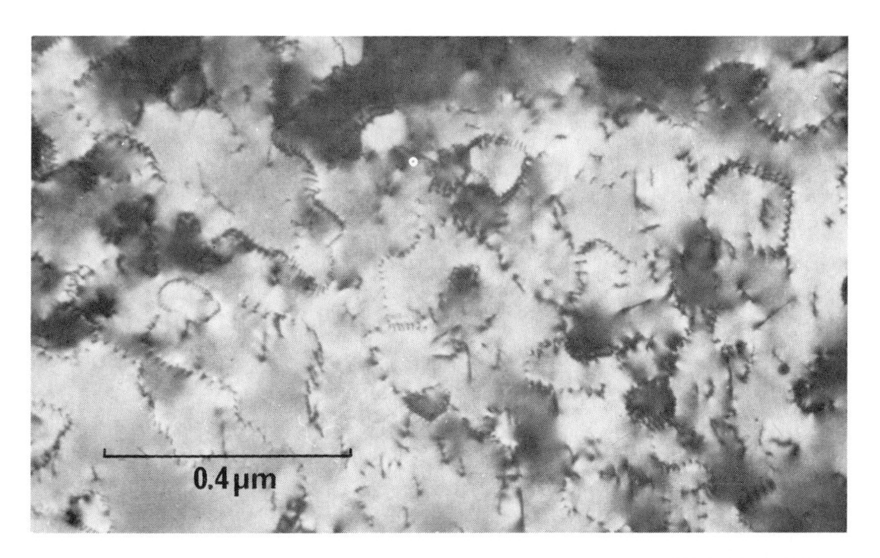

FIG. 9. Dislocation cells formed in a (111) film of PbTe grown on CaF_2 (D. J. Stirland, unpublished).

in this example, the excess of moiré fringes on one side of the hole shows that at least three dislocations must be formed when the hole is eliminated. A clear example of the influence of island misorientation on dislocation structure is shown in Fig. 9, where regions whose slight angular deviation from that of the surrounding lattice are seen to be bounded by dislocations. Indeed, this mosaic structure of epitaxial films is common and can also be revealed by X-ray methods (e.g., Holloway, 1966; Holloway *et al.*, 1966; Bobb *et al.*, 1966).

The influence of the angular misorientation of islands on dislocation formation can be fairly easily understood on the basis of the arguments put forward by Burgers (1940) and Read and Shockley (1950, 1952) to explain the mosaic structure of bulk crystals. The case of three small coalescing islands is illustrated in Fig. 10a and for two large islands in Fig. 10b. In the latter case, a wall of dislocations is produced, as first demonstrated for epitaxial films by Bassett (1960), and illustrated in Fig. 11 for the case of a Cd film on MoS_2. Unfortunately, a detailed statistical and geometrical analysis of the dependence of the total dislocation density arising from this effect on the spread of misorientation in a continuous film has not been made.

An additional and important mechanism that has been discussed in detail by Pashley (1964) evolves from the fact that nuclei are formed approximately at random on a substrate so that, even if they are *perfectly* aligned, when two

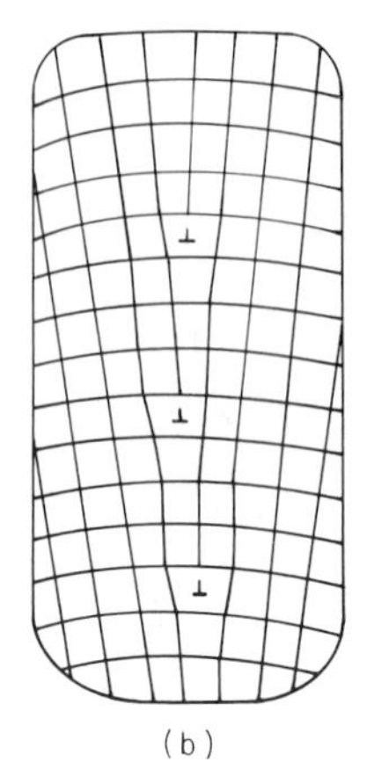

FIG. 10. Accommodation of rotational misfit: (a) between three small islands, and (b) between two large islands.

or more islands impinge their lattices will not match exactly. Thus a *translational* misfit will occur, which is shown schematically in Fig. 12. If the temperature is high enough, this misfit will be accommodated by the lattices distorting by the smallest displacement vector $\mathbf{\Delta}$ to minimize the elastic strain energy. In Fig. 12, the lattice points A and B would then coincide, rather than A and C or A and D. For three islands coalescing, we can define three displacement vectors $\mathbf{\Delta}_2{}^1$, $\mathbf{\Delta}_3{}^2$, and $\mathbf{\Delta}_1{}^3$ as shown in Fig. 13 and, if a closed path is described through the three islands, the linear arms of which start and end on lattice points, we must have

$$\mathbf{\Delta}_2{}^1 + \mathbf{\Delta}_3{}^2 + \mathbf{\Delta}_1{}^3 = \text{lattice vector} = \mathbf{S}$$

Let the islands coalesce, minimizing the elastic strain energy as described above; then by making a Burgers circuit around the hole (or line in which the three junctions meet) the Burgers vector of the dislocation formed is seen to equal **S**. These considerations lead to the conclusion that, for fcc crystals, Burgers vectors of incipient dislocations of the form $\langle\frac{1}{2}\frac{1}{2}0\rangle$, $\langle 100\rangle$, and $\langle 1\frac{1}{2}\frac{1}{2}\rangle$ are possible (Jacobs *et al.*, 1966); of course, strain-energy considerations would require the latter two types to dissociate eventually. Note that this argument allows the islands to be relatively displaced *normal* to the substrate, as well as parallel to it, so that the influence of substrate steps can be taken into account, as has been done more specifically by Mendelson (1967).

So far, the tacit assumption has been made that the deposit is elemental. Similar arguments can be made for compounds (e.g., PbSe or GaP), but additional considerations are needed. For example, suppose that we are dealing with an XY compound and that in Fig. 12 point A is occupied by an

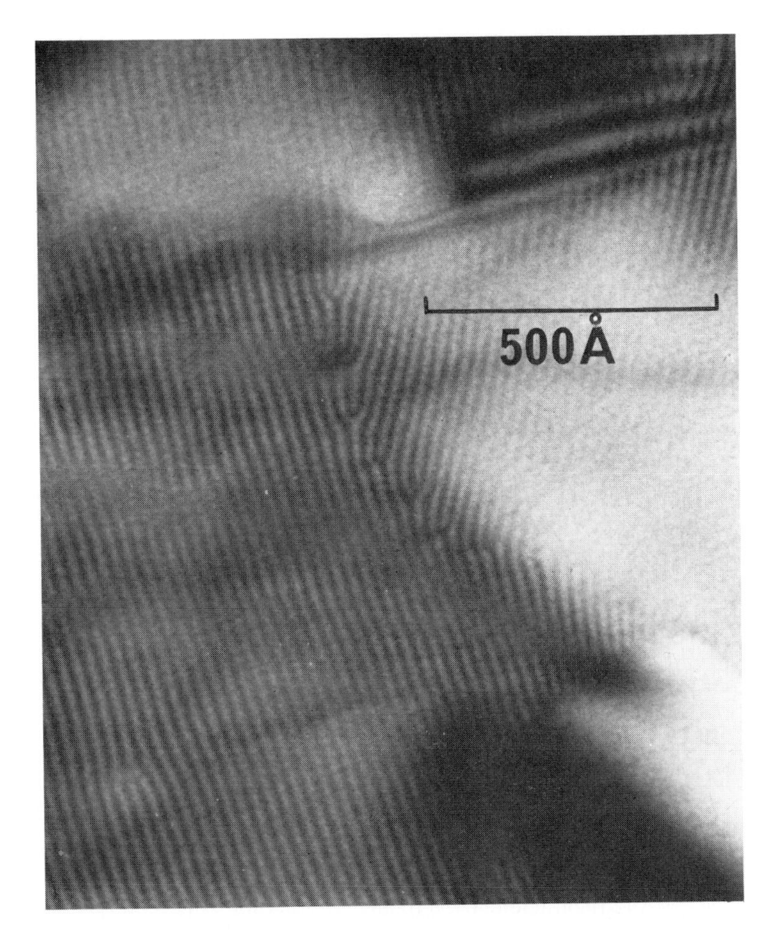

FIG. 11. Dislocation wall formed between misoriented (0001) Cd islands grown on MoS_2 (T. J. Law, unpublished).

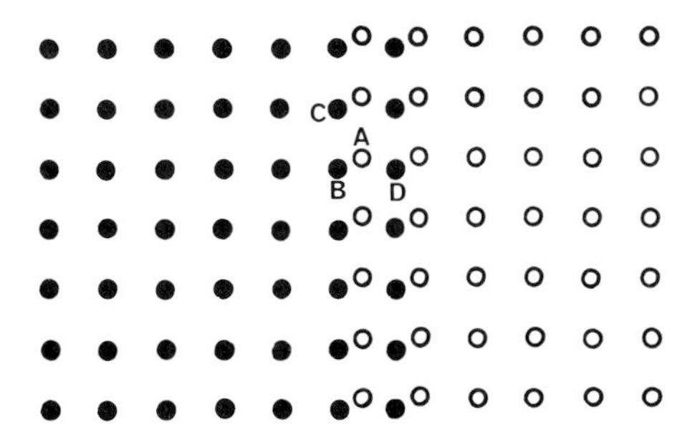

FIG. 12. Translational misfit between two parallel lattices (Jacobs *et al.*, 1966).

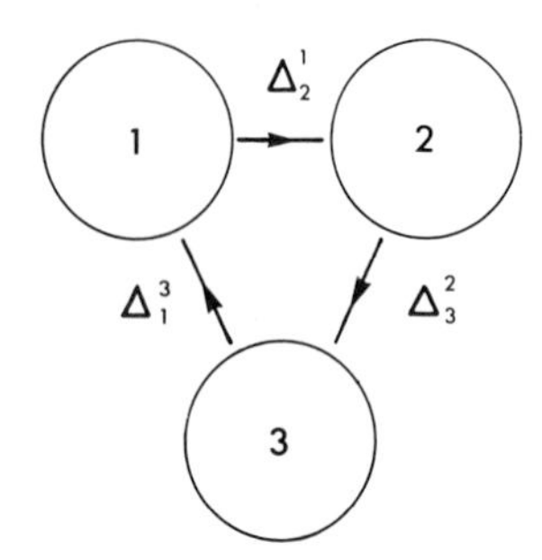

Fig. 13. Displacement misfits between three coalescing islands (Pashley, 1964, reproduced by permission from "Thin Films," p. 67, American Society for Metals).

X atom (or ion) and point B by a Y atom (ion). The smallest permissible value of **Δ** would then not be controlled by elastic considerations alone but also by bond energy and/or electrostatic factors, so that perhaps AD would be the chosen displacement. Detailed considerations of dislocation geometries in compounds are available (e.g., Hornstra, 1958; Holt, 1962; Hirth and Lothe, 1968) so that it is not important here to delve deeply into a variety of possibilities relevant to different crystal structures. The principle just outlined can be applied quite generally.

The next possibility for dislocation introduction is the aggregation of point defects to form dislocation loops. Because thin-film growth often occurs under highly nonequilibrium conditions it would be anticipated that excesses of vacancies or gas atoms would be trapped in a growing film. These point defects could aggregate to form dislocation loops that would subsequently grow and coalesce to form dislocations threading the foil. Although dislocation loops are occasionally observed in metallic epitaxial films, the existing experimental evidence is insufficient to warrant any strong weight being given to this mechanism in *elemental* deposits (except perhaps when growth occurs at very low temperatures). The situation is, however, not so clear-cut where compound deposits are concerned. Epitaxial films of compound semiconductors grown under high supersaturation are often highly imperfect, as judged from conductivity and Hall mobility measurements (Chopra, 1969). Electron-microscope observations show them to contain high dislocation densities, which in many cases must have arisen partly by the island-misfit mechanism discussed above. However, it is likely that nonstoichiometry of the growing film could lead to dislocations being produced by the mechanism under discussion.

The role of plastic deformation during film growth in generating dislocations might be expected to be highly significant, since it is well known that thin films are highly stressed during growth (Hoffman, 1966). Holes or island edges might provide sites at which dislocations are generated to relieve these internal stresses. Unfortunately, there is virtually no direct evidence in support of this mechanism, except in cases where the deposit initially grows pseudomorphically (see Chapter 8)

To summarize the foregoing arguments, it is noted that, of the dislocation-generation mechanisms that have been put forward, the "island-misfit" model has gathered strong experimental support especially from work involving in situ observations of epitaxial growth. Of the other possibilities, point-defect agglomeration is likely to be important in nonstoichiometric compounds, but there is a need for more experimental evidence on this aspect.

Most of the discussion so far has been concerned with the early stages of thin-film growth. The way that dislocation density changes during this regime has been described by Jacobs *et al.* (1966) for Au deposited on MoS_2 under conditions where the dislocations formed by coalescences are able to migrate fairly readily. Detailed analyses have not been made for situations where the latter condition is not fulfilled, although Stowell (1972) has discussed some of the implications. However, the variation of dislocation density with thickness in continuous epitaxial layers has not been intensively studied. It is anticipated that dislocations would migrate in thick films to lower the strain energy of the system, leading to annihilation of dislocations of opposite sign; Sloope and Tiller (1965) have demonstrated that this happens, but detailed considerations of this stage of epitaxial growth are generally lacking.

IV. Stacking Faults

A stacking fault is a planar defect across which the crystal has been displaced by a vector that is not a lattice translation vector. If a single stacking fault terminates within a perfect crystal, it must be bounded by a dislocation loop, the Burgers vector of which is also not a translation vector of the lattice; such dislocations are called "imperfect" or "partial." For example, in the fcc lattice, stacking faults occur on $\{111\}$ planes and are bounded by dislocations having Burgers vectors of the type $\frac{1}{6}\langle 112\rangle$ or $\frac{1}{3}\langle 111\rangle$. In the former case, the fault is produced by shear on a $\{111\}$ plane, whereas in the latter case removal or insertion of a layer of atoms is required so that the displacement is normal to the faulting plane. This is illustrated in Fig. 14, in which the conventional "abc" terminology representing the stacking sequence of atom planes in the fcc lattice is used. A fault made by the *removal* of a partial atomic layer is called "intrinsic" (Fig. 14a), and the *addition* of a layer of atoms yields an "extrinsic" stacking fault (Fig. 14b). From the viewpoint of the stacking sequence, an extrinsic fault in the fcc lattice is equivalent to two intrinsic faults on adjacent $\{111\}$ planes. Stacking faults can be created by the dissociation of perfect dislocations into two partial dislocations bounding a stacking fault; in this case, the sum of the Burgers vectors of the partials equals that of the perfect

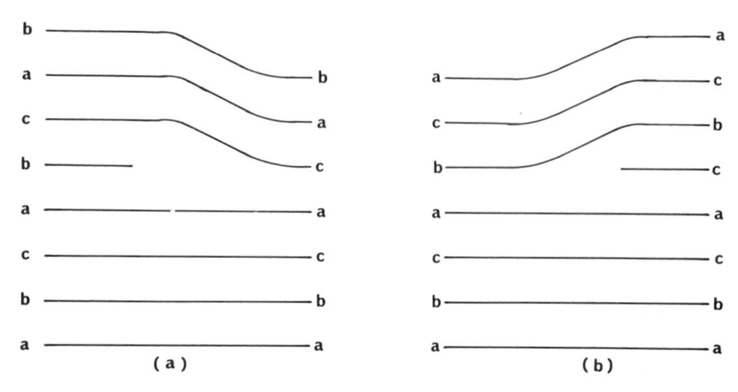

FIG. 14. Stacking sequences of (111) planes giving rise to: (a) intrinsic, and (b) extrinsic stacking faults.

dislocation. Further details about stacking faults and the geometrical configurations that are possible in different crystal lattices can be found in the standard texts on dislocation theory (Read, 1953; Cottrell, 1953; Friedel, 1956; Nabarro, 1967; Hirth and Lothe, 1968).

Stacking faults are made visible in transmission-electron-microscope images by virtue of the phase shift of magnitude $\alpha = 2\pi\mathbf{g}\cdot\mathbf{R}$, where $\mathbf{R}$ is the displacement vector of the stacking fault, which occurs across the fault (Whelan and Hirsch, 1957a, b); this gives rise to a series of interference fringes parallel to the crystal surface (Fig. 15a). In moiré images, stacking faults produce a displacement of the fringes across the fault. The magnitude of the displacement is given by $(\alpha D/2\pi)$, where D is the moiré spacing. In fcc crystals $\alpha = 0$ or $\pm(2\pi/3)$. Thus when $\alpha = 0$, no fringe displacement is seen (and, also, no fringes are seen in the normal transmission image), and when $\alpha = \pm(2\pi/3)$, the moiré fringes are displaced by $\pm(D/3)$ across the fault. This is illustrated in Fig. 15b. Further details of electron-microscope images of stacking faults can be found in the references cited earlier on transmission electron microscopy (Section III).

Early electron-microscope observations of epitaxial metal films revealed that stacking faults were often observed in very large numbers. Also, observations of Si layers grown on Si demonstrated that stacking faults were the predominant type of defect. Several possibilities have been suggested for the origin of these faults, most of which coincide with those that have already been discussed in relation to dislocations. Here, the two most likely explanations will be dealt with: (a) the accommodation of misfit between coalescing islands, and (b) the aggregation of point defects to form loops or tetrahedra of stacking faults.

Consider first the island-misfit model (Fig. 13). If the relative displacement

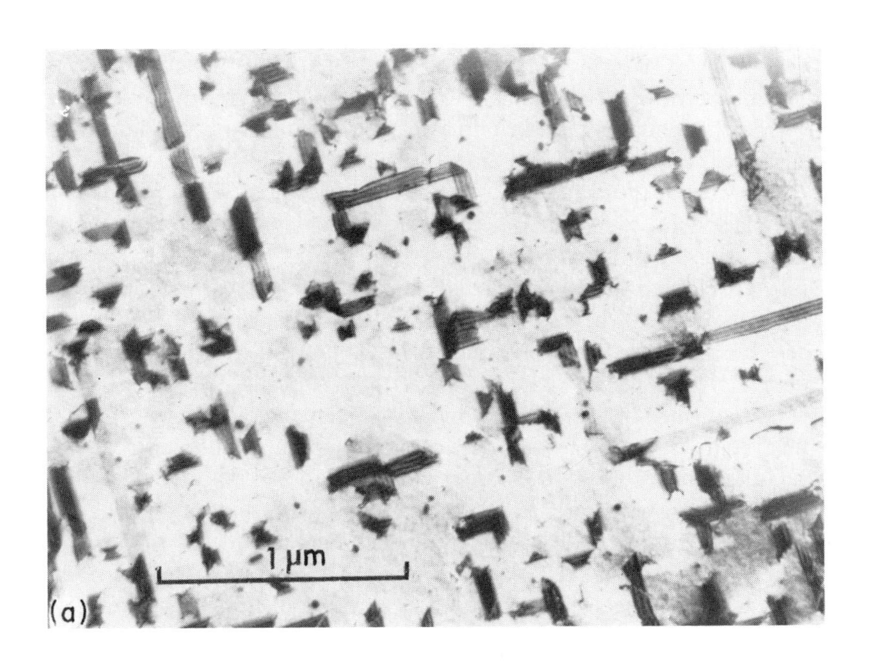

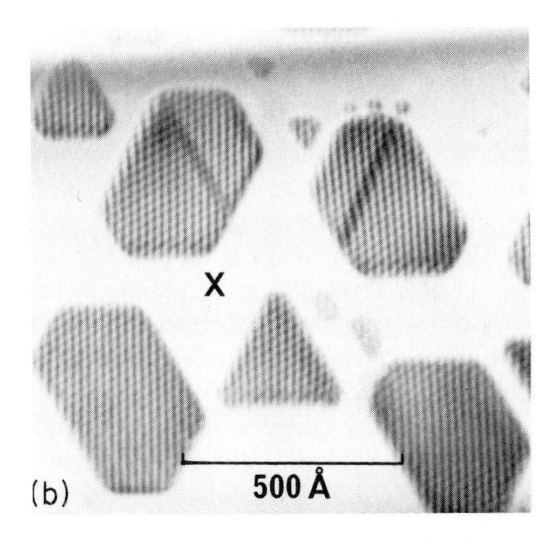

FIG. 15. (a) Stacking faults in Ag film grown on NaCl (Bassett and Pashley, 1959); (b) moiré images containing stacking faults. Note that in island × the fringes normal to the stacking fault are not displaced (Stowell, 1966).

between two agglomerating islands $\mathbf{\Delta}_i^j$ is equal to (or close to) an allowed displacement vector for a stacking fault (e.g., $\frac{1}{6}[211]$ in the fcc lattice), it is energetically favorable for this misfit to be accommodated mainly by a stacking fault, while the remainder is taken up by elastic strain in the compound island. This argument can be applied to more complex situations, such as when three islands coalesce. If the displacement between two islands is close to a stacking-fault displacement, say $\frac{1}{6}[211]$, the hole at the center of the islands contains an incipient dislocation with Burgers vector $(\sum \mathbf{\Delta}_i^j - \frac{1}{6}[211])$. If two or three of the $\mathbf{\Delta}_i^j$ are close to stacking-fault displacement vectors, two or three stacking faults may be created when the islands coalesce; Matthews (1962) has discussed the possible configurations in detail, so his arguments will not be repeated here.

The consequence of translational misfit between agglomerating islands can be illustrated very simply for the case of so-called autoepitaxy, where deposit and substrate are the same material. Booker and Stickler (1962) studied Si overgrowths on (111) Si substrates and postulated that the many stacking-fault "triangles" observed by both optical and electron microscopy originated from nuclei that grew in such a way that a stacking fault existed between them and the substrate. This is illustrated in Fig. 16a, where nuclei P, Q, and R grow in the correct relation to the substrate, but the first layer of nucleus S is incorrectly placed; in P, Q, and R the stacking sequence above the "a" substrate layer is (bcabc···), whereas in S it is (cabcabc···). Figures 16b–d show the deposit being built up layer by layer after the islands P, Q, R, and S have grown together, and it is evident that an outward-growing "triangular" fault is formed; it is a truncated tetrahedron. If the stacking fault at the substrate is extrinsic, those on the inclined planes are intrinsic and vice versa. More detailed considerations of these defects are given by Booker (1964, 1966) and Mendelson (1964a, b, 1967).

Operation of the island-misfit model was observed in situ by Jacobs *et al.* (1966) for the case of (111) Au crystals grown on MoS_2. In addition to stacking faults being formed as a result of the coalescence of islands, it was observed that they could be eliminated as well. One mechanism for the elimination of stacking faults is illustrated in Fig. 17. Here island X contains a stacking fault AB, and when islands X and Y coalesce a Shockley partial dislocation must be formed at B. This can then glide toward A and eliminate the stacking fault, also causing a displacement of part of island X relative to the substrate.

The precise origin of stacking faults in autoepitaxial deposits (e.g., Si) is not yet clarified; some type of defect must be responsible for the nuclei that are faulted relative to the substrate. Various suggestions such as oxygen contamination (Finch *et al.*, 1963) and metallic impurities (Pomerantz, 1967) have been proposed as initiation sites for the faulted nuclei. A recent detailed study of both the kinetics of Si surface structure formation and the variation

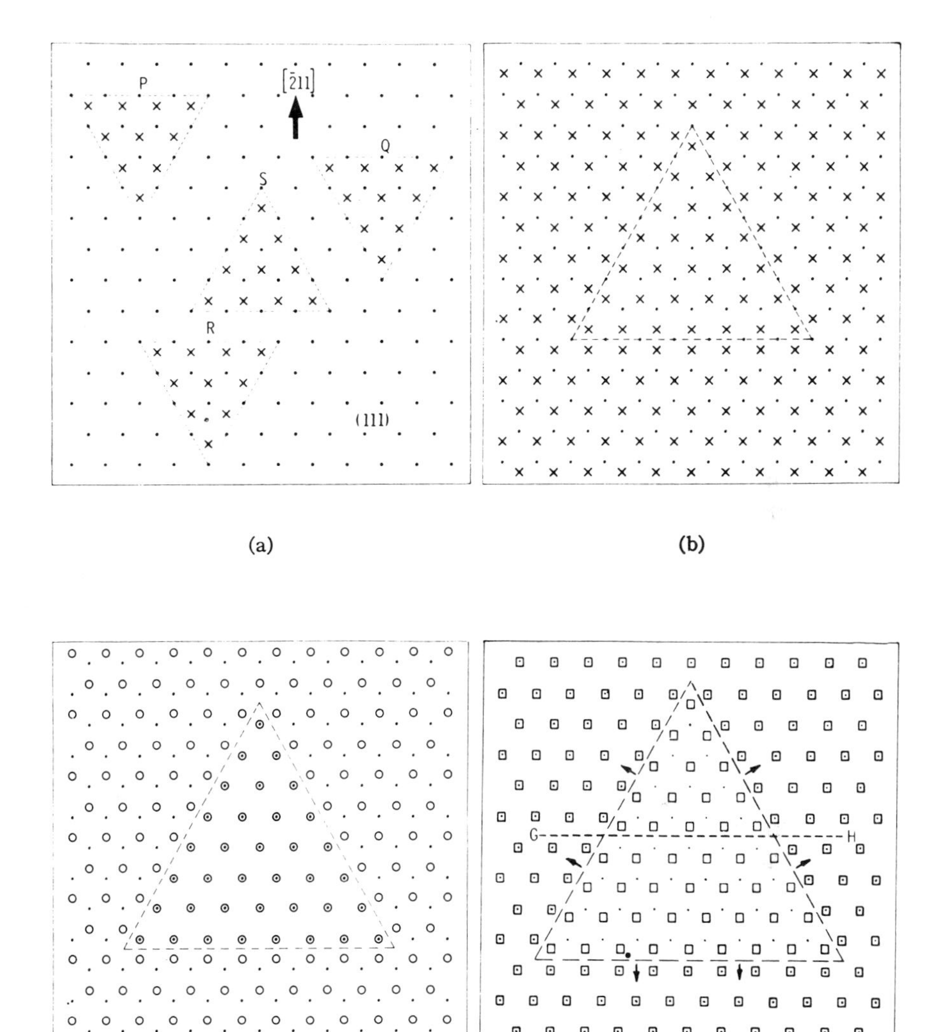

FIG. 16. Formation of stacking-fault tetrahedra due to translational misfit of coalescing islands (Booker and Stickler, 1962, reproduced by permission of the American Institute of Physics). The different atom layers are: $n = 0$ (·), substrate layer, type a positions; $n = 1$ (×), epitaxial layer; $n = 2$ (○), epitaxial layer; $n = 3$ (□), epitaxial layer. In part (a), P, Q, and R are type b positions and S is type c. In part (b), outside the triangle are type b positions and inside the triangle are type c. In part (c), outside the triangle are type c positions and inside are type a. In part (d), outside are type a and inside are type b.

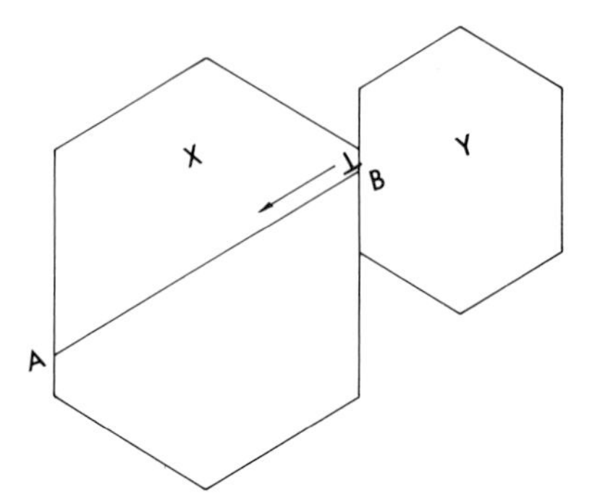

FIG. 17. Coalescence of islands X and Y. X contains a stacking fault AB and, on coalescence, a partial dislocation is formed at B that can glide out, so eliminating the fault.

of stacking-fault densities with temperature and deposition rate has led to another explanation (Thomas and Francombe, 1971). It was suggested that the surface structures are controlled by surface segregation of metallic impurities and that microprecipitates of these acted as nuclei for the stacking faults. The stacking-fault density was observed to vary exponentially with growth temperature (Fig. 18a) on both low-lifetime Si (high impurity content) and on high-lifetime material, but the densities on the former type of substrate were about an order of magnitude higher than those on the purer substrates. In addition, at a given growth temperature, the stacking-fault density was proportional to the deposition rate. These observations point to a triple correlation between impurity clusters (or patches) on the substrate, the nucleation kinetics of Si islands on these clusters, and the density of stacking faults that arise as a result of the island-mismatch model of Booker and Stickler (1962). The kinetic data have yet to be interpreted satisfactorily, and the fine details of the surface structural changes and their role in producing misfit between the nuclei need to be carefully examined. However, the bulk of the experimental data on stacking-fault production favors the island-misfit model.

Finally, consider the idea that stacking faults originate as a result of condensation of vacancies in a supersaturated film. In metal deposits on alkali halides, mica, MoS_2, etc., there is little experimental evidence to support this mechanism, but in Si–Si deposits the observation of tetrahedral stacking-fault configurations by Booker and Stickler (1962) led Jaccodine (1963) to suggest that a vacancy-condensation mechanism would be possible. [Stacking-fault tetrahedra have also been observed by Jaeger *et al.* (1968) in epitaxial Ag films grown on mica.] This has been examined quantitatively by Mendelson (1964b), who adopted a model similar to that used by Silcox and Hirsch (1959) for the nucleation of stacking-fault tetrahedra in quenched Au foils. The basis of the model is that a triangular Frank dislocation loop forms

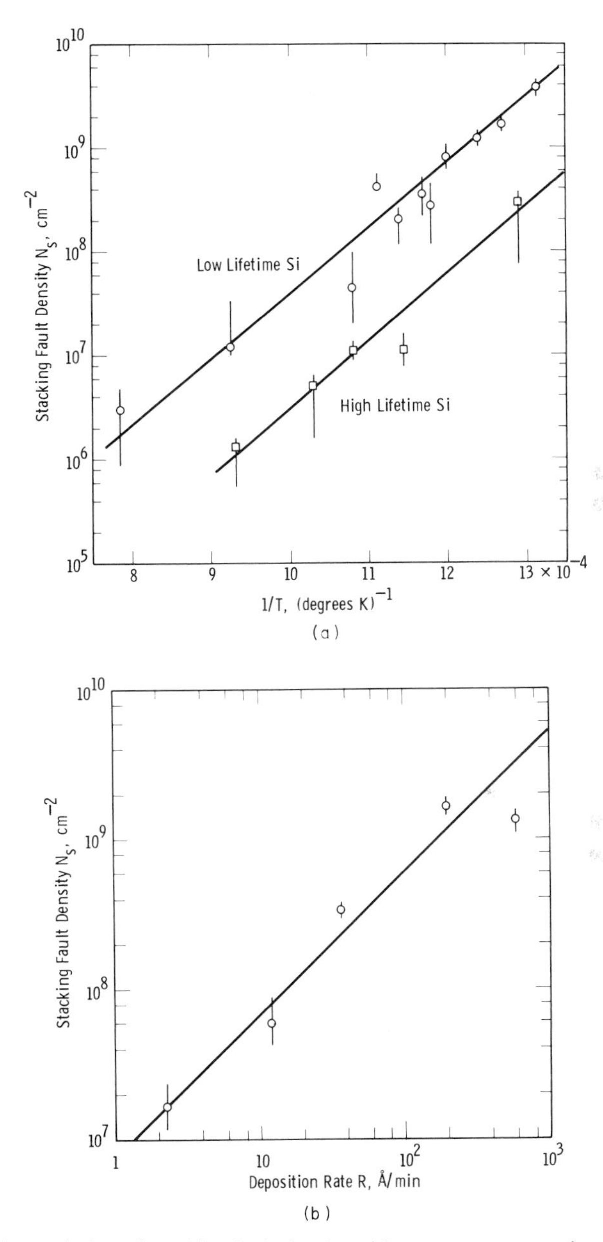

FIG. 18. (a) Variation of stacking-fault density with temperature at a deposition rate of 200 Å/min with $N_s = C_1 \exp E/kT$; (b) dependence of stacking-fault density on deposition rate at 520°C with $N_s = C_2 R^n$ (Thomas and Francombe, 1971, reproduced by permission of North-Holland Publ. Co.).

by vacancy condensation at the deposit–substrate interface. The $\frac{1}{3}\langle 111 \rangle$ Frank partials dissociate into glissile Shockley partials ($\frac{1}{6}\langle 112 \rangle$) plus sessile "stair rod" dislocations ($\frac{1}{6}\langle 110 \rangle$) that remain at the edges of the original triangle. The Shockley partials can then glide out on diverging $\{111\}$ planes to form a truncated tetrahedron of fault. There is a critical film thickness (or tetrahedron height) above which it is energetically unfavorable to nucleate this defect. This thickness is determined from the condition that the sum of the dislocation-line energy and the stacking-fault energy of the loop equals that of the truncated tetrahedron. Therefore, enough vacancies must be precipitated to form a vacancy disk before the film attains this critical thickness. Mendelson estimated the minimum rate of vacancy condensation to satisfy this criterion and the maximum possible rate of vacancy condensation, assuming the extreme condition that the vacancy content was that close to the melting point of Si; the growth temperature was 1250°C. These calculations, which are somewhat lengthy, will not be repeated here. It suffices to report that the minimum vacancy condensation rate greatly exceeded the most optimistic estimate of the maximum condensation rate, thus showing that the vacancy agglomeration mechanism is highly improbable for the case of Si deposits.

No similar calculations have been made for other materials but, since most of the evidence on stacking-fault generation points to the island-misfit mechanism being of major importance, it is safe to conclude that the vacancy-collapse mechanism is of minor significance except, perhaps, during the growth of nonstoichiometric compounds and during growth at low temperatures relative to the melting point of the deposit.

V. Twinning

The presence of microtwins in metal films grown on alkali halide substrates was noted in early studies of epitaxy using reflection electron diffraction (see Pashley, 1956), but the major part of our knowledge about these defects has been gathered from electron-microscope observations. The geometrical aspects of twinning in crystals have been well characterized and details of the crystallographic transformations that lead to twinning may be found in established texts (e.g., Hall, 1954). Because we shall be concerned here with growth twins in cubic crystals, the introductory remarks on twin geometry will be confined to these simple structures.

A twin is characterized by a reflection of atom positions across the twinning plane. In fcc crystals and those with diamond or sphalerite structures, the twinning plane is $\{111\}$, and one may use the "abc" stacking notation to represent the relative atom positions in matrix and twin. The stacking sequence of a (111) twin embedded in a matrix is shown in Fig. 19. Here the

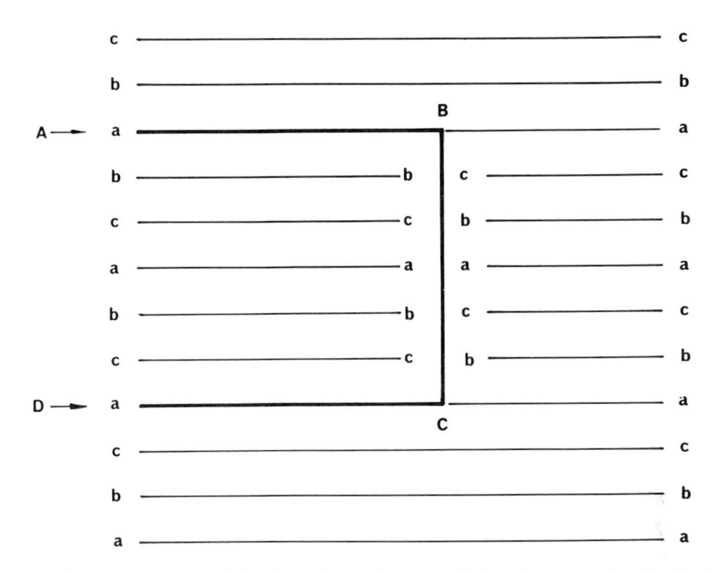

FIG. 19. Stacking of (111) planes in an fcc crystal showing an embedded twin.

(111) planes AB and DC are called "coherent" twin planes; nearest-neighbor atom distances are unchanged across these planes. Planes such as BC are called "incoherent" twin planes in view of the fact that the two lattices do not fit exactly at them, with the consequence that there is considerable atomic misfit. If a twin platelet terminates within a crystal, it must do so at an incoherent boundary; very elegant micrographs showing lattice fringes in twinned Si crystals have been published by Phillips (1972).

Because of the good atomic fit at a coherent twin boundary, the free energy of that interface γ_t is usually small compared with that of incoherent twin boundaries, which resemble high-angle grain boundaries. This factor will be seen to be important when the parameters that influence the growth or shrinkage of twins are discussed later.

In fcc crystals and those having the diamond or sphalerite structures, twins always occur on (111) planes. In (001) crystals, four sets of orthogonal twins are usually seen in epitaxial deposits. These can often be mistaken for thin stacking faults when viewed in an electron microscope because they give rise to fringes in the electron image that are similar to (but not identical with) those produced by stacking faults. Differentiation between stacking faults and very thin twins demands that detailed contrast experiments be carried out (see Amelinckx *et al.*, 1970, p. 257). However, thick twins can readily be distinguished by reference to the electron-diffraction pattern, which will exhibit extra reflections if twins are present. It can be shown that, if a crystal is twinned on its (hkl) plane, the (pqr) plane of the twin has indices $(p'q'r')$, with respect to the reciprocal lattice of the matrix, given by

$$(p'q'r') = (\bar{p}\bar{q}\bar{r}) + \left\{2\left|\frac{(pqr)}{(hkl)}\right|\cos\theta\right\}(hkl)$$

where $|(uvw)|$ is the length of the (uvw) reciprocal lattice vector and θ the angle between the normals to the (pqr) and (hkl) planes. Therefore, if $\{hkl\} \equiv \{111\}$, reflections from twins either coincide with those of the matrix or are displaced from them by vectors $\pm\frac{1}{3}(hkl)$. This allows the twins to be revealed by dark-field electron microscopy (Burbank and Heidenreich, 1960; Schoening and

Baltz, 1962; Pashley and Stowell, 1963). Bright- and dark-field electron images of twins in a (100) Au film are shown in Fig. 20.

It has already been demonstrated that, for both dislocations and stacking faults, the translational and rotational misfits between coalescing islands play an important part in the generation of these defects; it appears that in many cases twins originate as a consequence of these misfits. Burbank and Heidenreich (1960) first suggested that twins might form to accommodate translational misfit between coalescing islands, and, although twins have been observed by in situ electron-microscope methods to form when islands coalesce, the details of the way displacement misfits are manifested in microtwins are not at all clear.

At this point it is relevant to differentiate between *small* translational and rotational displacements occurring between epitaxially oriented islands and *gross* orientation differences between islands such as those that occur between (100) and (111) islands and multiply twinned particles in some fcc metal deposits on alkali halides (see Section VI for a detailed account of multiply twinned particles). In the former case, if two fcc islands with ideally planar {111} facets grow together and meet on a pair of *parallel* {111} facets, the microtwin formed cannot be greater than three {111} layers thick. Thicker twins could be produced if the joining facets contained steps. If the islands meet

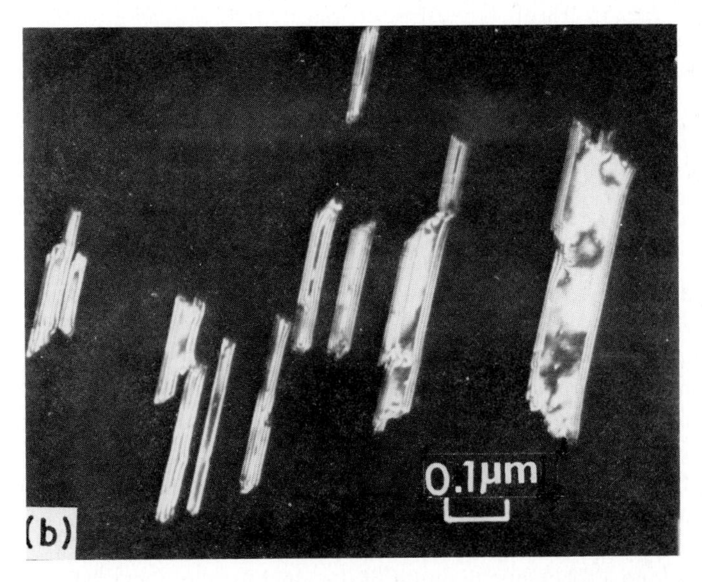

FIG. 20. Electron micrographs of microtwins in an epitaxial Au crystal: (a) bright-field, and (b) dark-field image from a twin reflection (Pashley and Stowell, 1963, reproduced by permission of Taylor and Francis Ltd.).

at a substrate surface step, a component of translational displacement normal to the substrate assists in the production of thicker twins, as has been discussed at length by Mendelson (1967). Matthews and Allinson (1964) have observed twins formed at substrate-surface steps in deposits of Ag on mica and on MoS_2. Thus, the mechanism by which thick (~ 50 Å or more) twins are produced in deposits that are well aligned with the substrate is not well understood. It was suggested by Hall and Thompson (1961) that twins arose as a result of accidental stacking faults on {111} faces of growing islands, but, since the bulk of the experimental evidence shows that twins are a consequence of island coalescences, this mechanism must be discounted.

An early model of twin formation was proposed by Matthews and Allinson (1963). They recognized that rotational misalignments were important and, stimulated by Bassett's (1960) observations of rotational realignment of coalescing islands, they suggested that *grossly* misoriented islands would rotate not into epitaxial positions but into twin relationships with the well-oriented islands. The way this reorientation process takes place was not specified. Jacobs *et al.* (1966) and Pashley and Stowell (1966) commented that such a process would most likely be a result of a recrystallization event. In this case, a high-energy interface formed between two coalescing islands might decompose into a coherent twin boundary and another high-angle grain boundary, which could migrate out of the composite island during the coalescence event. It appears that the model of Matthews and Allinson (1963) is relevant to growths of fcc metals on alkali halide substrates, especially since more recent work (Matthews, 1966b; Ino, 1966) has shown that grossly misoriented {111} islands and multiply twinned particles exist in the early stages of growth of these systems. Sato and Shinozaki (1971) have, in fact, observed multiply twinned particles transforming to twins during coalescence in deposits of Ag on MgO, although the twinning mechanism was not clarified.

Twin defects are also formed in semiconducting deposits such as Si, Ge, and GaAs grown both autoepitaxially and on foreign substrates. Holloway and co-workers (Holloway *et al.*, 1965; Holloway and Bobb, 1967a, b) have studied the occurrence of twins in GaAs grown on Ge, using both reflection electron diffraction and X-ray topography. Multiple twinning frequently occurs in this system, and the associated defects are seen as hillocks on the growing interface. Analogous, although geometrically different, features are seen in Si films grown on Si and are termed "tripyramids," the crystallographic characteristics of which have been extensively studied by Booker (1965). These are seen in the optical micrograph of Fig. 21 [the matrix orientation is (111)], and a schematic diagram of the ideal form of these defects is shown in Fig. 22. Booker (1965) demonstrated that the regions A, B, and C are each doubly twinned with respect to the substrate. At the defect–substrate

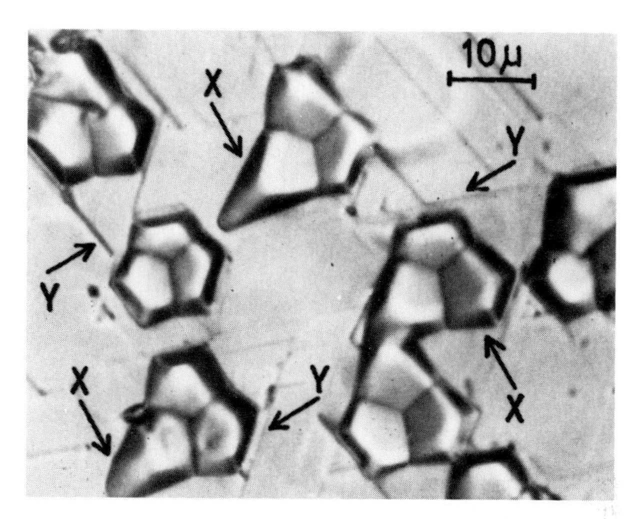

FIG. 21. Tripyramids X and associated defects Y in a (111) epitaxial Si layer (Booker, 1965, reproduced by permission of Taylor and Francis Ltd.).

interface, these regions do not meet in a point but are associated with a triangular defect that is twinned about the normal to the substrate; subsequently, Booker and Joyce (1966) showed this triangular defect to be associated with β-SiC contamination on the substrate, which causes the growth center to be in twin orientation. Thus A, B, and C are singly twinned relative to the Si growing on the SiC patch. The "associated defects" (Fig. 22) are singly

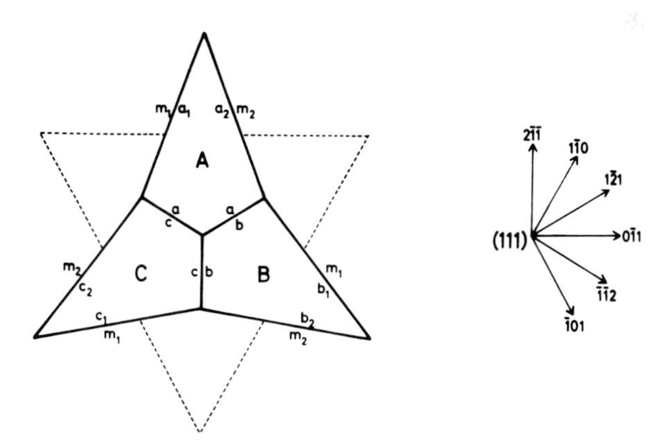

FIG. 22. Idealized form of the defects shown in Fig. 21. A tripyramid is shown by solid lines, associated defects by broken lines (Booker, 1965, reproduced by permission of Taylor and Francis Ltd.).

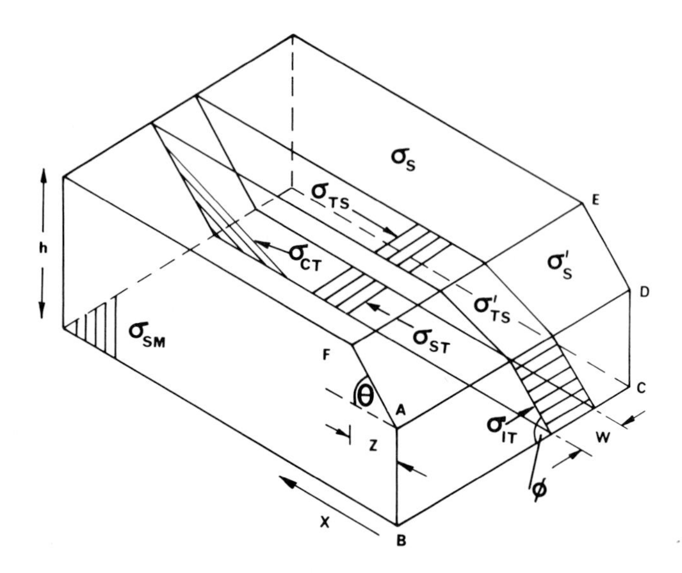

FIG. 23. Schematic diagram of a microtwin intersecting a surface groove (after Stowell and Law, 1969, reproduced by permission of Taylor and Francis Ltd.). (See text for details.)

twinned lamellae on inclined (111) matrix planes. It was suggested that these defects resulted from coalescences during growth of perfectly oriented nuclei with the twinned nucleus grown on the contamination patch.

So far we have been concerned with the introduction of twins. Early reflection electron-diffraction studies showed that twins became less prominent as deposits of fcc metals grew thicker than a few hundred angstroms. There are two main possibilities for this behavior, namely, that the twins are (i) "buried" by faster growing orientations, or (ii) completely removed from the deposit. Stowell and Law (1969), in a study of epitaxial Cu films grown on NaCl, showed the latter to be the case and proposed a mechanism by which the twins could be removed. This is illustrated in Fig. 23, which shows schematically a single twin embedded in a (001) matrix. An important observation in this work was that twins are prevented from shrinking in a continuous deposit by surface grooves that are remnants of the channels found in the final stages of discontinuous growth. The plane ABCD in Fig. 23 is a vertical section through a film from the bottom of such a groove (ADEF), which is normal to the twin plane. The width of the twin projected onto the deposit-substrate interface is w, the film thickness h, the inclination of the twin to the free surface ϕ, and the various surface free energies are shown as σ_i in the figure. In order for the twin to shrink, the incoherent twin boundary has to move out of the groove. Now, the driving force F per unit area of the incoherent twin boundary is given by

$$F = -(\sigma_{IT} \tan\theta/h) + (2\sigma_{CT} \csc\phi/w) + [(\sigma_S' - \sigma_{TS}') \sec\theta + (\sigma_{SM} - \sigma_{ST})]/h$$

where θ is the groove angle. This is zero for a critical value λ of h/w given by

$$\lambda = (\sin\phi/2\sigma_{CT})[\sigma_{IT} \tan\theta - (\sigma_S' - \sigma_{TS}') \sec\theta - (\sigma_{SM} - \sigma_{ST})]$$

so the expression for F may be rewritten as

$$F = 2\sigma_{CT} \csc\phi[(h/w) - \lambda]/h$$

Thus, when h/w exceeds the critical value λ, it will be energetically favorable for the twin to move out of the groove; once it has done so, there is no retarding force on it, and the twin can shrink to zero size. The incoherent twin boundary was assumed to migrate by a diffusion-controlled process akin to grain-boundary migration in bulk material, and Stowell and Law (1969) showed that both the energetic and kinetic analyses were in good agreement with the experimental observation that all the twins were eliminated at a thickness of 4000 Å.

Finally, in this section a special form of twinning is considered, which seems to be unique to epitaxial deposits. This occurs as a result of a commonly found effect known as "multiple positioning," whereby the deposit is oriented in several crystallographically equivalent ways on the substrate. When an fcc deposit is grown in (111) orientation on a (111) surface or on the basal plane of a hexagonal crystal, it is frequently found that two orientations are present, both of which have close-packed directions aligned in the interface. In terms of the abc stacking notation, one orientation is stacked Abcabc··· and the other is stacked Acbacb···, where A represents the atom positions in the substrate surface. These two orientations would be twins if stacked on top of each other, but they are nucleated side by side; when they grow together a special type of incoherent twin boundary is formed that is called a "double positioning" (DP) boundary and is illustrated in Fig. 24. In films of Ag, Cu, Au, Pb, etc., and in deposits of lead chalcogenides formed on substrates such as mica and MoS_2, these boundaries tend to be normal to the substrate and are close to {112} planes, although they frequently deviate from this orientation and have a stepped appearance, as in Fig. 25.

As these DP boundaries are incoherent twin boundaries, at which the atomic (or molecular) structure is disordered, they provide a source of high internal energy; the free energy of an incoherent twin boundary is $\sim\frac{2}{3}$ that of a high-angle grain boundary (Fullman, 1951). Similar incoherent twin boundaries can be found in fcc deposits having orientations other than (111) and in other crystal structures; the (111) orientation will be discussed here because it is the only one to be studied in any depth.

Although the initial nuclei in deposits of, say, Ag on mica are separated by $\sim$500 Å and the distribution of the two (111) orientations is approximately

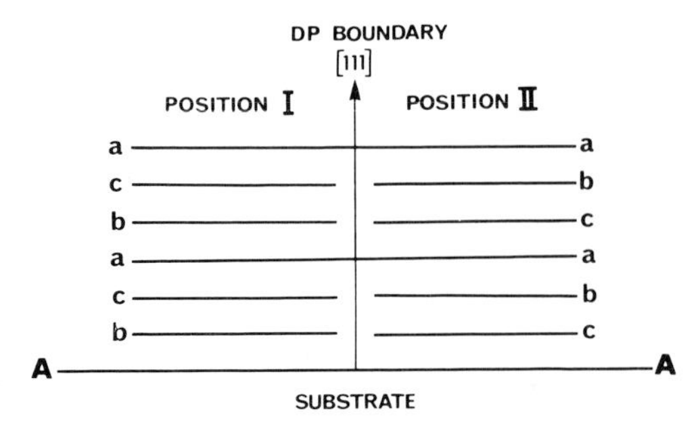

FIG. 24. Stacking of (111) planes at a double-positioning boundary (Jacobs and Stowell, 1965, reproduced by permission of Taylor and Francis Ltd.).

random, continuous films grown at ~200°C or higher contain domains of one orientation that are several microns across. This observation indicates that many of the DP boundaries have been lost during growth, and the manner in which this occurs has been illustrated by means of the in situ technique, mainly by Pashley and co-workers (Pashley, 1965; Jacobs *et al.*, 1966; Pashley and Stowell, 1966; Stowell and Law, 1966).

Double-positioning boundaries are formed when two islands coalesce. They are removed by a mechanism that is similar, if not identical, to that by which grain boundaries move in bulk crystals, that is, by glide normal to the boundary plane. This process has been observed directly in the electron microscope and is illustrated for Au on MoS_2 in Fig. 26; here the crystal has been tilted so that only one of the two (111) orientations is diffracting strongly and so appears dark in the image.

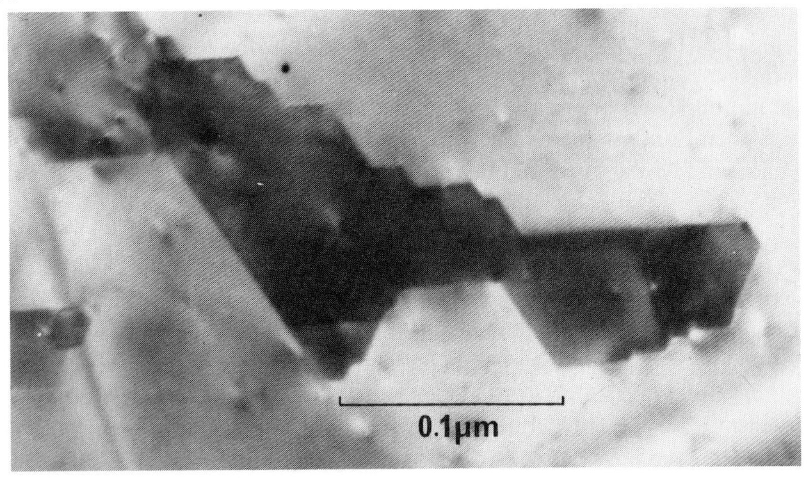

FIG. 25. Double-positioning domain in a (111) epitaxial gold film; in places, the DP boundary is stepped, alternating between different {112} planes (T. J. Law, unpublished).

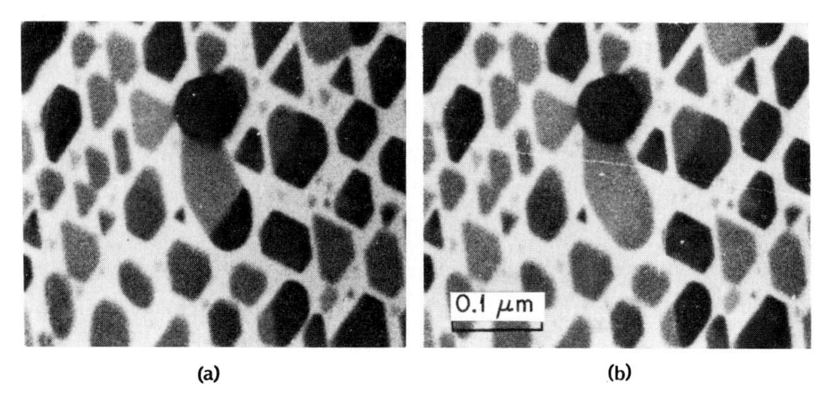

FIG. 26. Sequence showing the elimination of a DP boundary in a (111) Au deposit on MoS_2. In (a) the boundary between dark and light areas is a DP boundary that has migrated out of the island in (b) during the $\frac{1}{8}$ sec between exposures of the adjacent frames of the ciné film (Stowell and Law, 1966, reproduced by permission of Internationaler Buchversand GmbH).

What are the driving forces for this motion and by what mechanism does the boundary move? Partial answers to these questions were given by Stowell and Law (1966) for the system Au–MoS_2; this is unfortunately the only system that has been studied in detail. These authors estimated upper and lower limits for the times needed to remove a migrating DP boundary from small Au islands; usually, a boundary was eliminated in a time that corresponded to only a few frames of the motion picture films taken of these events. This was done for substrate temperatures between 350 and 475°C. It was suggested that the driving force was provided mainly by the free energy of the DP boundary. Knowing the island shape, the driving force can be calculated, and the basis of the calculation is illustrated in Fig. 27. Assume that the side faces of the hexagonal island shown are normal to the substrate and that all of the free surfaces have the same surface free energy. When the boundary moves to the right its area is decreased leading to a gain in free energy. If the

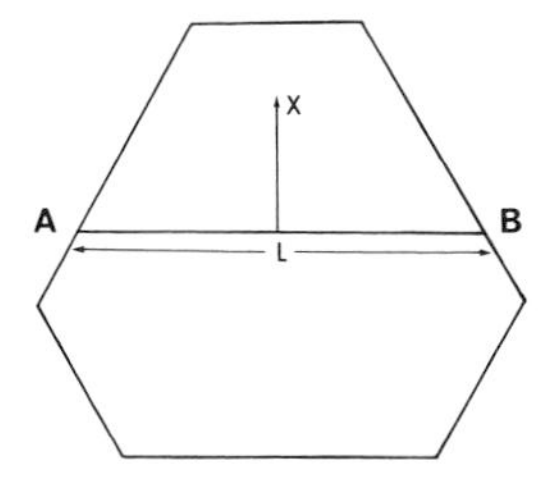

FIG. 27. Schematic illustration of a DP boundary (AB) in a hexagonal platelet.

surface free energy of the boundary is σ, its length $L(x)$, and the number of atoms per unit area of the boundary N, the driving force per atom of the boundary is

$$F = -[\sigma/NL(x)]\, dL(x)/dx$$

It is assumed that the boundary moves by a diffusion-controlled process (Turnbull, 1951) such that the velocity V is related to F by

$$V = MF$$

where M, the boundary mobility, is given by $M = D_{\mathrm{m}}/kT$, with D_{m} taking the form of a diffusion coefficient. So, if the sides of the island are straight, it may easily be shown that the time taken to eliminate the boundary is given by

$$t = \frac{kT}{D_{\mathrm{m}}} \frac{N}{\sigma} \frac{\bar{L}}{\Delta L} x^2$$

where $\bar{L}$ is the average boundary length during the time it migrates a distance x and ΔL the change in boundary length. Here, $\bar{L}$, ΔL, and x can be measured, N and σ can be estimated, and so upper and lower limits of D_{m} can be obtained for each event analyzed. An Arrhenius plot of these limits for a range of temperatures is shown in Fig. 28. From this it was deduced that the activation

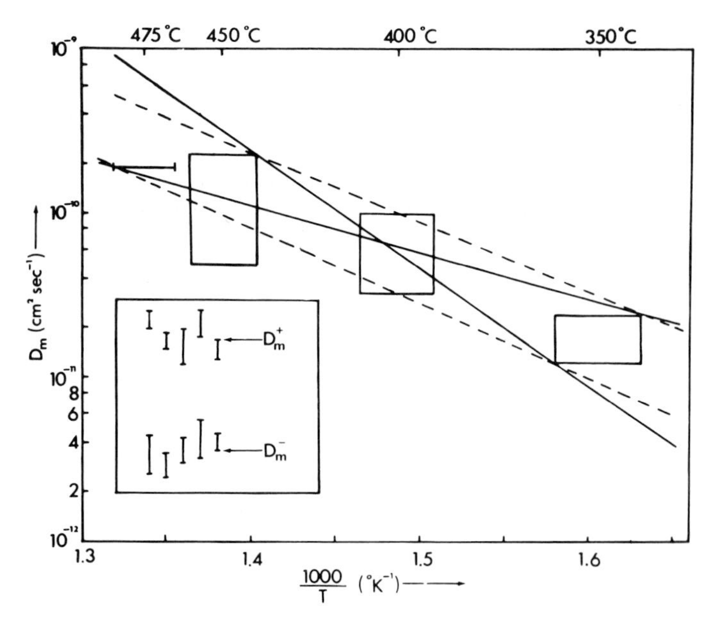

FIG. 28. Results of the analysis by Stowell and Law (1966) of DP boundary migration in epitaxial Au films. The effective diffusion coefficient D_{m} is plotted against $1/T$, and the boxes represent the limits of the experimental measurements (reproduced by permission of Internationaler Buchversand GmbH).

energy for the boundary migration process was between 13 and 33 kcal/mole, which is significantly less than that for volume diffusion in Au and is approximately what would be anticipated for grain-boundary migration in bulk crystals. The preexponential factor in D_m was also of a reasonable magnitude, so it can be concluded that a process similar to that obtaining in recrystallization of bulk crystals also obtains in these epitaxial thin films. Clearly it would be of great value to have similar measurements available on other crystals so that the applicability of these conclusions to the general field of epitaxial growth could be assessed. This is particularly relevant to the generalizations made by Stowell (1968, 1969) in his model describing structural changes in discontinuous and continuous film growth.

It was noted earlier in this section that DP boundaries are usually normal to the film in (111) deposits. However, this is not always the case, and the DP boundary sometimes contains steps that are parallel to the substrate; these parts of the boundary are therefore coherent twin interfaces, as shown schematically in Fig. 29. This type of configuration was first noticed by Matthews (1962), and its existence was convincingly demonstrated by Dickson *et al.* (1965) and Jacobs and Stowell (1965). The (111) portions of these boundaries are rarely perfectly planar; they frequently contain steps that may be interpreted as twinning dislocations if the step is only one interplanar spacing high; otherwise they are short segments of incoherent twin boundary. An example of a network of twinning dislocations in an epitaxial Au film is shown in Fig. 30. It is to be expected that these twinning dislocations would be eliminated as the deposit thickens because the free energy of the system can be decreased if the incoherent segments of the boundary move together to remove the (111) boundary areas (Fig. 29). This process would be similar to that involved in the complete elimination of twin lamellae discussed earlier in this section.

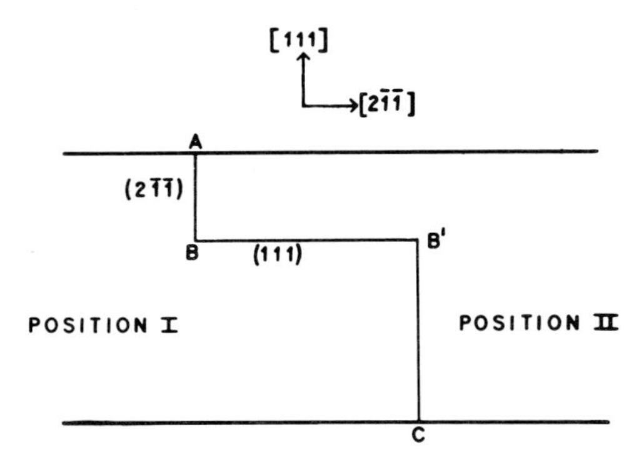

FIG. 29. Coherent DP boundary (BB′) in a (111) film (Dickson *et al.*, 1965, reproduced by permission of Taylor and Francis Ltd.).

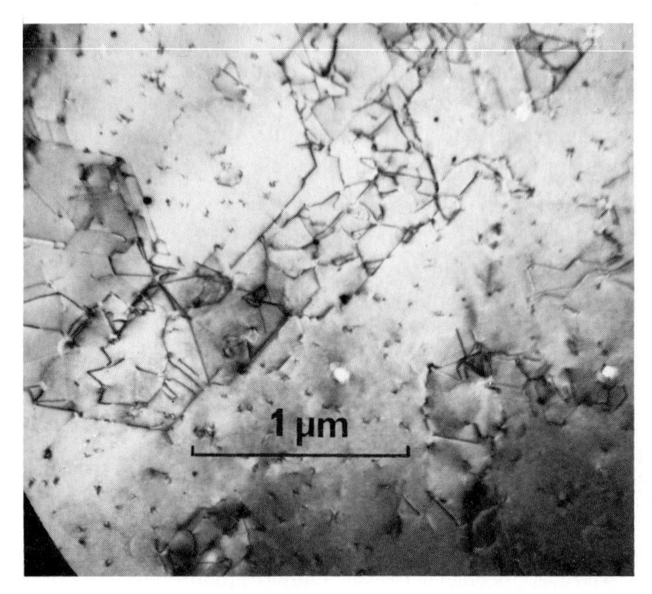

FIG. 30. Twinning dislocations at a coherent DP twin boundary in a (111) gold film grown on MoS_2 in UHV (T. J. Law, unpublished).

VI. Multiply Twinned Particles

Although multiply twinned particles (MTPs) should logically be treated in the section concerned with twin structures, these crystallographic features are sufficiently interesting and fundamentally important to be considered separately. We have already seen in the preceding section that in homoepitaxial Si and GaAs deposits multiply twinned regions are found that are apparently quite stable during growth of the films and that seem to be nucleated at substrate imperfections. The types of multiply twinned structures being considered here are different, in that they usually appear in the very early stages of the nucleation of metal deposits on substrates to which they are weakly bound (e.g., Au on KCl or NaCl). Allpress and Sanders (1964) and Mihama and Yasuda (1964) noticed anomalous (111) reflections in electron-diffraction patterns of thin (100) vapor deposits; the latter explained their result in terms of compound particles that are singly twinned. Later work by Ino (1966), Allpress and Sanders (1967), and others has shown this model to be incorrect. By careful analysis of the electron-diffraction patterns, supplemented by dark-field electron microscopy, the complex structures of these particles have been elucidated. In addition, these studies have led to interesting developments in the understanding of the initial formation of thin films.

Most of the illustrations in this section will refer to fcc metals grown on (100) surfaces of alkali halides, so it is relevant to consider first the main crystallite orientations that occur in these systems (Matthews, 1966b; Ino, 1966). In addition to the epitaxial (100) deposit, a variety of (111) crystals appear that are aligned with their close-packed directions parallel to low-index directions in the substrate surface. These are shown in Fig. 31, from which it is

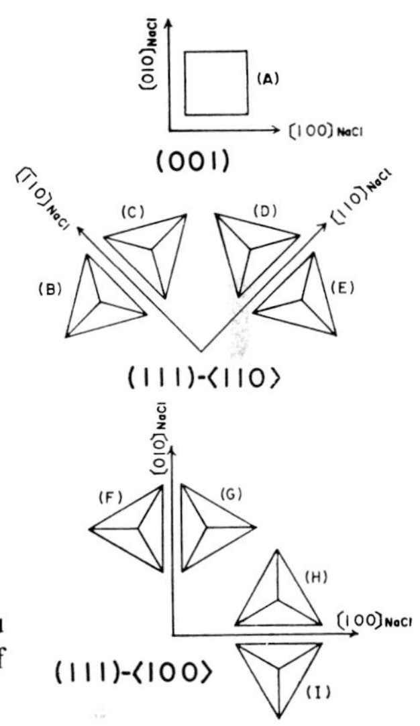

FIG. 31. Orientations of nuclei found in deposits of Au on NaCl in UHV (Ino, 1966, reproduced by permission of The Physical Society of Japan).

evident that there are eight principal orientations of the (111) nuclei. One set has a ⟨110⟩ deposit direction parallel to a ⟨110⟩ direction in the substrate. There are two ways to do this, and the number of orientations is doubled by double-positioning. Similarly, four orientations are obtained when the ⟨110⟩ deposit direction is parallel to ⟨100⟩ substrate directions. In (100) orientation, four {200} reflections are to be expected from the (100) nuclei, and all the (111) nuclei orientations in Fig. 31 give rise to a total of 24 {220} reflections; usually the 12 (220) reflections from the (111) nuclei having their ⟨110⟩ directions aligned with the substrate ⟨110⟩ directions are the strongest. The typical electron-diffraction pattern corresponding to the initial stage of growth of a Au deposit on NaCl at ~150°C in Fig. 32 shows the anomalous (111) reflections that arise from multiple twinning of the (111) nuclei.

Several different structures of MTPs have been successfully identified, and these are shown schematically in Figs. 33 and 34. These are the three most commonly found particle shapes and have pentagonal, hexagonal, and rhombic symmetry, as illustrated in these figures. Their twin structures were

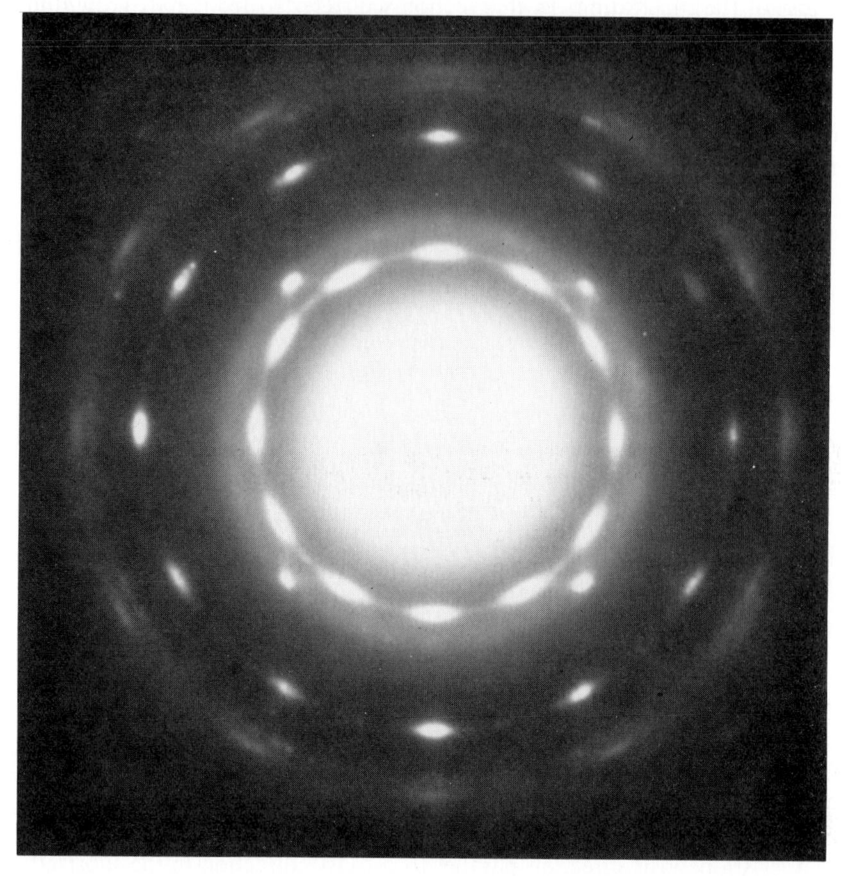

FIG. 32. Electron-diffraction pattern of Au deposit on NaCl, showing anomalous {111} reflections due to multiple twinning (D. J. Stirland, unpublished).

elegantly unravelled by Ino (1966) and Ino and Ogawa (1967). The first two of these can be described in terms of multiple twinning of the original (111) island of tetrahedral shape. In Fig. 33a, the tetrahedron OABC contains primary twins on each of its faces OAB and OAC, and these tetrahedral twins are again twinned on their faces OFA and OLA. These secondary twins do not match exactly, and a wedge OADPE remains to be filled. Direct observations of lattice planes in such a structure by Komoda (1967) have failed to detect this void, so it is concluded that the space is filled as a result of elastic strain of the whole structure, giving a coherent particle having pentagonal symmetry when viewed along the OA direction. Similarly, the hexagonal structure can be formed by placing primary twins on all three (111) faces of the nucleus and then putting secondary twins on the two upward-pointing faces of the primary twins. However, this structure is not capable of explaining all the contrast features, and Ino (1966) has proposed that the hexagonal particles contain in addition six tertiary twins, such as OSEF in Fig. 33b. This particle therefore contains a (111) nucleus and three primary, six secondary, and six tertiary twins. It will be noticed that the particle drawn in this idealized fashion contains numerous gaps, since the multiple twins do not fit exactly. As was the case for the pentagonal particle, these gaps are accommodated by overall elastic strain. Ino (1966) has pointed out that a more compact structure may be formed by filling in the hollow at the top of the structure shown in Fig. 33b with three quartic and one quintic twin, which after the gaps have been filled gives the icosahedron shown in Fig. 33c. This is more likely to be a

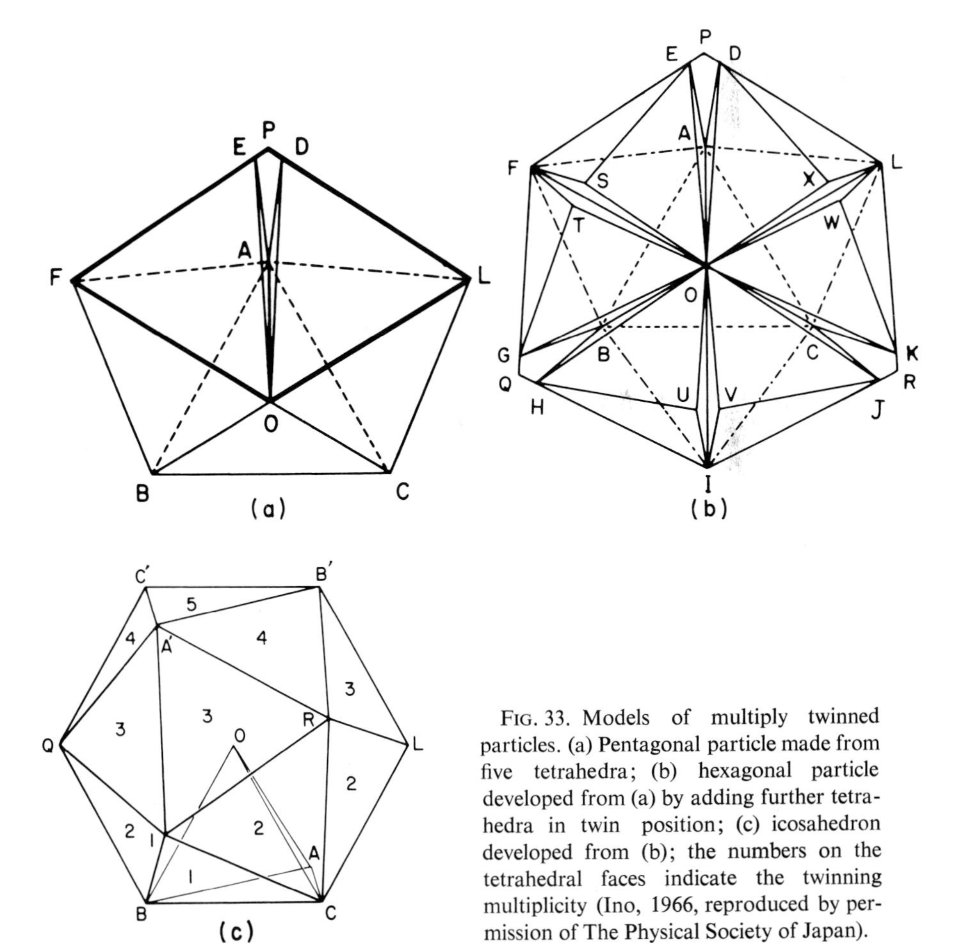

FIG. 33. Models of multiply twinned particles. (a) Pentagonal particle made from five tetrahedra; (b) hexagonal particle developed from (a) by adding further tetrahedra in twin position; (c) icosahedron developed from (b); the numbers on the tetrahedral faces indicate the twinning multiplicity (Ino, 1966, reproduced by permission of The Physical Society of Japan).

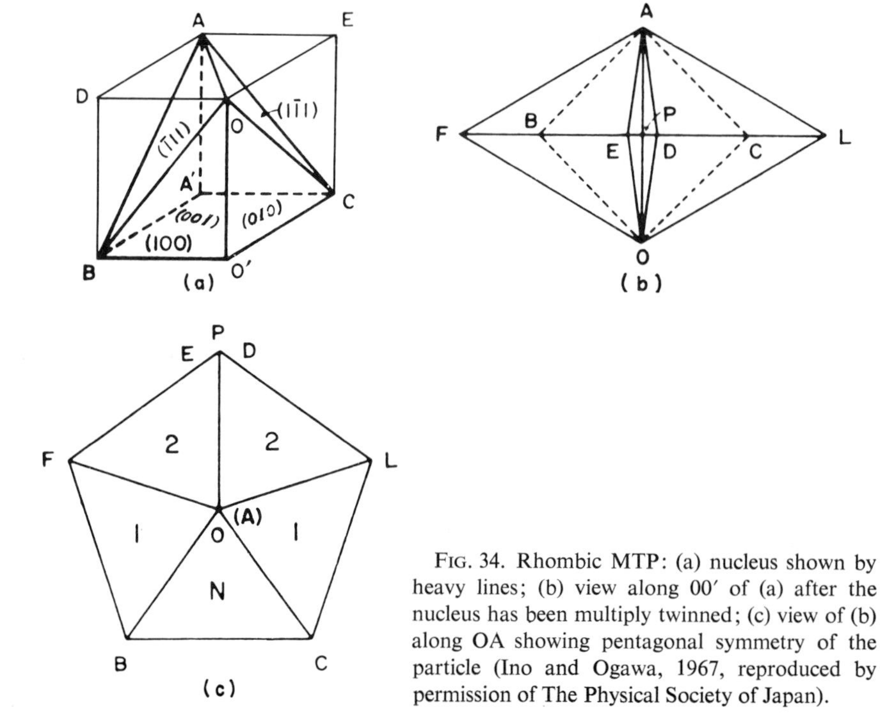

FIG. 34. Rhombic MTP: (a) nucleus shown by heavy lines; (b) view along 00′ of (a) after the nucleus has been multiply twinned; (c) view of (b) along OA showing pentagonal symmetry of the particle (Ino and Ogawa, 1967, reproduced by permission of The Physical Society of Japan).

realistic structure because the surface free energy is reduced by the extra twins.

The rhombic particles are more rarely seen and differ from those previously discussed in that they are formed around a nucleus in (100) orientation (Ino and Ogawa, 1967). In Fig. 34a a cubic island is truncated on two (111) planes OAB and OAC. Primary twins are erected on these faces and secondary twins placed on the top faces of the primary twins. This gives the shape seen in Fig. 34b when viewed along OO′, and when seen along O′A′ the pentagonal symmetry depicted in Fig. 34c is apparent. Once again, the twins are elastically strained to provide coherency. This model suffices to explain all the contrast features from these rhombic particles.

Although we have emphasized the work of Ino and Ogawa here, it is relevant that others have independently studied multiply twinned particles and have arrived at similar interpretations of the experimental results. For example, Allpress and Sanders (1967) studied MTPs in deposits of Au, Pd, and Ni on mica substrates cleaved in UHV; Gillet and Gillet (1969) and Stirland (1968) analyzed thin Au films on alkali halide substrates. MTPs are found not only in vapor deposits but also when metal vapors are condensed in gases (Kimoto and Nishida, 1967) and in electrodeposits (Schlötterer, 1964). They are also found in Ge films, and Mader (1971) has suggested that pentagonal structural units based on an extrapolation of the structure of MTPs form the basis for amorphous Ge.

The experimental observation of MTPs has led several workers to examine the stability of these structures and the possible reasons for their occurrence. It is now apparent that the bulk lattice structure is not necessarily the most favorable one energetically in very small particles, and we consider next the theoretical arguments that lend support to this statement.

The first serious attempt to explain why MTPs are more stable than crystallites that are untwinned was carried out by Fukano and Wayman (1969), who used the nearest-neighbor-bond approach to estimate surface and volume energies. From the number of atomic bonds in a particular atomic configuration, the ratio of the surface-to-volume energy per atom s/v was evaluated and used as a measure of particle stability. These authors analyzed the following particles: icosahedra, pentagonal decahedra, tetrahedra, octahedra, and pyramids, whose structures were built up from nuclei containing very small numbers of atoms so that the model structures put forward by Ino (1966), based on the stacking of tetrahedral twins, were not always reproduced. They pointed out that the minimum-sized nucleus for a pentagonal particle can be formed from the seven atoms shown in Fig. 35, which represents the most stable configuration of this number of atoms in free space. This nucleus can be built up into the particle shown in Fig. 36a, which contains 23 atoms and clearly exhibits fivefold symmetry. The atoms are not in perfect tetrahedral configurations, and the consequence of imposing maximum tetrahedral coordination is seen in Fig. 36b, from which the gaps previously noted can clearly be seen. Fukano and Wayman (1969) used models such as these in their calculations, which allowed for binding between a particle and a substrate in a rather simple and inaccurate manner, and which also included estimates of the effect of including gaps between tetrahedrally stacked atoms (cf Fig. 36b). The results of their calculations are displayed in Fig. 37 for the

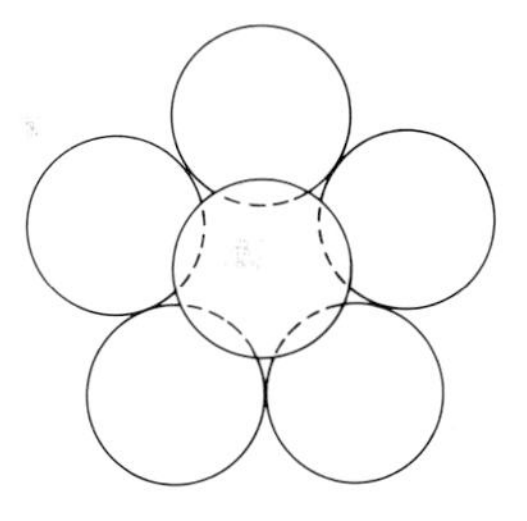

FIG. 35. Cluster of seven atoms that is the basic nucleus for a pentagonal particle. The seventh atom is at the center, below the plane of the page (Fukano and Wayman, 1969, reproduced by permission of American Institute of Physics).

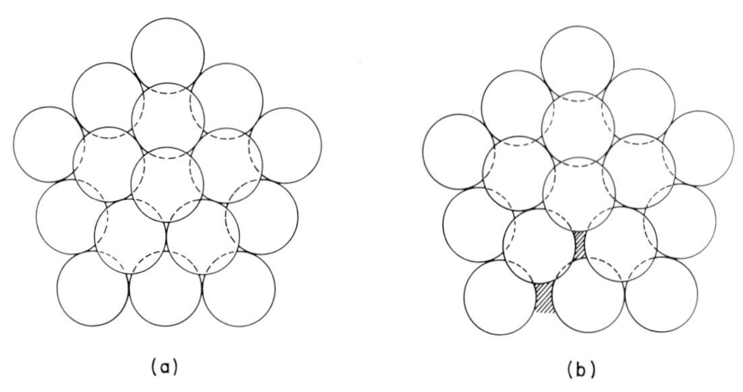

FIG. 36. (a) Pentagonal decahedron consisting of 23 atoms. Atoms are stacked above and below the plane of the outer ring of atoms. (b) Minimum-sized pentagonal decahedron formed from five twin crystals; the atoms have exact tetrahedral coordination, and the region of imperfect atomic contact is hatched (Fukano and Wayman, 1969, reproduced by permission of American Institute of Physics).

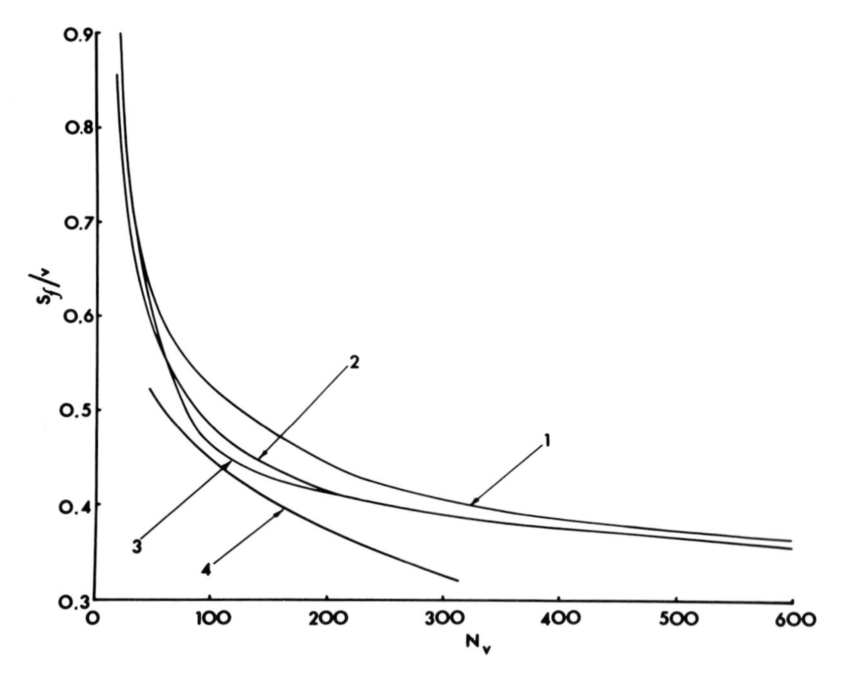

FIG. 37. Surface-to-volume energies (S_f/v) of (1) tetrahedron, (2) pentagonal decahedron, (3) octahedron, and (4) icosahedron, and (100) pyramid as a function of N_v the number of atoms in the particle (after Fukano and Wayman, 1969).

case of tetrahedral, pentagonal, octahedral, and icosahedral particles. Results for a (100)-based pyramid having inclined (111) faces are almost indistinguishable from those for the tetrahedron, and so are not shown on the figure. All the points in Fig. 37 refer to perfectly symmetrical configurations, and the lines joining these points do not accurately represent the ratio of surface-to-volume energy (indeed, s/v plots will have downward-pointing cusps at the positions of fully closed figures). These results show that tetrahedra and pyramidal (100) shapes are the least stable; the (111) octahedron is the most stable of the shapes constructed on the basis of the normal crystal structure. The pentagonal decahedron is more stable than the octahedron for small clusters ($N_v < 70$), and the icosahedron is the most stable of all the cluster geometries considered. However, Fukano and Wayman (1969) pointed out that, in large particles, the icosahedra will be highly strained internally, which would tend to reverse their conclusions about the relative stability of icosahedral and octahedral particles.

A more refined treatment of this problem was carried out by Allpress and Sanders (1970), who evaluated the interatomic interactions on a high-speed computer. Internal energies of various cluster configurations were calculated by assuming pairwise interactions using both Morse and Mie potentials. In the main part of this work only cluster configurations that correspond to complete shells of atoms were considered but, unlike the calculations of Fukano and Wayman (1969), they allowed the atom positions to be adjusted so that the minimum internal energy could be evaluated. The two different potentials used were:

$$\text{Morse potential:} \qquad E(r)/E_0 = 1 - \exp\{a(1 - r/r_0)\}^2 - 1$$

$$\text{Mie potential:} \qquad E(r)/E_0 = \{n(r_0/r)^m - m(r_0/r)^n\}/(m-n)$$

where $E(r)$ is the energy of atoms a distance r apart and has a minimum value E_0 at $r = r_0$, and the parameters a, m, and n took the following values:

$$\text{Morse potential:} \qquad a = 3, 4, 5, 6$$

$$\text{Mie potential:} \qquad m = 4.5, \quad m = 5, \quad m = 6$$

$$n = 6, 7, 8, \quad n = 7, 8, \quad n = 10, 12, 14$$

When accounting for atomic relaxation, all atoms were made to relax uniformly; further, the clusters were assumed to be free from external constraints, i.e., the influence of a substrate was not taken into account.

Five cluster configurations were evaluated, namely an fcc tetrahedron, an fcc octahedron, an fcc sphere, a pentagonal bipyramid, and an icosahedron. The pentagon was constructed from distorted tetrahedral units having dimensions shown in Fig. 38a, and the icosahedra were built up from the unit

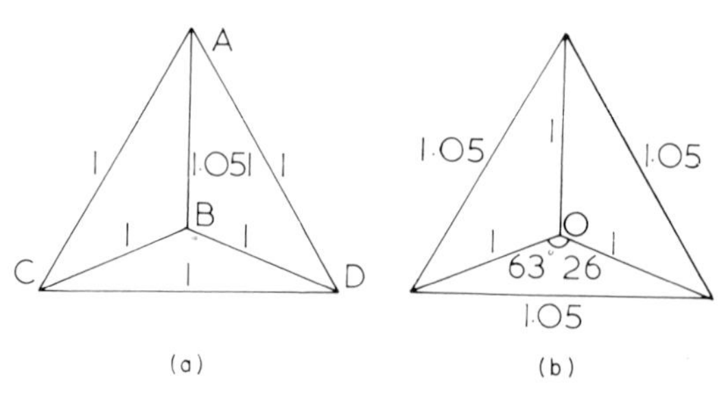

FIG. 38. Units on which MTPs were constructed by Allpress and Sanders (1970): (a) pentagonal bipyramid where AB is the fivefold axis; (b) icosahedron having 0 as center (reproduced by permission of Australian Institute of Physics).

shown in Fig. 38b. The relative stability of these clusters may be obtained from the plot of the computed energy per atom for the case of a Morse potential with $a = 4$ shown in Fig. 39. The results are similar to those of Fukano and Wayman (1969) showing that the icosahedron and the pentagon are the most stable structures for very small particles. In addition, the nearest-neighbor distances were evaluated for all the geometries examined. These calculations, although not taking the cluster–substrate interaction into account, emphasized that the most stable forms of small clusters are not what one would expect on

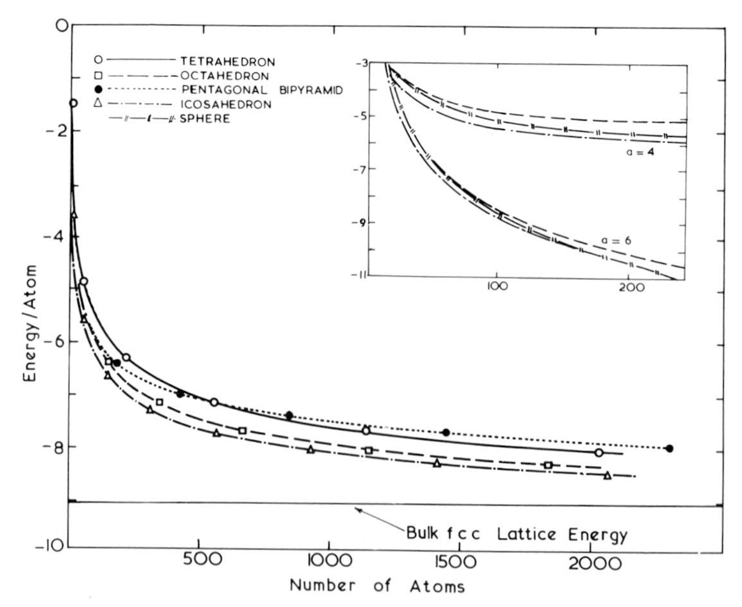

FIG. 39. Calculated energy per atom against number of cluster atoms for $a = 4$ (Morse potential). The inset shows results for a small number of atoms and also for $a = 6$ (Allpress and Sanders, 1970, reproduced by permission of Australian Institute of Physics).

the basis of the disposition of atoms found in the bulk lattice; they are, in essence, defect structures relative to the bulk space lattice.

It is worth noting here that considerable interest is currently being shown in the detailed structure of very small clusters. It is beyond the ambit of this review to go into great detail on this topic, and the interested reader is referred to several recent papers on this subject for further information (Romanowski, 1969; Burton, 1970, 1971; Dave and Abraham, 1971; Abraham and Dave, 1971) and in particular the extensive review by Hoare and Pal (1971).

The two sets of stability calculations with which we have just dealt were concerned with building up stable clusters in an atom-by-atom fashion; the structures produced are not exactly the same as those envisaged by Ino (1966) and Ino and Ogawa (1971). A different approach has been taken by Ino (1969) to assess the stability of multiply twinned particles, which employs the geometric models put forward by himself and Ogawa (Ogawa and Ino, 1967). This involves treating the structures as strained MTPs and employs macroscopic parameters such as surface free energy, twin boundary energy, and deposit–substrate adhesive energy. Ino (1969) evaluated the total free energies of normal (untwinned) particles U_n and multiply twinned particles U_m, expressed as

$$U_n = -U_c + U_s - U_a, \qquad U_m = -U_c + U_s - U_a + U_e + U_t$$

where U_c, U_s, and U_a are the cohesive energy, the surface energy, and the adhesive energy to the substrate, respectively; U_e is the elastic energy arising from the requirement to strain the MTPS to close the gaps between the stacked tetrahedral twins; and U_t is the free energy of the twin interfaces. He calculated the difference $U = U_m - U_n$ in order to assess the relative stability of clusters containing the same number of atoms. Tetrahedra, icosahedra, decahedra, a Wulff polyhedron composed of (111) and (100) faces, and a decahedron having both (111) and (100) faces were considered. The analysis was carried out with reference to perfect geometries (i.e., it did not allow for atoms that do not fit exactly into the assumed shapes). The elastic energy was evaluated using anisotropic elasticity theory, and the values for the energetic parameters were taken, for the most part, from the literature. The mathematical arguments are somewhat lengthy and will not be detailed here, but the major results of the calculations are summarized in the plot of $(U - U_c)/r$ vs r shown in Fig. 40; here r is the edge length of the polyhedron under consideration. It is worth noting that the terms involving the surface, adhesive, and twin boundary energies are proportional to r and the term involving the elastic energy is dependent on r^2. It was found that the total energy of the pentagonal decahedron is always greater than that of the Wulff polyhedron (and the decahedral Wulff polyhedron), so that the former particle is essentially unstable. For very small particles, the icosahedron is the most stable particle

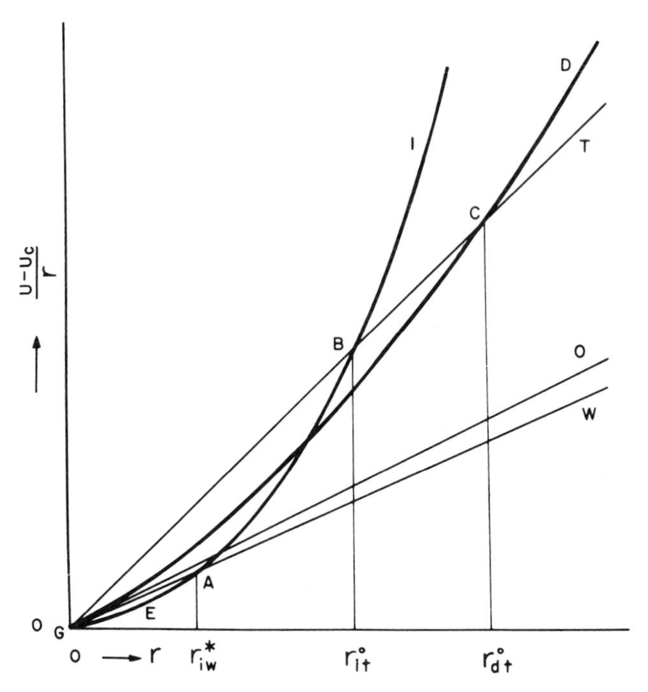

FIG. 40. Stability of icosahedral (I) and decahedral (D) MTPs compared with the tetrahedron (T), octahedron (O), and Wulff polyhedron (W) (Ino, 1969, reproduced by permission of The Physical Society of Japan).

but becomes unstable above a critical size given by r_{iw}. Ino (1969) calculated the critical dimensions indicated in Fig. 40 for both free particles and particles on a NaCl substrate.

It was argued that, because thin-film growth occurs under conditions that are far from equilibrium, the Wulff polyhedra would not be formed so that the particle-growth sequence would occur along the path GEABT in Fig. 40. The largest diameters observed for icosahedral particles of Au grown on NaCl are about 400 Å, which is in good agreement with the predicted value of 409 Å for $2r^0_{it}$. Thus far there seems to be good agreement between theory and experiment (Table I). However, it should be noted that Ino (1969) states that tetrahedra are observed in the growth of Au on NaCl; this is a suspect deduction, and it is more likely that the quasi-equilibrium shape is not the Wulff polyhedron discussed by Ino (1969) but a truncated tetrahedron, or an octahedron, as suggested by Fukano and Wayman (1969).

To summarize this discussion on MTPs, it is important to note that the observation and characterization of these fascinating structures has led to much valuable work relevant to the initial stages of formation of epitaxial

TABLE I

CALCULATED CRITICAL DIAMETERS $2r^*_{iw}$, $2r^0_{it}$, AND $2r^0_{dt}$ FOR STABLE AND QUASI-STABLE STATES OF ATOMIC CLUSTERS[a]

	$\gamma_a = 0$ (in free space)			$\gamma_a \neq 0$ (on NaCl)		
	$2r^*_{iw}$ [Å]	$2r^0_{it}$ [Å]	$2r^0_{dt}$ [Å]	$2r^*_{iw}$ [Å]	$2r^0_{it}$ [Å]	$2r^0_{dt}$ [Å]
Au	106.8	436.3	3961	102.1	409.4	3550
Ag	75.6	306.9	2905	69.8	273.5	2373
Cu	67.6	279.1	2833	63.2	259.2	2404
Ni	43.1	212.6	3385	39.0	189.4	2765
Pd	49.9	263.9	2447	44.8	234.6	1992
Pt	56.3	238.4	3784	51.9	213.4	3118
Pb	97.5	446.1	2258	89.2	398.5	1852
Al	29.3	486.1	2199	15.4	406.4	1602
Si	27.6	133.7	830	25.1	119.2	679
Ge	15.5	109.8	725	13.3	96.9	583
γ-Fe	44.0	200.6	1727	40.2	179.2	1417
β-Co	97.6	375.9	1574	90.9	337.8	1305

[a] After Ino, 1969.

deposits. Although there is controversy about the precise way in which these particles are grown (Gillet and Gillet, 1972) and about their energetics, it is evident that they are defect structures that can become energetically unstable as growth proceeds; this factor leads one to enquire further into the consequences of their existence. Energetically unstable MTPs may behave in a quite different manner to clusters that are stable (or less unstable). Indeed, it appears that MTPs are able to migrate as a whole on alkali halide substrates (Metois *et al.*, 1972).

VII. Defects in Alloys and Compounds

The object of this section is to describe some of the defects peculiar to alloys and compound structures, including superdislocations and planar defects such as antiphase domain boundaries (APBs). Epitaxial films of the structures considered here will, of course, contain normal dislocations and, where the crystallography is suitable, stacking faults and twins.

There exists a large number of ordered structures, too many to deal with in detail in this section, and the reader is referred to the excellent review by Marcinkowski (1963) for an account of the crystallography and defects found in some of the more common ordered materials. In addition, Amelinckx

(1972) deals with more recent advances in electron-microscope contrast effects from antiphase boundaries.

Consider first what is meant by an antiphase boundary. This is illustrated for a two-dimensional simple cubic compound of A and B atoms in Fig. 41. Within the region outlined by the dashed square, A and B atoms have been interchanged so that across the dashed boundary A atoms face A atoms. This interface is an APB and encloses an antiphase domain. In a real crystal, such as CuAuI, which is fc tetragonal and has complete layers of Cu and Au atoms alternating normal to the *c*-axis, APBs give rise to fringe patterns in an electron-microscope image, as in Fig. 42. Next consider the permissible glide displacements in Fig. 41. It is evident that the passage of a dislocation with Burgers vector $\frac{1}{2}[01]$ on an (01) plane creates an APB in its wake. In order not to create such a boundary, a dislocation must have a Burgers vector such as [01]; this is called a "superdislocation." A [01] superdislocation may dissociate into a $\frac{1}{2}[01]$ pair that bounds an APB, as in the upper part of Fig. 41. It is also evident from the lower part of this figure that, if an APB terminates within a single crystal, it does so at a dislocation.

Another type of defect in an ordered crystal is illustrated in Fig. 43. Here the structure is layered, and the boundary is a twin boundary; this is called an "order twin boundary" or a "coherent domain boundary" and is observed, for example, in CuAuI (Hunt and Pashley, 1963).

So far in this section we have been concerned with centrosymmetrical

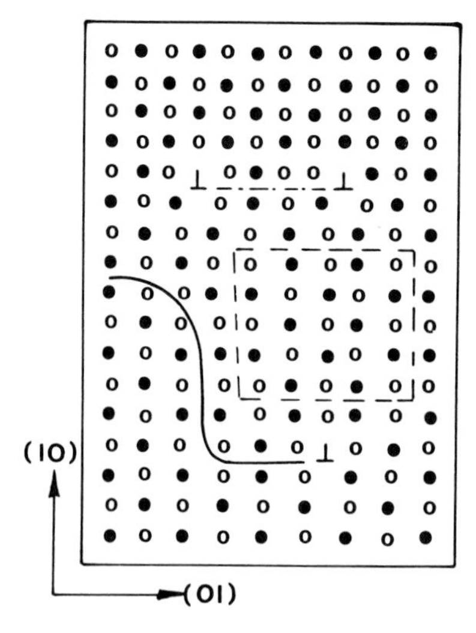

FIG. 41. Two-dimensional ordered lattice of A (●) and B(○) atoms showing an antiphase domain, APBs, and a superdislocation.

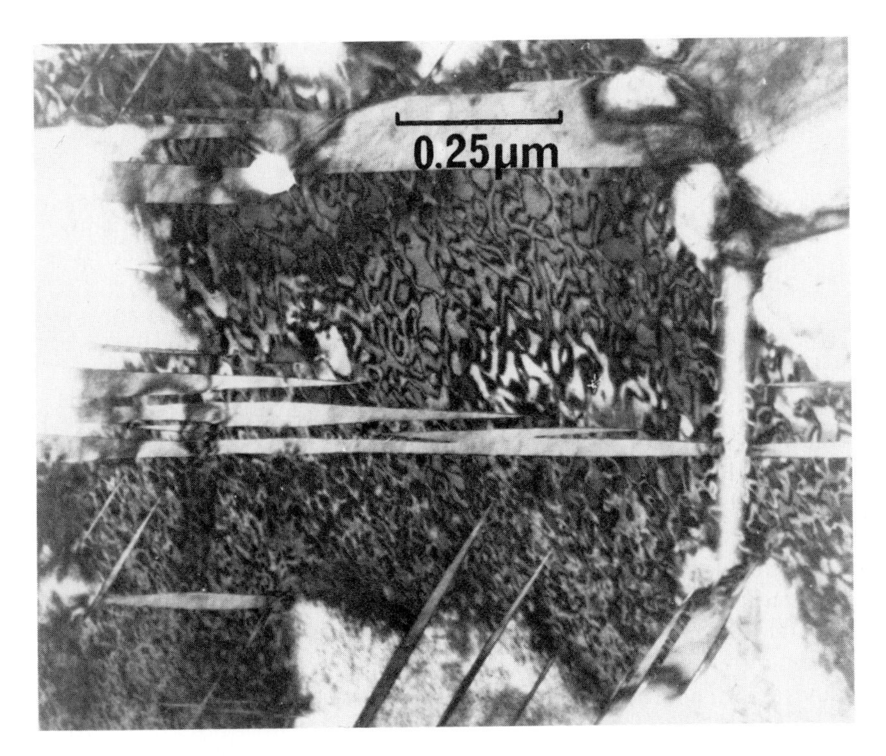

FIG. 42. Antiphase domain boundaries in a (111) film of CuAuI; the lenticular features are order twins on {101} planes.

structures. Additional factors evolve from consideration of defects in non-centrosymmetrical ordered crystals, such as compounds of group III–V and II–VI elements, which can solidify with the sphalerite structure; this gives an ordered superlattice in which interpenetrating fcc lattices of A and B atoms are displaced by $\frac{1}{4}a_0\langle 111\rangle$. The $(1\bar{1}0)$ section of the ordered sphalerite lattice

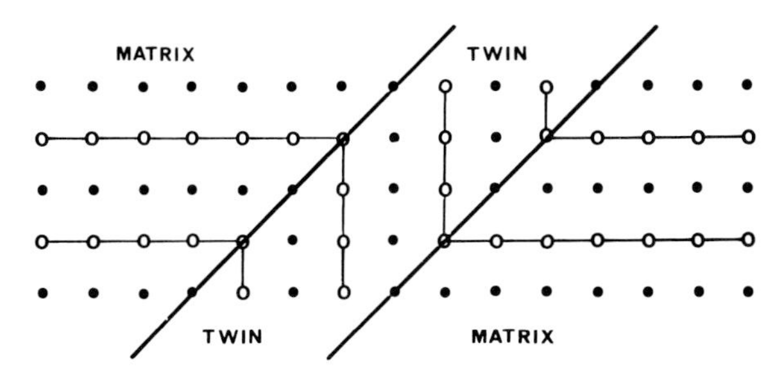

FIG. 43. Schematic diagram of an order twin for A (●) and B (○) atoms.

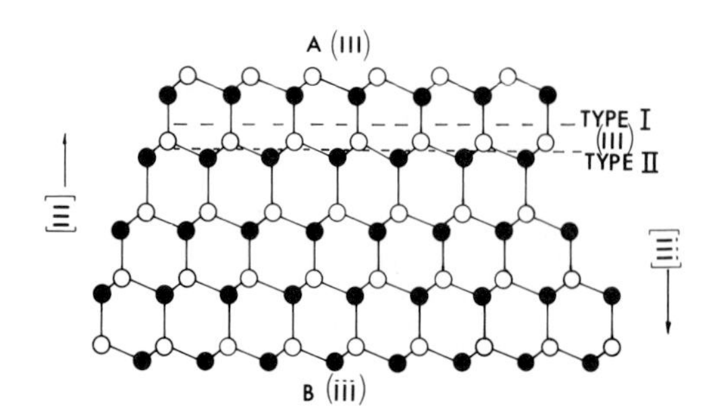

FIG. 44. Sphalerite structure [(1$\bar{1}$0) projection] showing two possible {111} surface types (Holt, 1969).

is shown in Fig. 44. Two distinct {111} surfaces can exist that contain either all A or all B atoms. Holt (1969) examined theoretically the defect structures in this lattice and showed that two different types of APB are possible. These are illustrated in Fig. 45. In Fig. 45a an APB is drawn at which there are equal numbers of wrong A–A and B–B bonds (a Type I APB), whereas in Figs. 45b and c respectively, only wrong B–B and A–A bonds exist (Type II APBs). The latter type of APB is important in that it represents an *excess* of one class of atom in the crystal; it may be viewed as a thin planar precipitate. In compounds grown under nonstoichiometric conditions, this type of defect may be prominent.

Holt (1969) also demonstrated that two different forms of twin are possible in the sphalerite lattice. These are shown in Figs. 46a and b, and from Fig. 46c it is evident that when inverted and upright twins join an APB must run into the twin boundary.

It is unfortunate that little experimental work has been performed to observe the defects just described in epitaxially grown ordered films, although extensive studies have been carried out on sequentially deposited films that have been subsequently interdiffused and then ordered (see, for example, Pashley and Presland, 1959; Sato, 1964; Hunt and Pashley, 1963). Mihama (1971) has detected APBs in crystallites of CuAu grown on NaCl, and Stoemenos and Vincent (1972) have studied order twins in annealed films of GeTe and GeTe–SnTe. In the latter case deposits crystallized in a noncentrosymmetrical rhombohedral phase with a very high defect structure in the as-deposited state. Annealing was necessary in order that the defects would be recognized. Multiple positioning of the initial nuclei leads to the possibility that order twins are formed to relieve internal strains when islands coalesce [analogous to the situation that obtains during ordering of bulk crystals (Tanner and Ashby, 1969)].

Finally, we deal with a mechanism of defect generation in alloyed films

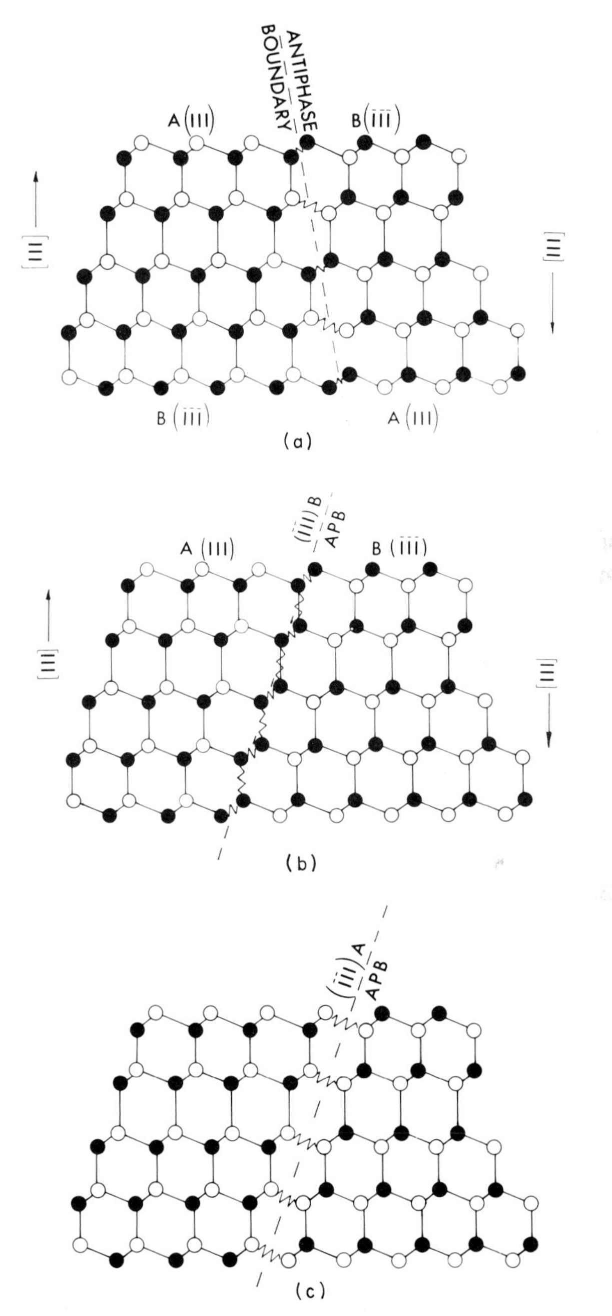

Fig. 45. Antiphase boundaries in the sphalerite structure [$(1\bar{1}0)$ projection]. Zig-zag lines denote wrong bonds. (a) Type I APB on $(\bar{1}\bar{1}3)$ plane; (b) type II APB on $(\bar{1}\bar{1}1)$ involving wrong B–B bonds only and an excess of B atoms; (c) type II APB involving wrong A–A bonds with an excess of A atoms (Holt, 1969).

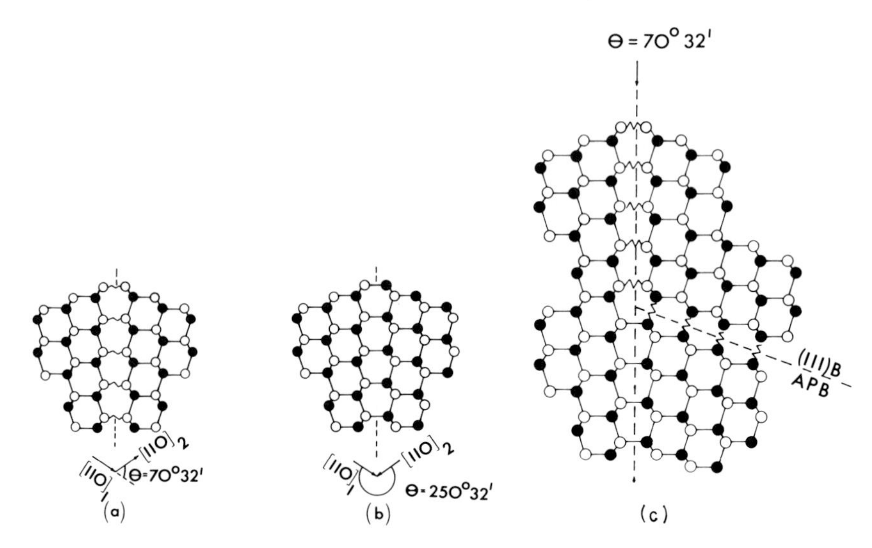

FIG. 46. Twin boundaries on {111} in the sphalerite lattice. Part (a) shows an inverted twin and (b) an upright twin. Where the two join (c), an APB is present (Holt, 1969).

that is particularly relevant to commercially important semiconducting epitaxial deposits. This mechanism depends upon a change in lattice parameter with solute content to produce the internal stresses that lead to defect generation; it operates in diffused layers and graded heterojunctions. Prussin (1961) first discussed this mechanism in detail in connection with dislocation generation in boron- and phosphorous-diffused Si wafers. He assumed that in a crystal containing a gradient in solute concentration C the strain components ε are linearly related to concentration, i.e., $\varepsilon = -\beta C$, where β is a constant. Consider a crystal containing a concentration gradient in the z-direction and consider an element lying between z and $z+dz$ in which the dislocation density is ρ. The dimensional change dx in an orthogonal direction is $\rho\alpha\, dz$, where α is the x-component of the Burgers vector and, if the dislocations compensate the lattice strain due to the solute, we have

$$dx = \beta(\partial C/\partial z)\, dz$$

So

$$\rho = (\beta/\alpha)\, \partial C/\partial z$$

This relates the dislocation density to the concentration gradient, and Prussin (1961) has treated in detail the case for diffused layers. Experimental support for this model was obtained by Washburn *et al.* (1964) for (100) Si wafers and by Joshi and Wilhelm (1965), Lawrence (1966), and Levine *et al.* (1967a) for (111) crystals. It evolves that the misfit dislocations move in from the diffusion-source surface, mainly by glide in (100) crystals but by nonconservative motion in other orientations. In addition precipitates are produced in diffused crystals (Levine *et al.*, 1967b).

A detailed analysis of dislocation generation by this mechanism during epitaxial growth is given by Abrahams *et al.* (1969), who investigated graded $GaAs_{1-x}P_x$ heterojunctions. The deposits were grown on (100) GaAs substrates and the phosphorous concentration was varied by changing the arsine and phosphine flow rates in the vapor streams. The authors noted that, because the deposit nucleates heterogeneously, the misfit dislocations should be *segmented* (Fig. 47) and will not initially form an ideal cross grid of continuous dislocations. In the case of a single misfit plane the linear dislocation density in the interface is $n_L = \Delta a/a^2$, where a is the mean lattice parameter and Δa the change in lattice parameter across the misfit plane. The density of segmented dislocations in the interface n_s is therefore $2n_L/L_{av}$, where L_{av} is the average total dislocation length. Since the segmented dislocations cannot end in the misfit plane, $n_I(=4n_L/L_{av})$ *inclined* dislocations must thread unit area of the interface.

When a linear concentration gradient exists, the density of segmented dislocations threading a plane normal to the interface is given by

$$n_A = (\Delta a/a^2)\, dC/dz$$

where now Δa is the difference in lattice parameter across the graded zone;

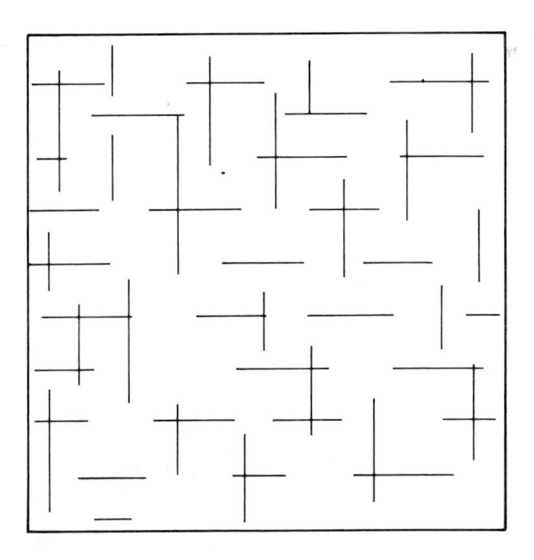

FIG. 47. Schematic diagram illustrating segmented misfit dislocations (Abrahams *et al.*, 1969, reproduced by permission of Chapman and Hall Ltd.).

correspondingly $n_s = 2n_A/L_{av}$. Again, there must be *inclined* dislocations, and Abrahams *et al.* (1969) argued that, once sufficient dislocations have been formed in the early stages of growth to accommodate the strain induced by the concentration gradient, no new inclined dislocations will be formed and n_I is related simply to n_A by

$$n_I = n_s n_A^{-1/2} \propto (dC/dz)^{1/2}$$

These relationships were verified experimentally. Of course, the misfit dislocations, which had Burgers vectors lying in the interface plane and at 45° to it, interacted to form partial networks, and the main configurations observed are seen in Fig. 48. This elegant work demonstrates not only how the misfit dislocations propagated through the graded region but also that dislocation configurations are dependent on the initial stage of strain accommodation.

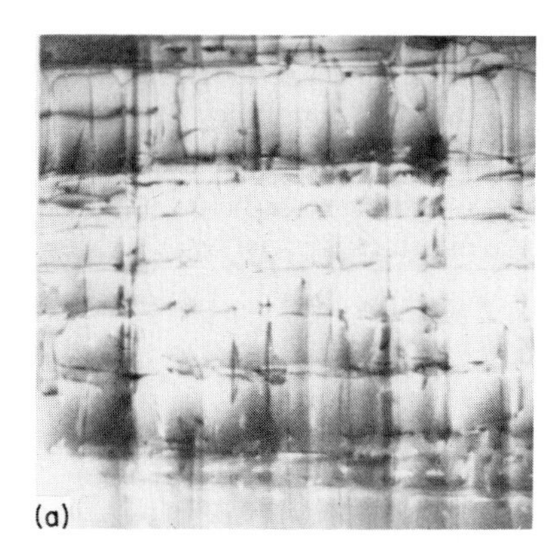

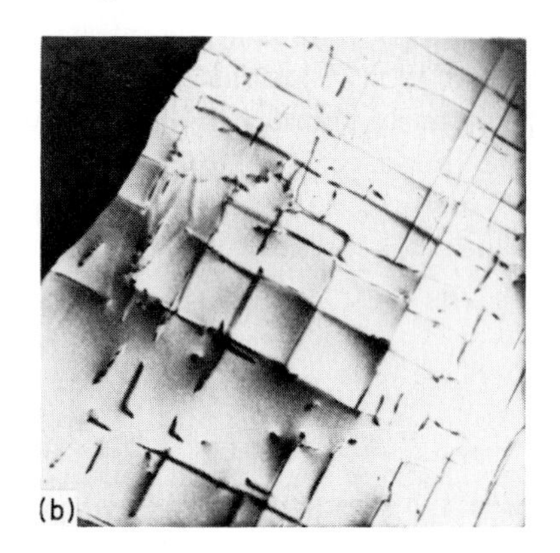

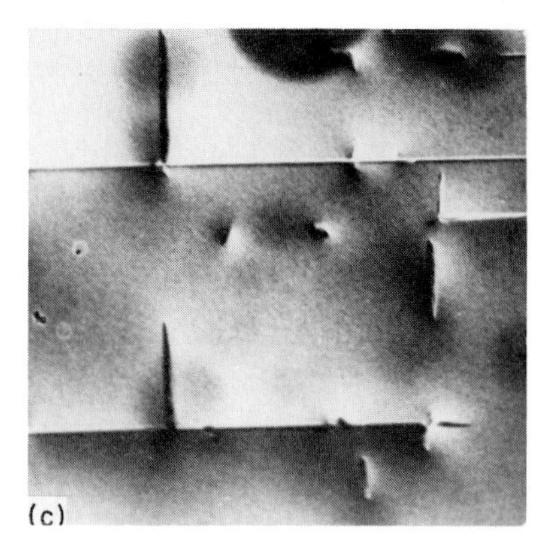

FIG. 48. Misfit-dislocation networks in graded $GaAs_{1-x}P_x$ heterojunctions grown with three different concentration gradients: (a) 5% P/μm; (b) 1.7% P/μm; (c) 0.21% P/μm (Abrahams *et al.*, 1969, reproduced by permission of Chapman and Hall Ltd.).

VIII. Conclusions

The main theme of this chapter has been to illustrate the means by which crystal defects are generated during the growth of epitaxial films. It is apparent that the constraints imposed by a substrate and the often highly nonequilibrium conditions under which epitaxial growth is obtained can both influence markedly the defect structure. In addition, it should be evident that much more fundamental research is needed before the roles of these factors can be completely understood. This survey has not been exhaustive, and interesting details relevant to specific deposit–substrate systems have been omitted. Also, because of the limited scope of this survey factors such as the influence of incompatibility of deposit–substrate thermal expansivities, high-energy radiation, and ion implantation have been ignored.

It can be concluded that the dominant source of imperfections in epitaxial films is the misfit between deposit and substrate (and any contaminant that may exist on the substrate). If the misfit is small, the defects arise mainly as a result of coherency loss between substrate and overgrowth (which is dealt with more thoroughly in Chapters 6–8). If the misfit is large, the defects arise predominantly from the lack of exact lattice registry between heterogeneously nucleated deposit particles. In intermediate situations, which have not been closely studied, both factors are expected to be important. Therefore, unless special heat treatments are employed following the growth process, it is unrealistic to expect to obtain a defect-free epitaxial film.

ACKNOWLEDGMENTS

The author is grateful to the many people who have provided illustrations for this article and to those who have assisted in its preparation. Special thanks are due to Dr. D. W. Pashley, F.R.S., for his encouragement and constructive comments.

This chapter is published by permission of the Chairman of Tube Investments Limited.

References

Abraham, F. F., and Dave, J. V. (1971). *J. Chem. Phys.* **55**, 1587.

Abrahams, M. S., Weisberg, L. R., Buiocchi, C. J., and Blanc, J. (1969). *J. Mater. Sci.* **4**, 223.

Allpress, J. G., and Sanders, J. V. (1964). *Phil. Mag.* **10**, 645.

Allpress, J. G., and Sanders, J. V. (1967). *Surface Sci.* **7**, 1.

Allpress, J. G., and Sanders, J. V. (1970). *Aust. J. Phys.* **23**, 23.

Amelinckx, S. (1964). The direct observation of dislocations. *Solid State Phys. Suppl. 6.*

Amelinckx, S. (1972). *Surface Sci.* **31**, 296.

Amelinckx, S., Gevers, R., Remaut, G., and Van Landuyt, J. (eds.) (1970). "Modern Diffraction and Imaging Techniques in Material Science." North-Holland Publ., Amsterdam.

Bassett, G. A. (1960). *Proc. Eur. Regional Conf. Electron Microsc., Delft* (A. L. Houwink and B. J. Spits, eds.), Vol. 1, p. 270. De Nederlandse Vereniging voor Electronmikrosc., Delft.

Bassett, G. A. (1964). *Proc. Int. Symp. Condensation Evaporation Solids* (E. Rutner, P. Goldfinger, and J. P. Hirth, eds.), p. 599. Gordon and Breach, New York.
Bassett, G. A., and Pashley, D. W. (1959). *J. Inst. Metals* **87**, 449.
Bassett, G. A., Menter, J. W., and Pashley, D. W. (1958). *Proc. Roy. Soc. London* **A246**, 345.
Bassett, G. A., Menter, J. W., and Pashley, D. W. (1959). *In* "Structure and Properties of Thin Films" (C. A. Neugebauer, J. B. Newkirk, and D. A. Vermilyea, eds.), p. 11. Wiley, New York.
Bauer, E. (1958). *Z. Kristallogr.* **110**, 395.
Bobb, L. C., Holloway, H., and Maxwell, K. H. (1966). *J. Phys. Chem. Solids* **27**, 1679.
Booker, G. R. (1964). *Discuss. Faraday Soc.* **38**, 298.
Booker, G. R. (1965). *Phil. Mag.* **11**, 1007.
Booker, G. R. (1966). *J. Appl. Phys.* **37**, 441.
Booker, G. R., and Joyce, B. A. (1966). *Phil. Mag.* **14**, 301.
Booker, G. R., and Stickler, R. (1962). *J. Appl. Phys.* **33**, 3281.
Burbank, R. D., and Heidenreich, R. D. (1960). *Phil. Mag.* **5**, 373.
Burgers, J. M. (1940). *Proc. Phys. Soc. (London)* **52**, 23.
Burton, J. J. (1970). *J. Chem. Phys.* **52**, 345.
Burton, J. J. (1971). *Surface Sci.* **26**, 1.
Chopra, K. L. (1969). "Thin Film Phenomena." McGraw-Hill, New York.
Coopersmith, B., Curzon, A. E., Kimoto, K., and Lisgarten, N. D. (1966). *In* "Basic Problems in Thin Film Physics" (R. Niedermayer and H. Mayer, eds.), p. 83. Vandenhoeck and Ruprecht, Gottingen.
Cottrell, A. H. (1953). "Dislocations and Plastic Flow in Crystals." Oxford Univ. Press, London and New York.
Dave, J. V., and Abraham, F. F. (1971). *Surface Sci.* **26**, 557.
Dickson, E. W., Jacobs, M. H., and Pashley, D. W. (1965). *Phil. Mag.* **11**, 575.
Finch, R. H., Queisser, H. J., Thomas, G., and Washburn, J. (1963). *J. Appl. Phys.* **34**, 406.
Frank, F. C., and Van der Merwe, J. H. (1949). *Proc. Roy. Soc. London* **A198**, 216.
Friedel, J. (1956). "Les Dislocations." Gauthiers-Villars, Paris. English ed. (1964). Pergamon, Oxford.
Fukano, Y., and Wayman, C. M. (1969). *J. Appl. Phys.* **40**, 1656.
Fullman, R. L. (1951). *J. Appl. Phys.* **22**, 448.
Gillet, E., and Gillet, M. (1969). *Thin Solid Films* **4**, 171.
Gillet, E., and Gillet, M. (1972). *J. Crystal Growth* **13/14**, 212.
Hall, E. O. (1954). "Twinning and Diffusionless Transformations in Metals." Butterworths, London and Washington, D.C.
Hall, M. J., and Thompson, M. W. (1961). *Brit. J. Appl. Phys.* **12**, 495.
Heidenreich, R. D. (1964). "Fundamentals of Transmission Electron Microscopy." Wiley (Interscience), New York.
Hirsch, P. B., Howie, A., Nicholson, R. B., Pashley, D. W., and Whelan, M. J. (1965). "Electron Microscopy of Thin Crystals." Butterworths, London and Washington, D.C.
Hirth, J. P., and Lothe, J. (1968). "Theory of Dislocations." McGraw-Hill, New York.
Hoare, M. R., and Pal, P. (1971). *Advan. Phys.* **20**, 161.
Hoffman, R. W. (1966). *In* "The Use of Thin Films in Physical Investigations" (J. C. Anderson, ed.), p. 261. Academic Press, New York.
Holloway, H. (1966). *In* "The Use of Thin Films in Physical Investigations" (J. C. Anderson, ed.) p. 111. Academic Press, New York.
Holloway, H., Wollmann, K., and Joseph, A. S. (1965). *Phil. Mag.* **11**, 263.
Holloway, H., Richards, J. L., Bobb, L. C., Perry, J., and Zimmerman, E. (1966). *J. Appl. Phys.* **37**, 4694.

Holloway, H., and Bobb, L. C. (1967a). *J. Appl. Phys.* **38**, 2711.
Holloway, H., and Bobb, L. C. (1967b). *J. Appl. Phys.* **38**, 2893.
Holt, D. B. (1962). *J. Phys. Chem. Solids* **23**, 1353.
Holt, D. B. (1969). *J. Phys. Chem. Solids* **30**, 1297.
Hornstra, J. (1958). *J. Phys. Chem. Solids* **5**, 129.
Hunt, A. M., and Pashley, D. W. (1963). *J. Aust. Inst. Met.* **8**, 61.
Ino, S. (1966). *J. Phys. Soc. Japan* **21**, 346.
Ino, S. (1969). *J. Phys. Soc. Japan* **27**, 941.
Ino, S., and Ogawa, S. (1967). *J. Phys. Soc. Japan* **22**, 1365.
Jaccodine, R. J. (1963). *Appl. Phys. Lett.* **2**, 201.
Jacobs, M. H., and Stowell, M. J. (1965). *Phil. Mag.* **11**, 591.
Jacobs, M. H., Pashley, D. W., and Stowell, M. J. (1966). *Phil. Mag.* **13**, 129.
Jaeger, H., Mercer, P. D., and Sherwood, R. G. (1968). *Surface Sci.* **11**, 265.
Jesser, W. A., and Matthews, J. W. (1967). *Phil. Mag.* **15**, 1097.
Jesser, W. A., and Matthews, J. W. (1968). *Acta Met.* **16**, 1307.
Joshi, M. L., and Wilhelm, F. (1965). *J. Electrochem. Soc.* **112**, 185.
Kimoto, K., and Nishida, I. (1967). *J. Phys. Soc. Japan* **22**, 940.
Komoda, T. (1967). *Jap. J. Appl. Phys.* **6**, 1047.
Lawrence, J. E. (1966). *J. Electrochem. Soc.* **113**, 819.
Levine, E., Washburn, J., and Thomas, G. (1967a). *J. Appl. Phys.* **38**, 81.
Levine, E., Washburn, J., and Thomas, G. (1967b). *J. Appl. Phys.* **38**, 87.
Mader, S. (1971). *J. Vac. Sci. Technol.* **8**, 247.
Marcinkowski, M. J. (1963). *In* "Electron Microscopy and Strength of Crystals" (G. Thomas and J. Washburn, eds.), p. 333. Wiley (Interscience), New York.
Matthews, J. W. (1959). *Phil. Mag.* **4**, 1017.
Matthews, J. W. (1962). *Phil. Mag.* **7**, 915.
Matthews, J. W. (1966a). *Phil. Mag.* **13**, 1207.
Matthews, J. W. (1966b). *J. Vac. Sci. Technol.* **3**, 133.
Matthews, J. W. (1967). *Phys. Thin Films* **4**, 137.
Matthews, J. W., and Allinson, D. L. (1963). *Phil. Mag.* **8**, 1283.
Matthews, J. W., and Allinson, D. L. (1964). *Phil. Mag.* **10**, 9.
Mendelson, S. (1964a). *In* "Single-Crystal Films" (M. H. Francombe and H. Sato, eds.), p. 251. Pergamon: Oxford.
Mendelson, S. (1964b). *J. Appl. Phys.* **35**, 1570.
Mendelson, S. (1967). *Surface Sci.* **6**, 233.
Metois, J. J., Gauch, M., Masson, A., and Kern, R. (1972). *Thin Solid Films* **11**, 205.
Mihama, K. (1971). *Proc. Int. Congr. Electron Microsc., 7th, Grenoble, 1970* p. 405. Soc. Fr. Microsc. Electron., Paris.
Mihama, K., and Yasuda, Y. (1964). Work reported by Ogawa and Ino (1971).
Nabarro, F. R. N. (1967). "Theory of Crystal Dislocations." Oxford Univ. Press, London and New York.
Ogawa, S., and Ino, S. (1971). *In* "Advances in Epitaxy and Endotaxy" (H. G. Schneider and V. Ruth, eds.), p. 183. VEB Deutscher Verlag für Grundstoffindustrie, Leipzig.
Pashley, D. W. (1956). *Advan. Phys.* **5**, 173.
Pashley, D. W. (1959). *Phil. Mag.* **4**, 324.
Pashley, D. W. (1964). *In* "Thin Films," p. 59. ASM, Metals Park, Ohio.
Pashley, D. W. (1965). *Advan. Phys.* **14**, 327.
Pashley, D. W., and Presland, A. E. B. (1959). *J. Inst. Metals* **87**, 419.
Pashley, D. W., and Stowell, M. J. (1962). *Proc. Int. Conf. Electron Microsc., 5th, Philadelphia, Pennsylvania* (S. S. Breese, ed.), paper GG–1. Academic Press: New York.

Pashley, D. W., and Stowell, M. J. (1963). *Phil. Mag.* **8**, 1605.
Pashley, D. W., and Stowell, M. J. (1966). *J. Vac. Sci. Technol.* **3**, 156.
Pashley, D. W., Stowell, M. J., Jacobs, M. H., and Law, T. J. (1964). *Phil. Mag.* **10**, 127.
Phillips, V. A. (1960). *Phil. Mag.* **5**, 571.
Phillips, V. A. (1972). *Acta Met.* **20**, 1143.
Pocza, J. F. (1967). *Proc. Colloq. Thin Films, 2nd* p. 93. Hungarian Acad. of Sci., Budapest.
Pomerantz, D. (1967). *J. Appl. Phys.* **38**, 5020.
Prussin, S. (1961). *J. Appl. Phys.* **32**, 1876.
Read, W. T. (1953). "Dislocations in Crystals." McGraw-Hill, New York.
Read, W. T., and Shockley, W. (1950). *Phys. Rev.* **78**, 275.
Read, W. T., and Shockley, W. (1952). *In* "Imperfections in Nearly Perfect Crystals" (W. Shockley, ed.), p. 352. Wiley, New York.
Romanowski, W. (1969). *Surface Sci.* **18**, 373.
Sato, H. (1964). *In* "Single-Crystal Films" (M. H. Francombe and H. Sato, eds.), p. 341. Pergamon, Oxford.
Sato, H., and Shinozaki, S. (1971). Private communication.
Sato, H., Shinozaki, S., and Cicotte, L. J. (1969). *J. Vac. Sci. Technol.* **6**, 62.
Schlötterer, H. (1964). *Z. Krist.* **119**, 321.
Schoening, F. R. L., and Baltz, A. (1962). *J. Appl. Phys.* **33**, 1442.
Silcox, J., and Hirsch, P. B. (1959). *Phil. Mag.* **4**, 72.
Sloope, B. W., and Tiller, C. O. (1965). *J. Appl. Phys.* **36**, 3174.
Stirland, D. J. (1968). *Thin Solid Films* **1**, 447.
Stoemenos, J., and Vincent, R. (1972). *Phys. Status Solidi (a)* **11**, 545.
Stowell, M. J. (1966). *In* "The Use of Thin Films in Physical Investigations" (J. C. Anderson, ed.), p. 131. Academic Press, New York.
Stowell, M. J. (1968). *Thin Films* **1**, 55.
Stowell, M. J. (1969). *In* "Molecular Processes on Solid Surfaces" (E. Drauglis, R. D. Gretz, and R. I. Jaffee, eds.), p. 461. McGraw-Hill: New York.
Stowell, M. J. (1972). *Thin Solid Films* **12**, 341.
Stowell, M. J., and Law, T. J. (1966). *Phys. Status Solidi* **16**, 117.
Stowell, M. J., and Law, T. J. (1969). *Phil. Mag.* **19**, 1257.
Tanner, L. E., and Ashby, M. F. (1969). *Phys. Status Solidi* **33**, 59.
Thomas, G. (1962). "Transmission Electron Microscopy of Metals." Wiley, New York.
Thomas, R. N., and Francombe, M. H. (1971). *Surface Sci.* **25**, 357.
Turnbull, D. (1951). *Trans. AIME* **191**, 3.
Washburn, J., Thomas, G., and Queisser, H. J. (1964). *J. Appl. Phys.* **35**, 1909.
Whelan, M. J., and Hirsch, P. B. (1957a). *Phil. Mag.* **2**, 1121.
Whelan, M. J., and Hirsch, P. B. (1957b). *Phil. Mag.* **2**, 1303.
Yagi, K., Kobayashi, K., and Honjo, G. (1969). *J. Appl. Phys.* **40**, 3857.

CHAPTER 6

ENERGY OF INTERFACES BETWEEN CRYSTALS

J. H. van der Merwe

Department of Physics
University of Pretoria
Pretoria, South Africa

C. A. B. Ball

Department of Applied Mathematics
University of Port Elizabeth
Port Elizabeth, South Africa

I. Introduction

A. Model

An epitaxial bicrystal consists of a single-crystal substrate bonded to a single-crystal overgrowth at a common interface, the crystal lattice of the overgrowth having a definite orientation with respect to the lattice of the substrate. The phenomenon of epitaxy has been reviewed by various authors, e.g., Chopra (1969), Pashley (1965), Royer (1928), Schneider (1969), and the theory of epitaxial bicrystals by van der Merwe (1964, 1966). This chapter serves as a brief introduction to the theory concerned with the structure of epitaxial bicrystals.

The theory is restricted to bicrystals of the following type: The substrate is thick and may be treated mathematically as semiinfinite. The surface onto which the overgrowth is bonded is atomically flat and large in extent. The overgrowth crystal is homogeneously thick and two dimensional, i.e., the lateral dimensions are large compared to the thickness. The transition from the substrate to the overgrowth is sharp, and hence the interface is clearly defined between two adjacent planes of atoms. The interfacial atomic arrays of the overgrowth and substrate are rectangular and are parallel to each other.

Epitaxial bicrystals in which interdiffusion takes place at the interface have been considered by Vermaak and van der Merwe (1964, 1965) and the case of three-dimensional epitaxial island overgrowths by Jesser and Kuhlmann-Wilsdorf (1967), Cabrera (1964), Niedermayer (1968), Szostak and Moliére (1966), Jesser and van der Merwe (1971), and others.

The forces between the two crystals may be resolved into tangential and normal components with respect to the interface. These forces are assumed to be restricted to atoms in the planes immediately adjoining the interface. Due to the periodic nature of the crystals it is clear that the tangential force on an interfacial atom is periodic in terms of its lateral displacment relative to atoms on the other side of the interface. This force is generally expressed in terms of a periodic interfacial potential or potential-energy density, but the optimum form is not known. It will certainly depend on the actual bicrystal system considered. In view of the mathematical complexity involved, the choice of a realistic model is very limited.

In the simple case of a monolayer on a rigid substrate with the lattice parameters of substrate and overgrowth equal in one direction and different in the perpendicular direction in the interface, a sinusoidal model for the interfacial potential of an overgrowth atom was proposed by Frenkel and Kontorowa (1938), which may be written as

$$V = \tfrac{1}{2}W[1-\cos(2\pi x/a)] \tag{1}$$

where a is the spacing of the substrate atoms and x the displacement of an overgrowth atom from a potential minimum site in the direction in which the lattice parameters are different .The amplitude W will be large for strong bonding and small for weak bonding (van der Merwe, 1964). The tangential force on the overgrowth atom will be

$$F = -\partial V/\partial x = -(\pi/a)W\sin(2\pi x/a) \tag{2}$$

In the more general case of thicker overgrowths, it is convenient to work with shear stresses at the interface. These are expressed in terms of a periodic interfacial potential-energy density. The sinusoidal model may be written in the form

$$V = (\mu c^2/4\pi^2 d)[1-\cos(2\pi U/c)] \tag{3}$$

where the shear stress $p_{zx}(0,x)$ at the interface

$$p_{zx}(0,x) = \partial V/\partial U = (\mu c/2\pi d)\sin(2\pi U/c) \tag{4}$$

corresponds to the so-called "Peierls–Nabarro model" (van der Merwe, 1950). Here μ is an interfacial shear modulus, U the displacement from a potential minimum, d the separation of interfacial atomic planes, and c the period. It will be seen that this reduces to Hooke's law for small displacements.

The sinusoidal model may be considered as the first approximation of a Fourier series. A refined approximation with two harmonic terms has been used by the authors (Ball and van der Merwe, 1970). In addition, a model consisting of a series of parabolic arcs has been introduced by van der Merwe (1963a, b) to demonstrate that the forces acting normal to the interface contribute a neglegible amount to the energy of the system. Figure 1 shows the three different forms of the potential-energy densities mentioned.

The interactions within the single crystals are taken into account by considering the crystals to be isotropic elastic media (Nabarro, 1947; van der

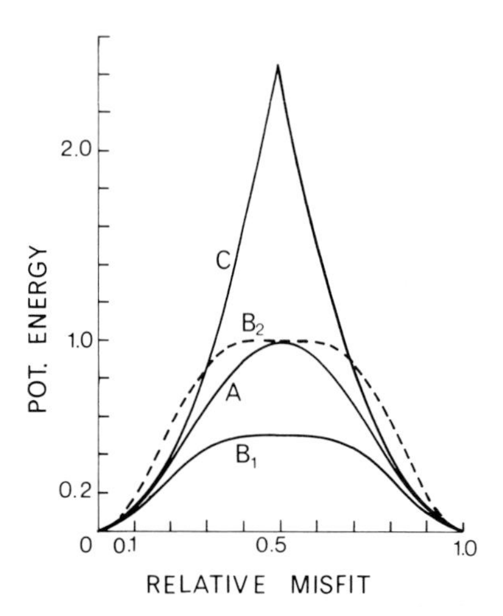

FIG. 1. Illustrations of different representations of interfacial potentials. Curve A: sinusoidal, Eq. (1) or (3); B: Fourier approximation containing first two harmonic terms where amplitudes are chosen such that B_1 osculates with A at 0 and B_2 has the same overall amplitude as A; C: parabolic model whose arcs osculate with A in the troughs (van der Merwe, 1964).

Merwe 1950). It is assumed that the elastic moduli of films are the same as those of bulk materials. This may not be valid for very thin films.

B. EQUILIBRIUM CRITERION

The purpose of this work is to calculate the equilibrium configuration of a given substrate–overgrowth system. It is assumed that when the bicrystal is formed, the deposition rate of the overgrowth is slow enough to ensure that the system is in quasi-equilibrium at all times. This would imply sufficiently high temperatures so that the deposited atoms will move to equilibrium positions. It is also assumed that the bicrystal is large enough to be treated as a thermodynamic system. If the deposition takes place at constant temperature and pressure, the system will be in equilibrium when the Gibbs free energy is a minimum. For the system under consideration this condition may be approximated by a minimum energy requirement

$$E = \text{minimum} \tag{5}$$

where E is the total energy (van der Merwe, 1966). The total energy of a bicrystal will in general depend on a number of parameters. These include the misfit, the overall strain, the thickness of the film, and the bonding both at the interface and in the crystals.

The term "misfit" is used to describe the disregistry between the substrate and overgrowth arrays of atoms at the interface. If a and b are the natural lattice parameters of substrate and overgrowth, respectively, then the misfit may be defined as

$$1/P = (b-a)/a \tag{6}$$

Normally there will be two different misfit parameters associated with the two perpendicular directions of the rectangular arrays of atoms at the interface.

Under the combined actions of the crystal and interfacial forces, the interfacial atoms assume a configuration that has rightly been named an "array of misfit dislocations" (Frank and van der Merwe, 1949a).

If the overgrowth has finite thickness, an overall lateral strain of the correct magnitude, and in a sense that reduces the mismatch, will reduce the energy. The overall strain is defined by

$$e = (\bar{b}-b_0)/b_0 \tag{7}$$

where $\bar{b}$ and b_0 are, respectively, the strained and unstrained average lattice parameters of the overgrowth. If there is misfit in the two perpendicular directions, there will be two overall strain parameters.

The bonding is taken into account in the shear moduli and Poisson's ratios of the two crystals and of the interface.

For a given bicrystal system the natural lattice parameters, bonding, and thickness are specified, and the energy is minimized with respect to the strains in order to obtain the equilibrium configuration.

II. Thin Films

A. Model

1. *Assumptions and Reference System*

In this section we consider (van der Merwe, 1970) an overgrowth so thin that the strain gradient in the film normal to the interface may be neglected. This is essentially correct for a monolayer and may be a reasonable approximation for films of a few atomic layers. It is further assumed that the substrate may be regarded as rigid, but it has been shown that this assumption introduces significant errors when the film is not in the pseudomorphic state (Ball, 1970) and is thicker than one atomic layer.

A set of cartesian axes is chosen with origin in the interface at a point where an overgrowth atom is in a potential minimum of the substrate forces. The x- and y-axes are in the interface along the perpendicular rows of atoms passing through the origin. The rectangular interfacial atomic meshes have natural side lengths a_i in the substrate and b_i in the overgrowth ($i = x, y$). Each potential trough and corresponding overgrowth atom may be designated by the pair of integers (n, m) indicating the nth position from the origin along the x-axis and the mth position along the y-axis, respectively. The positions of the atoms may thus be written as

$$x_{nm} = a_x(n+\xi_{nm}), \qquad y_{nm} = a_y(m+\eta_{nm}) \tag{8}$$

where $(a_x\xi_{nm}, a_y\eta_{nm})$ is the displacement of the (n, m)th atom from the position of the (n, m)th trough.

2. *Strains, Stresses, and Energy*

The linear and shear strains in the film are given by

$$e_x = (x_{n+1,m} - x_{nm} - b_x)/b_x \tag{9}$$

$$e_{xy} = [(y_{n+1,m} - y_{nm})/b_x] + [(x_{n,m+1} - x_{nm})/b_y] \tag{10}$$

For an unstrained film the vernier of mismatch p_i may be written

$$p_i = P_i b_i = (P_i+1)a_i, \qquad i = x, y \tag{11}$$

When the overgrowth is subjected to overall strains $\bar{e}_i$, so that the average lattice parameters change from b_i to $\bar{b}_i$, it follows that

$$\bar{e}_i = (\bar{b}_i - b_i)/b_i = (\bar{P}_i^{-1} - P_i^{-1})\alpha_i \tag{12}$$

where

$$1/\bar{P}_i = (\bar{b}_i - a_i)/a_i \qquad \text{and} \qquad 1/P_i = (b_i - a_i)/a_i \tag{13}$$

are, respectively, the misfits with and without overall strain, and

$$\alpha_i = a_i/b_i \tag{14}$$

The lateral and shear forces in the film plane and the strain-energy density per interfacial atom may be written (Timoshenko, 1934; van der Merwe, 1970)

$$T_x = 2\mu t(e_x + \sigma e_y)/(1-\sigma), \qquad T_{xy} = \mu t e_{xy} \tag{15}$$

$$\varepsilon = \tfrac{1}{2}\mu\Omega[2(e_x{}^2 + e_y{}^2 + 2\sigma e_x e_y)/(1-\sigma) + e_{xy}^2] \tag{16}$$

where μ, σ, and t represent, respectively, the shear modulus, Poisson's ratio, and thickness of the film. The volume per interfacial overgrowth atom is given by

$$\Omega = t b_x b_y \tag{17}$$

The total energy E of a film consisting of NM overgrowth atoms may be obtained by summing the potential energy

$$V_{nm} = \tfrac{1}{2}W_x[1 - \cos(2\pi x_{nm}/a_x)] + \tfrac{1}{2}W_y[1 - \cos(2\pi y_{nm}/a_y)] \tag{18}$$

and the elastic energy ε_{nm} in (16), giving

$$E = \sum_{n,m} [V_{nm}(\xi_{nm}, \eta_{nm}) + \varepsilon_{nm}(\xi_{nm}, \eta_{nm})] \tag{19}$$

B. Equilibrium Configuration

The equilibrium position of the (n, m)th atom will be given by

$$\partial E/\partial \xi_{nm} = \partial E/\partial \eta_{nm} = 0 \tag{20}$$

If the displacements vary slowly from atom to atom, n and m may be treated as continuous variables, and if in addition solutions of the kind

$$\xi(n, m) = \xi(n), \qquad \eta(n, m) = \eta(m) \tag{21}$$

are considered, then the equilibrium equations become

$$0 = \pi W_x \sin 2\pi\xi - 2\mu\Omega\alpha_x{}^2 [(1-\sigma)]^{-1} \, d^2\xi/dn^2 \tag{22}$$

with an analogous equation for the y-direction. Higher-order derivatives are neglected.

Equation (22) may be written as

$$d^2\xi/dn^2 = (\pi/2l_x{}^2) \sin 2\pi\xi \tag{23}$$

where

$$l_x{}^2 = \mu\Omega\alpha_x{}^2/W_x(1-\sigma) \tag{24}$$

The first and second integrals of Eq. (23) are, respectively,

$$d\xi/dn = (1-k_x{}^2 \cos^2 \pi\xi)^{1/2}/k_x l_x \tag{25}$$

and

$$\pi n/k_x l_x = F[k_x, \pi(\xi - \tfrac{1}{2})] \tag{26}$$

where

$$k_x = [1 + l_x{}^2 e_{1x}^2]^{-1/2}; \qquad e_{1x} = d\xi/dn \quad \text{at} \quad \xi = 0 \tag{27}$$

$F[k, \phi]$ is the incomplete elliptic integral of the first kind, and $n = 0$, where $\xi = \frac{1}{2}$.

The dependence of $\xi(n)$, illustrated in Fig. 2, may be interpreted as a sequence of dislocations, the so-called "misfit" or "interfacial dislocations" located at half-integer values of ξ and spaced at intervals

$$\bar{P}_x = 2l_x k_x K(k_x)/\pi \qquad \text{in units of } \bar{b}_x \tag{28}$$

as may be deduced from Eq. (26), with $K(k)$ the complete elliptic integral of the first kind.

When dislocations are completely or practically nonexistent,

$$k_x = 1, \qquad \bar{P}_x = \infty \tag{29}$$

This configuration in which the film perfectly matches the substrate is referred to as the "coherent" or "pseudomorphic" one.

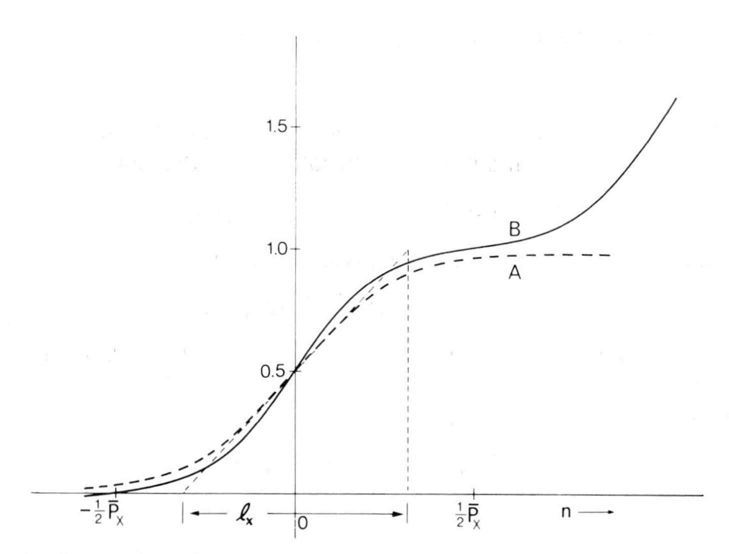

FIG. 2. Curves of $\xi(n)$ from Eq. (26); B for a sequence of dislocations located at $n = 0$, $\pm\bar{P}_x$, $\pm 2\bar{P}_x, \ldots$, and A for a single dislocation at $n = 0$. The length l_x is a measure of the width of a single dislocation (Frank and van der Merwe, 1949a).

C. Strains, Stresses, Work, and Energy

Since $\xi(n)$ and $\eta(m)$ are known functions, the strains e_{xy}, e_x, and e_y are also known, i.e.,

$$e_{xy} = 0, \qquad e_x = \alpha_x[d\xi/dn - (1/P_x)] = \alpha_x[(1-k_x{}^2\cos^2\pi\xi)^{1/2}/(k_x l_x) - P_x^{-1}] \tag{30}$$

with an analogous expression for e_y. Hence the stresses and energies in Eqs. (15) and (16) are also determined.

The work done in generating a unit length of dislocation by pulling a straight free boundary lying parallel to the y-axis over a potential crest is given by

$$\begin{aligned} w_x &= \int_0^1 \bar{T}_x a_x \, d\xi \\ &= 2\mu t a_x[2\alpha_x E(k_x)/\pi k_x l_x - (\alpha_x/P_x) + \sigma\bar{e}_y]/(1-\sigma) \end{aligned} \tag{31}$$

where $\bar{T}_x$ and $\bar{e}_y$ are the averages in the interval $0 \leqslant m \leqslant \bar{P}_y$.

The total energy E is approximated by the replacing the summation in Eq. (19) by an integral. The average energy per atom thus obtained is

$$\bar{\varepsilon} = \frac{E}{NM} = \sum_i W_i \left[\frac{4E(k_i)l_i}{\pi k_i \bar{P}_i} - \frac{2l_i^2}{\bar{P}_i P_i} + \frac{l_i^2}{P_i^2} + 1 - k_i^{-2} \right]$$

$$+ 2\sigma \prod_i W_i^{1/2} \left(\frac{l_i}{\bar{P}_i} - \frac{l_i}{P_i} \right), \qquad i = x, y \tag{32}$$

where $E(k)$ is the complete elliptic integral of the second kind. It is of interest that the last term of Eq. (32) is the contribution due to the Poisson phenomenon. This term may be written as

$$\varepsilon_{\text{p}} = 2\sigma \prod_i W_i^{1/2} l_i \bar{e}_i / \alpha_i \tag{33}$$

D. Stability of an Island with Free Boundaries

1. *Metastability*

In this subsection the equilibrium of a rectangular island consisting of NM atoms with sides parallel to the x- and y-axes is considered. It is assumed that no forces act on the perimeter and hence

$$T_n = T_{nr} = 0 \tag{34}$$

where T_n and T_{nr} are the normal and shear forces on the boundary of the island. If the parameters n and m are considered continuous variables, Eq. (34) becomes

$$T_x = [2\mu t/(1-\sigma)][e_x(\xi_0) + \sigma e_y(\eta)] \tag{35}$$

and

$$T_{xy}(\xi_0, \eta) = 0 \tag{36}$$

for the boundary at $x = \xi_0$. It is evident from (35) that it is not possible to choose ξ_0 so that T_x vanishes for all η, showing that the simple solutions are not exact for this case. However, it follows from simple arguments, and Saint Venant's principle, that misfit dislocations will not be fed in or emitted spontaneously when

$$e_x(\tfrac{1}{2}) + \sigma e_y(0) > 0, \qquad e_x(0) + \sigma e_y(\tfrac{1}{2}) < 0 \tag{37}$$

respectively. Conditions (37) reduce to

$$[(1-k_x^2)^{1/2}/k_x l_x] + [\sigma\alpha_y(1/k_y l_y - 1/P_y)]/\alpha_x$$

$$< 1/P_x < (1/k_x l_x) + \sigma\alpha_y[(1-k_y^2)^{1/2}/k_y l_y - (1/P_x)]/\alpha_x \tag{38}$$

when the necessary substitutions are made. It also becomes

$$\sigma/(1+\sigma) < l/P < 1/(1+\sigma) \tag{39}$$

for the coherent configuration in the special case of quadratic symmetry, i.e.,

$$a_x = a_y \qquad \text{and} \qquad b_x = b_y$$

2. *Stability*

The configurations discussed in Section II, D,1 are normally metastable, and of these the one of lowest energy is stable. For large islands this is equivalent to the condition that no work is required to generate a misfit dislocation. It follows from Eq. (31) that this condition is satisfied if

$$1/P_x = [2E(k_x)/\pi k_x l_x] + \sigma(\bar{e}_y/\alpha_x) \tag{40}$$

which reduces to

$$1/P_x = (2/\pi l_x) - (\sigma\alpha_y/\alpha_x P_y) \tag{41}$$

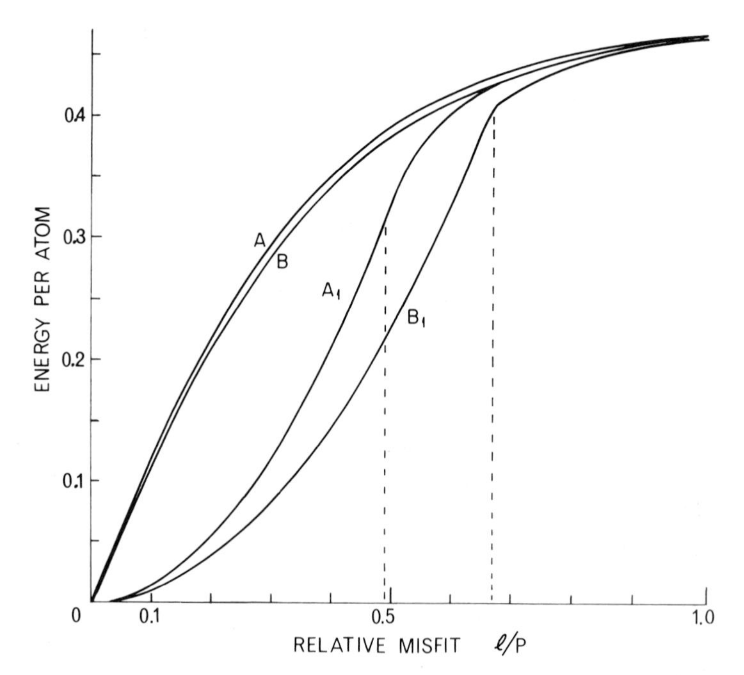

FIG. 3. Mean energy per atom $\bar{\varepsilon}$, in units of W, versus relative misfit l/P, when $a_x = a_y$, $b_x = b_y$. A and B correspond, respectively, to $l/\bar{P} = l/P$, i.e., $\bar{b} = b$, and lowest energy. A_1 and B_1 are the corresponding curves due to the one-dimensional model of Frank and van der Merwe (1949a, b; van der Merwe, 1970).

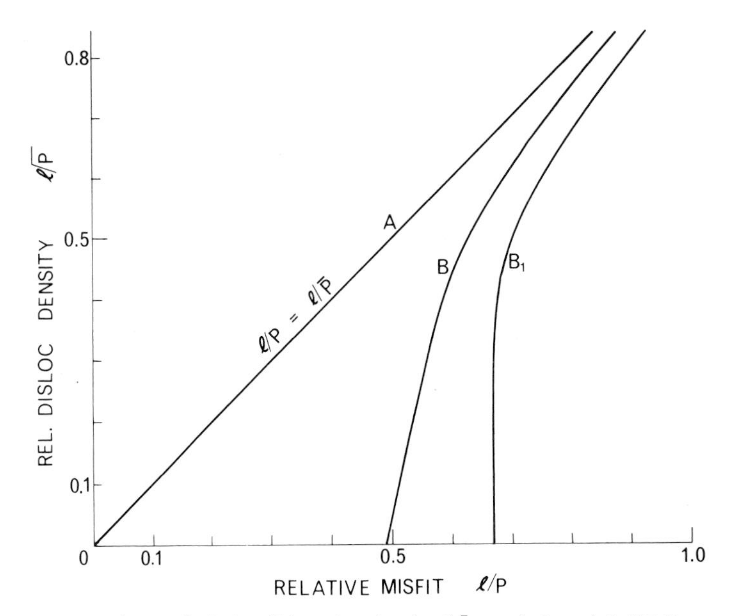

FIG. 4. Dependence of relative dislocation density $l/\bar{P}$ on relative misfit l/P. B corresponds to lowest energy when $a_x = a_y$, $b_x = b_y$, and B_1 to the corresponding curve deduced from the one-dimensional model. A represents the case $\bar{b} = b$ (van der Merwe, 1970).

for the coherent configuration and further to

$$1/P = 2/[\pi l(1+\sigma)] \tag{42}$$

for quadratic symmetry.

The dependence of $\bar{\varepsilon}$, the energy per atom, on relative misfit in the case of quadratic symmetry is illustrated in Fig. 3, and the dependence of relative dislocation density $l/\bar{P}$ on relative misfit l/P is shown in Fig. 4. The results are compared with those of Frank and van der Merwe (1949a) for a one-dimensional case. It will be seen that theory predicts a limiting misfit, below which a film will be stable in the coherent state.

E. Discussion

These results strictly apply to monolayers but have been extrapolated to multiple films for which it is a rather poor approximation because the elastic properties in both overgrowth and substrate are not properly taken into account (Ball, 1970).

The model given in this section correctly includes the contribution to the total energy due to the Poisson phenomenon. This was not taken into account in the one-dimensional model of Frank and van der Merwe (1949a, b) and is responsible for the differences exhibited in Figs. 3 and 4.

The calculations have been applied to a refined model (Frank and van der Merwe, 1949c) in which a second harmonic term is included in the potential in Eq. (18). This refinement increases the limiting misfit by about 15% if the overall amplitude is kept fixed.

The consequence of assuming that m and n are continuous and the tacit assumption that P_i in Eqs. (11) are integers is to produce a smooth energy curve that otherwise would have cusps in positions where the lattice parameters are in the ratio of small integers (Fletcher, 1964; du Plessis and van der Merwe, 1965).

It must be noted here that the islands discussed in this section are considered so large that the generation or escape of a single dislocation alters the energy of the system only by an infinitesmal amount.

III. Semiinfinite Overgrowths

A. Model

In this section a system consisting of an overgrowth B that is effectively infinitely thick on a substrate A that is also semiinfinite is considered (van der Merwe, 1950, 1963a, b). It will be assumed that the energies associated with the two perpendicular arrays of misfit dislocations may be calculated independently and added to obtain the total energy, and hence one considers misfit in the x-direction only. The relevant approximation is exact when the interfacial interaction consists of terms in which a given term is a function of only x or y. The axes are chosen so that the origin is on a dislocation line in the interface and the z-axis normal to the interface, with $z > 0$ corresponding to crystal B and $z < 0$ to crystal A (see Fig. 5). The problem is thus one of plane strain.

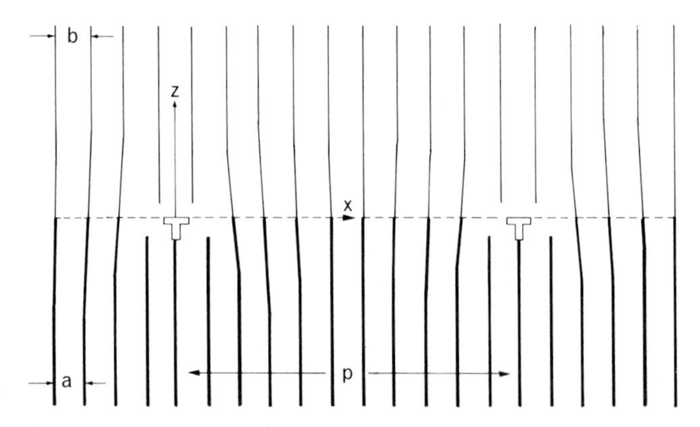

Fig. 5. Diagrammatic representation of the distortions due to the mismatching interface of two epitaxial crystals A and B with one-dimensional misfit. The distortions are interpreted as (misfit) dislocations of spacing p defined in Eq. (43).

In order to facilitate the mathematical description of the system, a reference lattice C with lattice parameter c defined by

$$p = (P+\tfrac{1}{2})c = Pb = (P+1)a \tag{43}$$

is introduced, where b and a are, respectively, the lattice parameters of B and A in the x-direction. Then p will be the spacing of misfit dislocations and P, as before, is assumed to be an integer. One may imagine the bicrystal to have been produced by taking a crystal C and then by some process extending the crystal half above a certain plane so that its lattice parameter increases from c to b and contracting the lower half so that its lattice parameter decreases from c to a. The displacement of an overgrowth atom from the corresponding substrate atom may then be written

$$U(x) = (c/2) + (cx/p) + u_b(x) - u_a(x) \tag{44}$$

where the term cx/p is a result of the fact that the displacement U increases by c in a vernier period p of x, $c/2$ indicates that the origin of axes is chosen on a dislocation line, and $u_a(x)$ and $u_b(x)$ are the elastic displacements of the corresponding substrate and overgrowth atoms, respectively. In this case it is convenient to use the ratio

$$c/p = (b-a)/\tfrac{1}{2}(a+b) \tag{45}$$

obtained from (43), as a measure of the misfit.

The Peierls–Nabarro model for the interfacial potential-energy density [Eq. (3)] will be used in the calculations given here, so that the shear stress $p_{zx}(0, x)$ at the interface will be given by Eq. (4). It will be assumed (van der Merwe, 1963) that the normal forces at the interface are negligible, i.e.,

$$p_{zz}(0, x) = 0 \tag{46}$$

From the periodic nature of the atomic configuration it will be seen that the stresses and strains will be periodic in x with period p, e.g.,

$$p_{zz}(z, x+p) = p_{zz}(z, x) \tag{47}$$

The shear stress will be antisymmetric in x about the plane $x = 0$, i.e.,

$$p_{zx}(z, x) = -p_{zx}(z, -x) \tag{48}$$

and the stresses and strains will vanish at sufficiently large distances from the interface, e.g.,

$$p_{xx}(z, x) = 0 \qquad \text{at} \quad z = \pm\infty \tag{49}$$

The stresses acting on the interface will be continuous in going from one crystal to the other, i.e.,

$$p_{zz}(0-, x) = p_{zz}(0+, x), \qquad p_{zx}(0-, x) = p_{zx}(0+, x) \tag{50}$$

B. GOVERNING EQUATION

Problems in plane strain are most easily solved (Timoshenko, 1934) using a stress function χ that satisfies the equation

$$\nabla^4 \chi = 0 \tag{51}$$

with the stresses defined by

$$p_{xx} = \partial^2\chi/\partial z^2, \qquad p_{zz} = \partial^2\chi/\partial x^2, \qquad p_{zx} = -\partial^2\chi/\partial x\,\partial z \tag{52}$$

Also, the Hookean relations reduce to the form

$$2\mu e_{xx} = (1-\sigma)\,p_{xx} - \sigma p_{zz} \tag{53}$$

A suitable solution of Eq. (51) satisfying all the boundary conditions except the Peierls–Nabarro force is (van der Merwe, 1950)

$$\chi = -(c\lambda p/2\pi^2)\,Z \sum_{n=1}^{\infty} A_n n^{-1} e^{\pm nZ} \cos nX \tag{54}$$

where

$$Z = 2\pi z/p, \qquad X = 2\pi x/p \tag{55}$$

and

$$\lambda^{-1} = [(1-\sigma_a)/\mu_a] + [(1-\sigma_b)/\mu_b] = \lambda_a^{-1} + \lambda_b^{-1} \tag{56}$$

μ_a and μ_b are the shear moduli of A and B, respectively, and σ_a and σ_b the corresponding Poisson's ratios. The positive index in (54) refers to crystal A and the negative index to crystal B. From Eqs. (52–(54) it follows that

$$p_{xx} = -(2c\lambda/p) \sum A_n (nZ \pm 2)\, e^{\pm nZ} \cos nX \tag{57}$$

$$p_{zz} = (2c\lambda/p) \sum A_n nZ e^{\pm nZ} \cos nX \tag{58}$$

$$p_{zx} = -(2c\lambda/p) \sum A_n (1 \pm nZ)\, e^{\pm nZ} \sin nX \tag{59}$$

$$p_{zx}(0,x) = -(2c\lambda/p) \sum A_n \sin nX \tag{60}$$

and

$$U/c = \psi(X) = \tfrac{1}{2} + (X/2\pi) + \pi^{-1} \sum A_n n^{-1} \sin nX \tag{61}$$

where u_a and u_b in Eq. (44) have been obtained by integrating Eq. (53).

The solution given in Eq. (54) will be the desired stress function provided that $p_{zx}(0,x)$ and U satisfy Eq. (4). This will be the case when ψ satisfies (van der Merwe, 1950) the relation

$$2\pi\left[\frac{d\psi}{dX} - \frac{1}{2\pi}\right] = -\frac{1}{2\pi\beta}\int_0^{\pi} [\sin 2\pi\psi(X+t) - \sin 2\pi\psi(X-t)] \cot \tfrac{1}{2}t \, dt \tag{62}$$

where

$$\beta = 2\pi d\lambda/\mu p \tag{63}$$

C. Solution

The solution of Eq. (62), obtained by an iterative method (van der Merwe, 1950), is

$$\psi(X) = \tfrac{1}{2} + \pi^{-1} \arctan\{[(1+\beta^{-2})^{1/2} + \beta^{-1}] \tan\tfrac{1}{2}X\} \tag{64}$$

where the coefficients A_n in Eq. (54) are given by

$$A_n = A^n = [(1+\beta^2)^{1/2} - \beta]^n \tag{65}$$

The stresses, strains, and displacements are now known infinite series and may be calculated. The displacement function $\psi(X)$ is illustrated in Fig. 6.

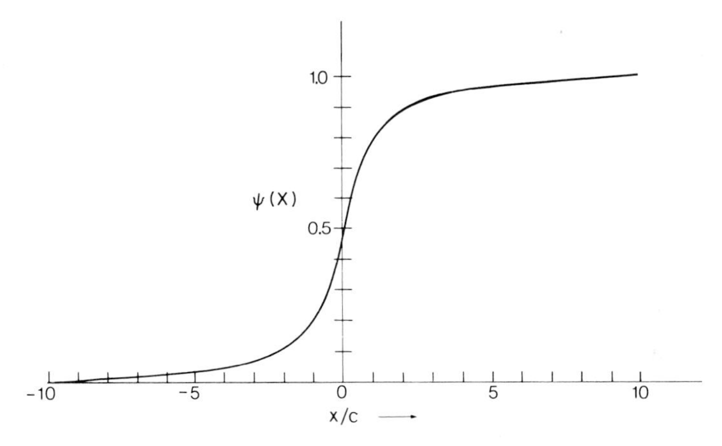

FIG. 6. Variation of the relative displacement $\psi(x) = U(x)/c$ corresponding to a dislocation with vernier of mismatch $P = 20$. Other values assumed are: $\mu_a = \mu_b = \mu_0$, $\sigma_a = \sigma_b = \sigma = 0.3$ (van der Merwe, 1964).

D. Energy

The average potential energy of misfit per unit area is calculated, using Eqs. (4) and (64):

$$\begin{aligned} E_{\mathrm{m}} &= (1/p)\int_{-p/2}^{p/2} (\mu c^2/4\pi^2 d)(1-\cos 2\pi\psi)\, dx \\ &= (1/p)\int_{-p/2}^{p/2} dx \int_0^U dU\, \partial V/\partial U = (\mu c^2/4\pi^2 d)[1-A] \end{aligned} \tag{66}$$

The elastic strain energy in the crystal halves may be obtained by integrating the strain-energy density or by calculating the work done by the interfacial stress in the displacements u_a and u_b according to the relation

$$E_e = \tfrac{1}{2}\int \mathbf{T}\cdot\mathbf{s}\,dA \tag{67}$$

where **s** is the displacement of the surface element dA, when acted on by the interfacial stress **T**. Here the first method will be used as it gives information on the distribution of the strain energy.

The average strain energy per unit area of interface of that part of the overgrowth that extends from the interface to a distance h from it is given by

$$E_e^{\ b}(h) = (1/p)\int_0^h dz \int_{-p/2}^{p/2} dx\,(1/4\mu_b)\,[(1-\sigma_b)(p_{xx}^2+p_{zz}^2) - 2\sigma_b\, p_{xx}\, p_{zz} + 2p_{zx}^2] \tag{68}$$

when plane strain is assumed. The integration is most easily performed using Eqs. (57)–(60) and (65). One finds

$$E_e^{\ b}(h) = -(1-\sigma_b)\frac{c^2\lambda\mu\beta}{4\pi^2\mu_b d}\left[\ln\left(\frac{1-A^2}{1-A^2e^{-2H}}\right) + \frac{HA^2e^{-2H}(H-1+A^2e^{-2H})}{(1-\sigma_b)(1-A^2e^{-2H})^2}\right] \tag{69}$$

where

$$H = 2\pi h/p \tag{70}$$

The total strain energy in the overgrowth is obtained by letting $H \to \infty$:

$$E_e^{\ b} = -[(1-\sigma_b)\,c^2\lambda\mu\beta/4\pi^2\mu_b d]\,\ln(1-A^2) \tag{71}$$

An analogous expression may be written for the strain energy in the substrate crystal.

Finally, the average energy per unit area of the interface will be

$$E = E_m + E_e^{\ a} + E_e^{\ b} = \frac{\mu c^2}{4\pi^2 d}[1-A-\beta\ln(1-A^2)] \tag{72}$$

In Fig. 7 the energies E_m, $E_e^{\ a}+E_e^{\ b}$, and E are plotted as functions of β.

E. Discussion

It is easily seen that the energy in (72) tends monotonically to

$$\tfrac{1}{2}[\mu c^2/2\pi^2 d] \tag{73}$$

[which is one-half the overall potential-energy amplitude of V in Eq. (3)] when β becomes large. Physically, this is equivalent to any one or a combination

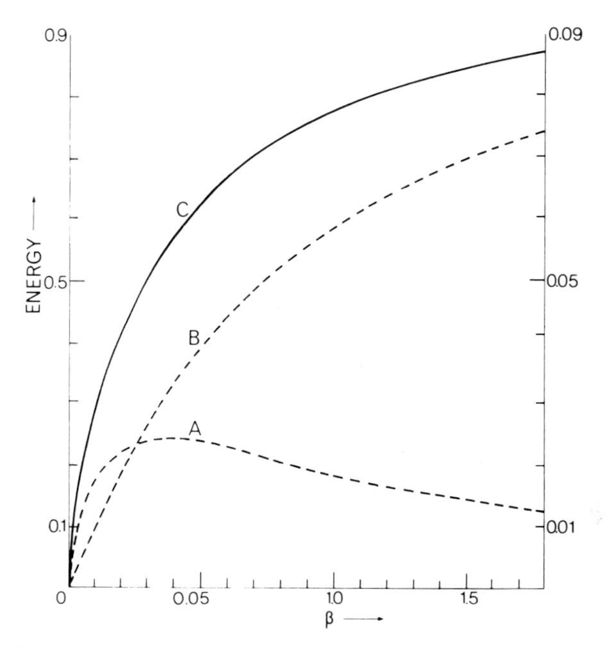

FIG. 7. Variation of mean energy per unit area (units $\mu c/4\pi^2$) with β in Eq. (63). A, B, and C represent, respectively, the strain energy $E_e^a + E_e^b$ defined by Eq. (71), the potential energy of misfit E_m in Eq. (66), and the total energy defined in Eq. (72). The assumptions $\mu_a = \mu_b = \mu_0$, $\sigma_a = \sigma_b = \sigma = 0.3$ have been made in obtaining the graphs (van der Merwe, 1950).

of the following cases: (i) both crystals are rigid, (ii) the interfacial interaction vanishes, and (iii) the misfit is large; as may be seen from Eq. (63).

It follows from Eqs. (69) and (71) that, of the strain energy in the overgrowth, only the fraction

$$R(h) = \left[\ln(1 - A^2 e^{-2H}) - \frac{HA^2 e^{-2H}(H - 1 + A^2 e^{-2H})}{(1-\sigma_b)(1 - A^2 e^{-2H})^2}\right] \Big/ \ln(1 - A^2)$$
$$\simeq -\frac{[H(H-1)/(1-\sigma_b) + 1]A^2 e^{-2H}}{\ln(1 - A^2)} \tag{74}$$

is stored beyond the plane $z = h$. The dependence of R on h/p, in the special case $\mu_a = \mu_b = \mu$, $\sigma_a = \sigma_b = \sigma$, is illustrated in Fig. 8 for $d/p \simeq c/p = 2$ and 20%. The value of R is seen to drop very rapidly with increasing h/p, and this property of the strain energy is used to justify the assumption that the energy in an overgrowth of finite thickness may be approximated by that of an infinite overgrowth in cases where the thickness exceeds half the dislocation spacing.

It is of interest that the strain energy is distributed between crystals A and B in the ratio

$$E_e^a/E_e^b = (1-\sigma_a)\,\mu_b/\mu_a(1-\sigma_b) \tag{75}$$

A basic requirement of the model is that Hooke's law be satisfied. The maximum strains occur close to a dislocation, and it may be shown (van der Merwe, 1950) that there will be a small region around a dislocation where the model is bad and overestimates the elastic strain and energy.

Calculations have also been done using a refined interfacial potential containing two harmonic terms (Ball and van der Merwe, 1970), which is believed to yield more accurate results. The inclusion of the second harmonic term introduces considerable mathematical difficulties, however, and only approximate solutions were obtained. The difference between the results obtained using the two models for interfacial potential depends on the relative amplitudes of the two harmonic terms in the refined model.

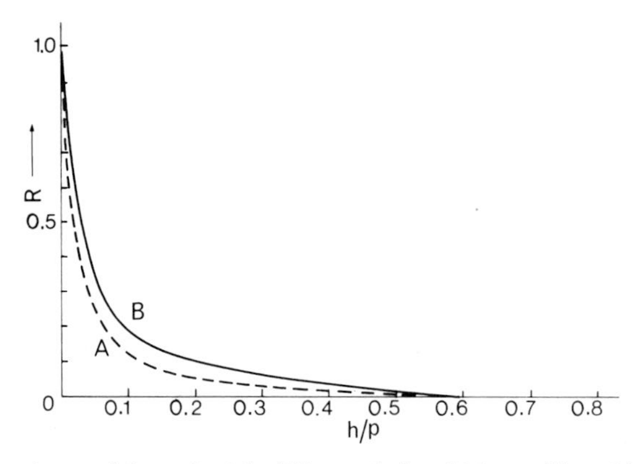

FIG. 8. Dependence of the ratio R in (74) on relative thickness h/p, taking $\mu_a = \mu_b = \mu_0$, $\sigma_a = \sigma_b = \sigma = 0.3$. Curves A and B correspond to $d/p \simeq c/p$ having the values 2 and 20%, respectively (van der Merwe, 1963b).

IV. Misfit Dislocation Energy in Overgrowths of Finite Thickness

A. INTRODUCTION

In Sections II and III the two extreme cases of epitaxial bicrystals—a monolayer overgrowth and an infinitely thick overgrowth—have been considered. There is, however, an intermediate region of film thicknesses of a few atomic layers for which neither of these approaches gives a reasonable approximation for the dislocation energy, and it is precisely this region that is of importance in the study of the growth of epitaxial layers. In this section various attempts to find expressions for the misfit dislocation energy of a layer of arbitrary thickness t are discussed. Two of these are simple extrapolations from the extreme cases.

The model is the same as that given in Section III, A, except that the boundary condition given in Eq. (49) is replaced by

$$0 = p_{zz}(z, x) = p_{zx}(z, x) \qquad \text{at} \quad z = t \tag{76}$$

$$0 = p_{zz}(z, x) = p_{zx}(z, x) \qquad \text{at} \quad z = -\infty \tag{77}$$

B. Exact Solution

Van der Merwe (1963a, b) analyzed a parabolic interfacial potential

$$V = \mu(U-nc)^2/2d, \qquad (n-\tfrac{1}{2})\,c < U < (n+\tfrac{1}{2})\,c, \qquad n = 0, \pm 1, \ldots \tag{78}$$

and the refinement

$$p_{zz}(0, x) = 2\mu W(x)/(1-2\sigma)d \tag{79}$$

representing a normal interfacial stress induced by relative displacement W of the adjoining surfaces of the two crystals at the interface. The calculated expression for the energy is in the form of an infinite series

$$E_{\text{d}} = \frac{-c}{2p}\sum_{n=1}^{\infty}(-1)^n\left(\frac{2\pi n}{p}B_n + C_n\right) \tag{80}$$

where the coefficients B_n and C_n are rather complicated functions of the crystal parameters. Since, in addition, the series in Eq. (80) converges very slowly and the interfacial potential (78) is unrealistic anyhow, it is of little interest here. The result showing that, except for small misfits, the energy of a monolayer differs at most by about 8% from that of an infinite crystal is worth mentioning however.

C. Various Extrapolations

1. *Monolayer Extrapolation*

It was pointed out in Section II that the model for a monolayer could be extrapolated by letting t in Eq. (17) be a film thickness of one or more atomic layers, but that appreciable errors are introduced as a consequence of the assumptions that the substrate is rigid and that the strain gradient in the film normal to the interface is negligible.

2. *Model of Ball*

Ball (1970) retained the assumption that the strain gradient in the film is negligible but incorporated the elasticity of the substrate. Although the parabolic model [Eq. (78)] was used, an approximate correction for the overestimation of the energy was made by introducing a scale factor τ that was used to introduce an effective shear modulus μ^*, defined by

$$\mu^* = \mu\tau \tag{81}$$

in terms of μ in Eq. (78). The value of τ was chosen so that the slope of the energy curve for a monolayer on a rigid substrate at zero misfit matches that obtained by using the more realistic Peierls–Nabarro model under the same conditions. One finds

$$\tau = 64/\pi^4 \tag{82}$$

The assumption is made that the scale factor yields reasonable matching at larger misfits and for thicker overgrowths. Evidence in favor of this is shown in Fig. 9, where the curves B and A correspond to the misfit dislocation

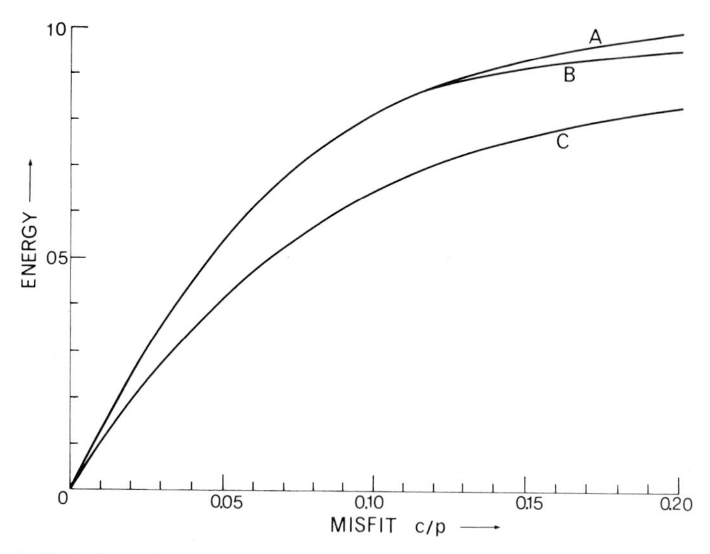

FIG. 9. Variation of mean misfit dislocation energy E_d (units $\mu c/4\pi^2$) of a monolayer with (one-dimensional) misfit c/p. Curve A: sinusoidal interfacial potential (Section II); B, C: scaled energy $E_d{}^*$ in Eq. (83), B for a rigid ($\mu_a = \infty$) and C for an elastic substrate. The curves have been obtained taking $\mu_a = \mu_b = \mu_0$ and $\sigma_a = \sigma_b = \sigma = 0.3$ (Ball, 1970).

energy of a monolayer on a rigid substrate using the Peierls–Nabarro and scaled energies, respectively.

The misfit dislocation energy of a film on an elastic substrate obtained in the analysis is

$$E_{\mathrm{d}}^* = (\mu^* c/4\pi^2) \sum_{n=1}^{\infty} [n^2 + (n/\beta^*) + \tfrac{1}{2}\gamma^*]^{-1} \tag{83}$$

where

$$\beta^* = 2\pi c\mu_a/[\mu^* p(1-\sigma_a)], \qquad \gamma^* = \mu^* p^2/[4\pi^2 \mu_b(1+\sigma_b)\, tc] \tag{84}$$

The series in Eq. (83) may be expressed in terms of diagamma functions. A curve of the elastic energy of a monolayer on an elastic substrate is shown in Fig. 9, curve C. It will be seen that the additional relaxation introduced by allowing for elastic displacements is significant even for a monolayer. While this model is considered to be the best for a monolayer, the analysis (Ball, 1970) shows that it cannot be expected to yield realistic results for overgrowths consisting of three or more atomic layers.

3. *Extrapolation of Semiinfinite Model*

It has been shown (van der Merwe, 1963a, b) [Eq. (74)] that the mean misfit dislocation energy per unit interfacial area E_{d} of an overgrowth of thickness t can be approximated to within an error of 2% by that of a semi-infinite overgrowth when

$$t \geqslant \tfrac{1}{2}p \tag{85}$$

The corresponding expression for the misfit dislocation energy will be Eq. (72).

4. *Refined Extrapolation*

The extrapolation in Section 3 has been refined in the following manner (van der Merwe and van der Berg, 1972). Calculations show that the effect of nearby dislocations or free surfaces on a given dislocation can be approximated in terms of the concept of screening. Thus the strain field of a misfit dislocation is assumed to have an effective range $\frac{1}{2}q$ defined by

$$\tfrac{1}{2}q = \begin{cases} \tfrac{1}{2}p, & \tfrac{1}{2}p \leqslant t \quad (86) \\ t, & \tfrac{1}{2}p \geqslant t \quad (87) \end{cases}$$

when the film thickness is t. This has been justified by using (74).

Relations (86) and (87) are illustrated by the straight-line sections AB and BC in Fig. 10 for c/q as a function of c/p.

The considerations that led to Eq. (85) and the results obtained from the parabolic model suggest that the straight line CB should be introduced more gradually. When a simple parabolic transition OB is chosen, Eqs. (86) and (87) can be written conveniently in the refined form:

$$c/q = \begin{cases} (c/4t) + (tc/p^2), & c/p \leqslant c/2t \\ c/p, & c/p \geqslant c/2t \end{cases} \tag{88}$$

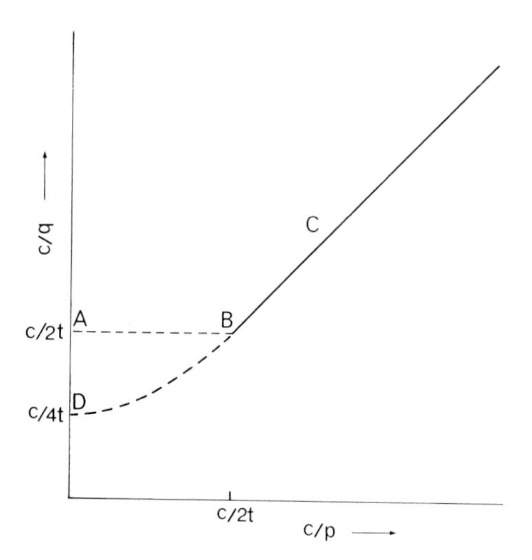

FIG. 10. Representation of the effective range of the strain field of misfit dislocations near free surfaces, Eqs. (86)–(88): ABC the conventional approximation and DBC the refined approximation (van der Merwe and van der Berg, 1972).

By the analyses that yielded the result (72), the energy per unit length of dislocation, when the overgrowth is infinite, is given by $pE_{\text{d}}[\beta(c/p)]$, or by $qE_{\text{d}}[\beta(c/q)]$ in terms of q. The mean energy per unit interfacial area is accordingly given by

$$E_{\text{d}}' = (q/c)\, E_{\text{d}}[\beta(c/q)]\,(c/p) \tag{89}$$

where c/q is defined in Eq. (88) and $E_{\text{d}}(\beta)$ in Eq. (72).

D. First Approximation

In this approximation (van der Merwe and van der Berg, 1972) the exact solution of the displacement function $\psi(x)$ given in Eq. (64) for the semi-infinite case is used as a first approximation for the displacement function for a finite overgrowth.

Stress functions satisfying Eq. (51) and the boundary conditions Eqs. (46)–(48), (50), and (76) are

$$\chi_a = z \sum_{n=1}^{\infty} C_n^{\;a} e^{mz} \cos mx \tag{90}$$

for the substrate A and

$$\chi_b = \sum_{n=1}^{\infty} [zC_n^{\;b} \cosh mz + (B_n^{\;b} + zD_n^{\;b}) \sinh mz] \cos mx \tag{91}$$

for the overgrowth B, where

$$m = 2\pi n/p \tag{92}$$

The coefficients B, C, and D are to be determined. From Eqs. (52), (90), and (91) it follows that

$$p_{xx}^a = \sum (2+mz) m C_n{}^a e^{mz} \cos mx \tag{93}$$

$$p_{zx}^a = \sum (1+mz) m C_n{}^a e^{mz} \sin mx \tag{94}$$

$$p_{zz}^a = -z \sum m^2 C_n{}^a e^{mz} \cos mx \tag{95}$$

$$p_{xx}^b = \sum \{(2D_n{}^b + mzC_n{}^b) m \cosh mz + (2C_n{}^b + mB_n{}^b + mzD_n{}^b) m \sinh mz\} \cos mx \tag{96}$$

$$p_{zx}^b = \sum \{(C_n{}^b + mB_n{}^b + mzD_n{}^b) m \cosh mz + (D_n{}^b + mzC_n{}^b) m \sinh mz\} \sin mx \tag{97}$$

$$p_{zz}^b = -\sum \{m^2 z C_n{}^b \cosh mz + m^2 (B_n{}^b + zD_n{}^b) \sinh mz\} \cos mx \tag{98}$$

After substituting (93)–(98) in (53) and letting $z = 0$, one obtains by integration

$$\mu_a u_a(x) = (1-\sigma_a) \sum C_n{}^a \sin mx \tag{99}$$

$$\mu_b u_b(x) = (1-\sigma_b) \sum D_n{}^b \sin mx \tag{100}$$

where the integration constants are zero.

The relative displacement U of interfacial atoms in B with respect to the corresponding atoms in A is given by

$$U = (cx/p) + u_b - u_a \tag{101}$$

where, in this case, the origin of axes is chosen at a point halfway between two interfacial dislocations. It follows from Eqs. (99) and (100) that

$$u_b - u_a = \sum [(D_n{}^b/\lambda_b) - (C_n{}^a/\lambda_a)] \sin mx \tag{102}$$

where λ_a and λ_b are defined in Eq. (56).

The essence of this approximation is that the term $u_b - u_a$ is identified with the corresponding development

$$u_b - u_a = (c/\pi) \sum_{n=1}^{\infty} A^n n^{-1} \sin mx \tag{103}$$

obtained for the semiinfinite case.

By substituting Eqs. (93)–(98) into (50), and taking into account Eqs. (102) and (103), the following relations for the unknown coefficients are obtained:

$$mB_n{}^b + C_n{}^b - C_n{}^a = 0 \tag{104}$$

$$C_n^{\,b} \cosh mt + (B_n^{\,b} + D_n^{\,b}) \sinh mt = 0 \tag{105}$$

$$(mB_n^{\,b} + C_n^{\,b} + mtD_n^{\,b}) \cosh mt + (mtC_n^{\,b} + D_n^{\,b}) \sinh mt = 0 \tag{106}$$

$$D_n^{\,b} \lambda_b^{-1} - C_n^{\,b} \lambda_a^{-1} = cA^n/\pi n \tag{107}$$

The solutions of Eqs. (104)–(107) are

$$B_n^{\,b} = 2c\lambda t^2 A^n/p\Delta \tag{108}$$

$$C_n^{\,b} = -2c\lambda \sinh^2 mt\, A^n/mp\Delta \tag{109}$$

$$C_n^{\,a} = 2c\lambda (m^2t^2 - \sinh^2 mt)\, A^n/mp\Delta \tag{110}$$

$$D_n^{\,b} = 2c\lambda (\sinh mt \cosh mt - mt)\, A^n/mp\Delta \tag{111}$$

$$\Delta = [\lambda(\sinh mt \cosh mt - mt)/\lambda_b] + [\lambda(\sinh^2 mt - m^2t^2)/\lambda_a] \tag{112}$$

The strain energy per unit area of interface is most easily obtained from the relation

$$E_e^{\,a} = (1/2p) \int_{-p/2}^{p/2} p_{zx}^a(0, x)\, u_a(x)\, dx \tag{113}$$

with a similar expression for the overgrowth. The interfacial potential energy is

$$E_m = (1/p) \int_{-p/2}^{p/2} dx \int_{-p/2}^{x} d\xi\, p_{zx}(0, \xi)\, \partial U/\partial \xi \tag{114}$$

in accordance with Eq. (66). One obtains on substitution and integration

$$E_m = (\mu c^2/4\pi^2 d)[1 - A] \tag{115}$$

and

$$E_e = E_e^{\,a} + E_e^{\,b} \tag{116}$$

$$= \frac{\mu c^2}{4\pi^2 d\lambda} \beta \sum \frac{(\sinh^2 mt - m^2t^2)\, A^{2n}/n}{[(\sinh mt \cosh mt - mt)/\lambda_b] + [(\sinh^2 mt - m^2t^2)/\lambda_a]} \tag{117}$$

The misfit dislocation energy per unit area E_d thus becomes

$$E_d = E_e + E_m \tag{118}$$

E. Discussion

Various approximate expressions for the misfit dislocation energy E_d of an epitaxial film of thickness t have been calculated above. The results are compared in Figs. 11a and 11b for a monolayer and a fivefold layer, respectively. The curves were obtained by setting $\mu_a = \mu_b = \mu$ and $\sigma_a = \sigma_b = \sigma = \frac{1}{3}$.

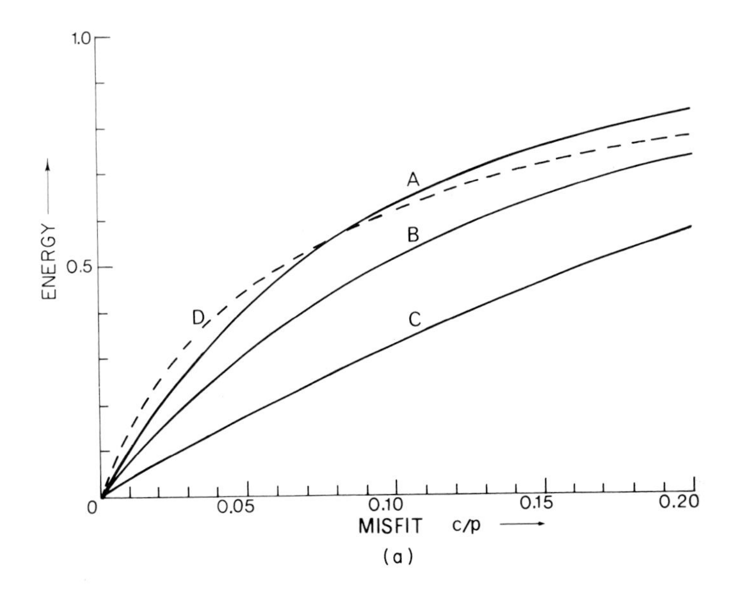

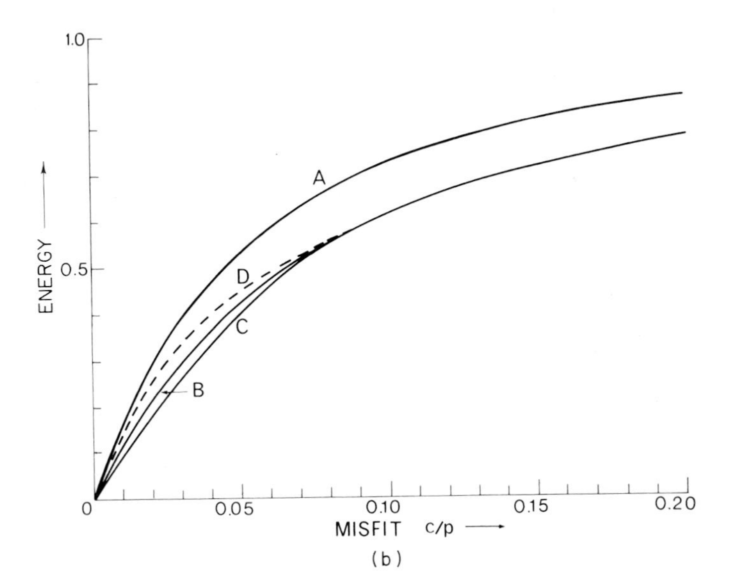

FIG. 11. Comparison of the dependence of misfit dislocation energy E_d (units $\mu c/4\pi^2$) on misfit c/p for different models. Curves A, B, and C correspond, respectively, to the model of Ball (83), the first approximation (118), and the refined extrapolation (89). Curve D is for a semiinfinite overgrowth and substrate, Eq. (72), with sinusoidal potential. (a) Monolayer; (b) fivefold layer (van der Merwe and van der Berg, 1972).

In each figure curves A, B, and C correspond to the model of Ball, the first approximation, and the refined extrapolation, respectively.

The model of Ball is certainly the more accurate one for a monolayer as it represents most accurately the properties of the film. In comparison, B lies somewhat too low, while C is far too low. However for thicker films of three or more layers, the model of Ball is certainly poor and will overestimate the strain energy. The refined extrapolation is definitely a poor approximation for fivefold and thinner films, but should be comparable to the first approximation for thicker films. This approximation is the easiest to work with in numerical calculations and will be useful in obtaining quick estimates, but should be used with caution. The first approximation will have the greatest overall accuracy but has the disadvantage that the series expression converges very slowly for small values of misfit and thickness.

V. Stability of Growing Epitaxial Crystals

A. Introduction

The energy of an epitaxial bicrystal consisting of a semiinfinite substrate and a growing two-dimensional overgrowth will depend on the film thickness t and the homogeneous strains $\bar{e}$ [See Eq. (7)]. The stable configuration is assumed to be the one of minimum energy, and in this section the strains $\bar{e}_{\mathrm{m}}$ that minimize the energy are calculated for given film thickness.

The additional assumption is made that the energies associated with the two perpendicular arrays of misfit dislocations may be calculated independently and added to obtain the total dislocation energy. Van der Merwe (1970) has shown for the monolayer case that there will be a small interaction contribution of the order of 2%. The energy associated with the overall strains $\bar{e}$ may be added to the dislocation energies to obtain the energy of the superimposed configuration (van der Merwe, 1963; Jesser and Kuhlmann-Wilsdorf, 1967).

B. Extrapolations

1. *Monolayer Extrapolation*

In Section II, D, 2, the stability of a film of thickness t on a rigid substrate was discussed on the assumption that the elastic strains in the film do not vary normal to the interface. If $P_{i(n)}$, $i = x, y$, are the numbers of atoms per dislocation for stable films of n atomic layers, then it may be shown, using the substitution

$$t = nb_z \tag{119}$$

where b_z is the thickness per layer, that

$$1/P_{i(n)} = 1/n^{1/2} P_{i(1)} \tag{120}$$

The dependence of the overall strains on thickness may be derived from Eq. (40), but as previously mentioned in Section II, E, it is a rather poor approximation.

2. *Thick-Film Extrapolation*

By using the approximate criterion that a coherent film is unstable when the dislocation energy for vanishing strain is less than the strain energy needed for coherency, van der Merwe (1963) deduced an expression for the limiting misfit at which the coherent configuration becomes unstable.

The energy of overall strain per unit area of interface may be written (Jesser and Kuhlmann-Wilsdorf, 1967)

$$E_{\mathrm{H}} = [\mu_b t/(1-\sigma_b)](\bar{e}_x{}^2 + \bar{e}_y{}^2 + 2\sigma_b \bar{e}_x \bar{e}_y) \tag{121}$$

If quadratic symmetry is assumed (i.e., $a_x = a_y = a$, $b_x = b_y = b$, $\bar{e}_x = \bar{e}_y = \bar{e}$) and the film is strained to coherency, then the overall strain energy will be

$$E_{\mathrm{H}}(e_0) = 2\mu_a t(1+\sigma_b) e_0{}^2/(1-\sigma_b) \tag{122}$$

where $e_0 = (a-b)/b$ is the strain required for coherency.

If the dislocation energy is assumed to be that of a semiinfinite overgrowth as given by Eq. (72), then one may obtain the following relation between limiting misfit ρ_c and t:

$$\begin{aligned} \ln \rho_c &+ 2\pi(1+\sigma)(1+\mu_b/\mu_a)\, t\rho_c/b \\ &+ \ln\{4\pi\mu_a/[(1-\sigma)(1+\mu_a/\mu_b)\,\mu_0\, e]\} = 0 \end{aligned} \tag{123}$$

where in the last term $e = 2.718\cdots$, and in general $\rho = c/p$.

Once the coherent configuration has become unstable the strain $\bar{e}_{\mathrm{m}}$, which minimizes the energy, is very small compared to the strain e_0 for coherency. The energy may thus be approximated in terms of the average strain $\bar{e}$ by an expression of the form

$$E = (\mu_0\, a/4\pi^2)(A - B\bar{e} + C\bar{e}^2) \tag{124}$$

where A, B, and C are constants determined by Eqs. (72) and (121). It follows that

$$\bar{e}_{\mathrm{m}} \cong [(1-\sigma)\,\mu_0\, b/8\pi^2(1-\sigma)\,\mu_b\, t]\, \beta \ln[2\beta(1+\beta^2)^{1/2} - 2\beta^2] \tag{125}$$

where

$$\beta = 2\pi\mu_b\,\rho/\{(1-\sigma)(1+\mu_b/\mu_a)\,\mu_0(1+\rho)\} \tag{126}$$

and terms in powers of ρ exceeding unity are neglected. In Eqs. (123), (125), and (126) the approximation $\sigma_a = \sigma_b = \sigma$ has been used. The results given by these equations may be expected to differ from those of the exact solution of the model by less than 10%, except possibly in a small region about the limiting strain.

C. Energy Minimization

1. *The Total Energy*

The total energy ε_{T} of an epitaxial overgrowth of finite thickness on a semiinfinite substrate may be written in the form (Ball, 1970):

$$\varepsilon_{\mathrm{T}} = E_{\mathrm{H}}\,S_0 + (E_{\mathrm{d}}{}^x + E_{\mathrm{d}}{}^y)\,S + \varepsilon_{\mathrm{f}} + \varepsilon_{\mathrm{ad}} \tag{127}$$

where S_0 is the interfacial area when the strains $\bar{e}_x$ and $\bar{e}_y$ vanish and S the interfacial area with nonzero overall strain; E_{H} is the energy of overall strain given by Eq. (121); and $E_{\mathrm{d}}{}^x$, $E_{\mathrm{d}}{}^y$ are the dislocation energies associated with the misfits in the x- and y-directions, respecitvely. The term ε_{f} is the total energy of the free surface of the film, which is here assumed to be independent of strains (thus ignoring the strain dependence predicted by Drechsler and Nicholas, 1969); $\varepsilon_{\mathrm{ad}}$ is the total energy of adhesion of the film onto the substrate when it is strained to coherency and is by definition independent of strain.

For simplicity we consider only the case of quadratic symmetry, although the considerations may be extended to the more general case of unequal lattice parameters. It then follows that

$$S = (1+\bar{e}^2)\,S_0 \tag{128}$$

and the minimizing strain $\bar{e}_{\mathrm{m}}$ will be the strain for which the quantity

$$E_{\mathrm{T}} = \varepsilon_{\mathrm{T}}/S_0 = E_{\mathrm{H}} + 2(1+\bar{e}^2)\,E_{\mathrm{d}} \tag{129}$$

is a minimum.

2. *Minimizing Strain*

For quadratic symmetry ε_{T} will be a minimum for the strain $\bar{e}_{\mathrm{m}}$ defined by

$$\partial E_{\mathrm{T}}/\partial \bar{e} = 0, \qquad \bar{e} = \bar{e}_{\mathrm{m}} \tag{130}$$

It must be remembered in the expression for E_{d} in Eq. (129) that the strain $\bar{e}$ is not zero and Eq. (43) must be replaced by

$$\bar{p} = (\bar{P}+\tfrac{1}{2})\,\bar{c} = \bar{P}\bar{b} = (\bar{P}+1)\,a \tag{131}$$

It is easily shown that

$$\mu = \mu_0(1+\tfrac{1}{2}\rho)$$

We define

$$\rho = \frac{c}{p} = \frac{-e_0}{1+\tfrac{1}{2}e_0}, \qquad \bar{\rho} = \frac{\bar{c}}{\bar{p}} = \frac{\bar{e}-e_0}{1+\tfrac{1}{2}(\bar{e}+e_0)} \tag{132}$$

and ρ and $\bar{\rho}$ may be used as measures of the misfit for unstrained and homogeneously strained overgrowths, respectively.

Equation (130) is of the form

$$f(e_0, \bar{e}_m, t) = 0 \tag{133}$$

from which one may obtain

$$\bar{e}_m = \bar{e}_m(e_0, t) \tag{134}$$

and then calculate the minimum value of E_T:

$$E_{T,\min} = E_T[e_0, \bar{e}_m(e_0,t), t] = E_{T,\min}(e_0, t) \tag{135}$$

The dependence of $E_{T,\min}$ on e_0 (or ρ) may be calculated for different values of t. A typical curve is shown in Fig. 12, where curve A is the energy per unit

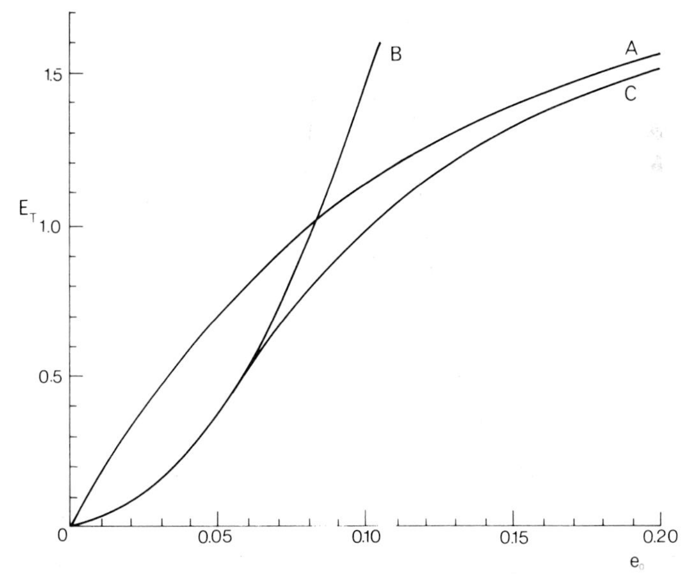

FIG. 12. Curves of $E_T(e_0)$ in units of $\mu_0 a/4\pi^2$ for a monolayer overgrowth. A: $\bar{e}=0$ (all misfit accommodated by misfit dislocations); B: $\bar{e}=e_0$ (coherent configuration); C: $\bar{e}=\bar{e}_{\min}$ (minimum energy configuration). Elastic constants: $\mu_0=\mu_a=\mu_b$, $\sigma_a=\sigma_b=0.3$.

unstrained area if all the misfit of a monolayer film is accommodated by dislocations, curve B is the energy per unit area if all the misfit is accommodated by overall strain e_0, and curve C shows $E_{\mathrm{T,min}}$. Here it is assumed that $\sigma_a = \sigma_b = 0.3$. Equations (115)–(118) were used to calculate E_{d}, these being considered the most generally suitable. Alternatively the dependence of $E_{\mathrm{T,min}}$ on t may be calculated for different values of e_0 (or ρ).

Of particular interest are growing films whose initial atomic layers are deposited in the coherent state, i.e.,

$$\partial E_{\mathrm{T}}/\partial |\bar{e}| \leqslant 0, \qquad \text{for} \quad |\bar{e}| = |e_0| \tag{136}$$

The derivative in (136) increases with thickness t and a limiting thickness t_c is reached at which

$$\partial E_{\mathrm{T}}/\partial |\bar{e}| = 0, \qquad \text{for} \quad |\bar{e}| = |e_0| \tag{137}$$

Above this thickness the minimum-energy configuration will be one that includes misfit dislocations. Equation (137) is of the form

$$f(t_c, e_0) = 0 \tag{138}$$

which may be solved for limiting thickness t_c. Figure 13 shows curves of limiting

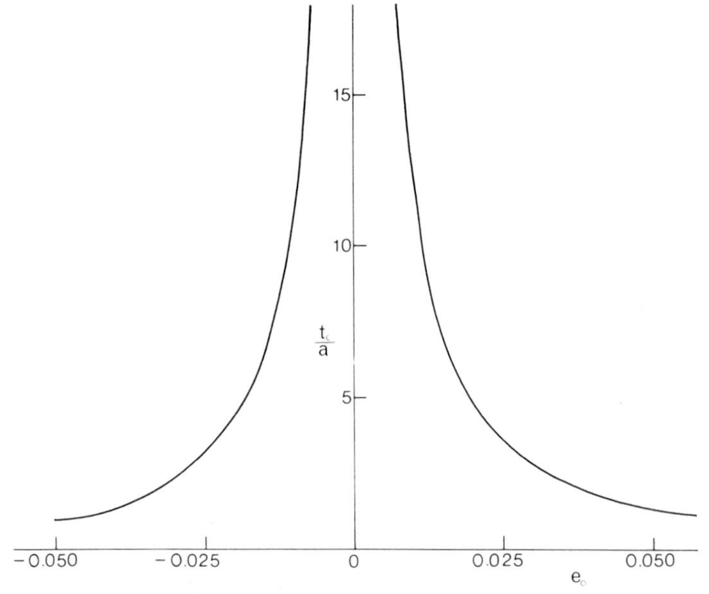

FIG. 13. Curves of limiting thickness $t_c(e_0)$ for positive and negative values of e_0. Elastic constants: $\mu_0 = \mu_a = \mu_b$, $\sigma_a = \sigma_b = 0.3$.

thickness as a function of e_0, using Eqs. (115)–(118) for E_d and $\mu_0 = \mu_a = \mu_b$, $\sigma_a = \sigma_b = 0.3$.

VI. Small Epitaxial Islands

A. Introduction

In the previous sections the epitaxial overgrowth was assumed to be large. Thus the discrete nature of the absorption or emission of a misfit dislocation and of other properties, related to the finite size of the island, have been ignored. The strain field emanating from an island into the substrate (Cabrera, 1964), the shape anisotropy (Szostak *et al.*, 1966; Niedermayer, 1968), the scatter about the ideal axial orientation (Jesser and Kuhlmann-Wilsdorf, 1967), the saw-tooth behavior of average strain (Vincent, 1969; Jesser and van der Merwe, 1971, 1972), and the apparent fluctuation in axial orientation of a growing island (Matthews, 1972) are interesting examples of cases that have been dealt with before.

The saw-tooth behavior of average strain in a small growing island will be briefly considered with the view of demonstrating some of the principles involved. For this purpose it suffices to use the simple model of a growing one-dimensional system (Frank and van der Merwe, 1949a, b), represented by a chain of atoms connected by elastic springs of natural length b and acted on by a sinusoidal substrate potential of wavelength a. For the purpose of the discussion it is assumed that $b > a$.

The first atom that is deposited will, apart from thermal vibrations, settle at the bottom of a potential trough. When the next one is added in a neighboring trough there will be an adjustment of positions whereby the atoms will, under the thrust of the connecting spring, ride up symmetrically on the far slopes of the neighboring potential troughs. When a third one is added the adjusted configuration will again be symmetrical, but the atoms at the free ends have now ridden higher up. The readjustment will continue when more atoms are added, until a stage is reached in which the atoms at the free ends ride over the potential crest. This configuration is unstable and will trigger a spontaneous extension of the chain in which the atoms at one end will slide down the hill and up some distance on the other side to introduce a misfit dislocation. In this discrete process the strain will change by a finite amount. If there are no frictional forces, the entire chain will also experience a sideways shift to form a new symmetrical configuration containing one misfit dislocation. It should be noted that the spontaneous instability referred to above is preceded by a metastable range of chain lengths. These and other properties of the system will be dealt with quantitively below.

B. Calculations

The governing equation for the one-dimensional model (Frank and van der Merwe, 1949a, b; Niedermayer, 1968), is analogous to (23), only l_x is replaced by

$$l = (\mu a^2/2W)^{1/2} \tag{139}$$

The solution of (23) can be written in terms of elliptic functions

$$\pi n/kl = F(k, \phi) - F(k, \phi_m) \tag{140}$$

When solved for ξ it is of the form

$$\xi_m(n) = \tfrac{1}{2} + \pi^{-1} am[(\pi n/kl) + F(k, \phi_m)] \tag{141}$$

where

$$\phi = \pi[\xi_m(n) - \tfrac{1}{2}], \qquad \phi_m = \pi[\xi_m(0) - \tfrac{1}{2}] \tag{142}$$

m is the number of dislocations and the origin of n is chosen at the midpoint of the symmetrical configuration. This choice of origin implies that

$$\xi_m(0) = 0, \tfrac{1}{2} \qquad \text{for} \quad m \text{ even, odd} \tag{143}$$

The displacement of the right-hand free end ($n = \frac{1}{2}N_m$) of a chain of $N_m + 1$ atoms can be written in the form

$$\xi_m(\tfrac{1}{2}N_m) = \tfrac{1}{2}(m+1) + \xi_m{}^1 + (\phi_m/\pi) \tag{144}$$

where $0 \leqslant \xi_m{}^1 \leqslant \frac{1}{2}$.

One may define a difference $\Delta\Phi_m$ by

$$\Delta\Phi_m \equiv \Phi_m - \phi_m = \pi[\xi_m(\tfrac{1}{2}N_m) - \tfrac{1}{2}] - \phi_m = \tfrac{1}{2}\pi(m+1) - \chi_m \tag{145}$$

where

$$\chi_m = \pi(\tfrac{1}{2} - \xi_m{}^1), \qquad \tfrac{1}{2}\pi \geqslant \chi_m \geqslant 0 \tag{146}$$

The applied force at the end of the chain can be expressed in terms of a tension such as T_x in (35). The condition that T_x vanishes at the end ($n = \frac{1}{2}N_m$) of the chain determines the values of $\xi_m{}^1$ and χ_m by means of the equation

$$0 = \mu t e_x = \pi t \alpha[(d\xi/dn) - (1/P)], \qquad n = \tfrac{1}{2}N_m$$

where $d\xi/dn$ is of form (25). It follows that

$$\chi_m = \arcsin(k^{-1} - l^2 P^{-2})^{1/2}, \qquad 0 \leqslant \chi_m \leqslant \tfrac{1}{2}\pi \tag{147}$$

χ_m will be real and meaningful provided

$$s \leqslant 1/k \leqslant (1 + l^2/P^2)^{1/2} \tag{148}$$

where s is the larger of 1 and l/P.

The number N_m is obtained from (140):

$$\pi N_m/2kl \equiv \Delta F_m = F(k, \Phi_m) - F(k, \phi_m) = (m+1)\, K(k) - F(k, \chi_m) \tag{149}$$

where the latter result follows from (145) and the properties of elliptic integrals.

The average energy per atom ε_m of the chain (Niedermayer, 1968) may be calculated using the methods of Section II, C:

$$\varepsilon_m = W\{(2\,\Delta E_m/k^2) - (2l/kP)\,\Delta\Phi_m + (1-k^{-2}+l^2/P^2)\,\Delta F_m\}/\Delta F_m \tag{150}$$

where

$$\Delta E_m \equiv E(k, \Phi_m) - E(k, \phi_m) = (m+1)\, E(k) - E(k, \chi_m) \tag{151}$$

Strictly speaking ε_m is the mean energy per bond in a row of N_m+1 atoms.

The dependence of ε_m on N_m and N_m/l is shown in Fig. 14 for $l/P = 0.8$ and the value $l = 7$, estimated previously (Frank and van der Merwe, 1949a, b). The numbers allotted to the curves designate the number of corresponding dislocations m.

The average strain in a row is given by

$$\bar{e} = (x - x_0{}^1)/x_0{}^1 \tag{152}$$

where x and $x_0{}^1 = N_m b$ denote, respectively, the actual and unstrained

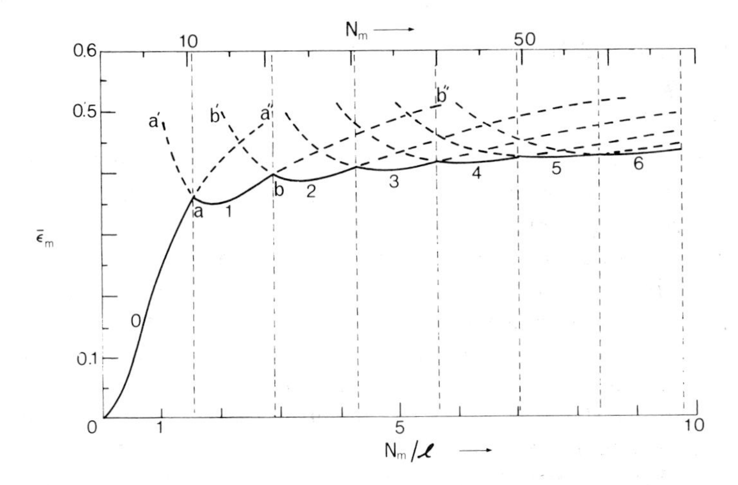

FIG. 14. Curves of the average energy per atom $\bar{\varepsilon}_m$ in units of W versus number of atoms N_m or relative number N_m/l for $l = 7$, $b > a$, and $l/p = 0.8$. The curves $m = 0, 1, 2 \ldots$ correspond to a chain containing $0, 1, 2 \ldots$ misfit dislocations. The solid line represents lowest energy configurations (Jesser and van der Merwe, 1971).

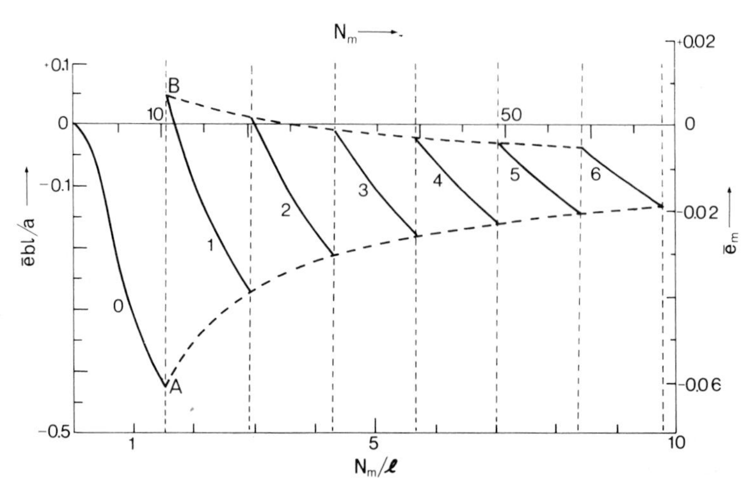

FIG. 15. Saw-tooth behavior of average strain $\bar{e}$ (in units of bl/a on left-hand scale) versus number of atoms in chain N_m or relative number N_m/l for $l = 7$, $l/P = 0.8$, and $b > a$. The curves $m = 0, 1, 2\ldots$ correspond to a chain containing $0, 1, 2, \ldots$ misfit dislocations (Jesser and van der Merwe, 1971).

lengths of the row of $N_m + 1$ atoms. It follows from the foregoing that

$$x = aN_m + 2a[\xi_m(\tfrac{1}{2}N_m) - \tfrac{1}{2} - \phi_m/\pi] \tag{153}$$

Hence

$$\bar{e} = a[2\,\Delta\Phi_m(l/\pi N_m) - (l/P)]/lb \tag{154}$$

The dependence $\bar{e}(N_m)$ is plotted in Fig. 15 for the case $l/P = 0.8$ and $l = 7$.

C. Discussion

Consider the building up of a chain by the addition of consecutive atoms as set out in Section A. The first one that goes down settles in a potential trough and has zero energy represented by the lowest point 0 on curve 0 in Fig. 14. As more atoms are added, the average energy per atom, consisting of potential energy of misfit and strain energy in the springs, increases in discrete steps along curve 0. When the point a″ is surpassed, instability sets in and a misfit dislocation is introduced spontaneously as described in Section A. However, the present considerations show that already at the point a the configuration has become metastable in that the state of lowest energy, represented by curve 1, is one with one dislocation. If more atoms are added to this configuration it becomes metastable at b, where lowest energy requires a transition to curve 2. Spontaneous instability would set in at point b″.

The solid-line curve represents the variation of lowest energy as atoms are added and more dislocations introduced. If on the other hand atoms are desorbed, starting with a stable configuration on curve 1, it becomes metastable again at a and complete instability sets in at a′, after which dislocations will be emitted spontaneously.

The intersections of the curves depend among others on the value of the parameter l/P. When $l/P \leqslant 2/\pi$ curve 0 drops to a position where it is everywhere below curves 1, 2, 3 .... This shows that the coherent configuration is then stable, whatever the value of N_m as has been shown in Section II.

Consider the average strain $\bar{e}$ plotted in Fig. 15. The curve 0A demonstrates the increase in compressive strain with increasing number of atoms. Point A corresponds to a in Fig. 14. When an additional atom is added, lowest energy requires the introduction of a dislocation—a transition to curve 1 in Fig. 14. This is manifested in $\bar{e}$ by a jump from A on curve 0 to B on curve 1 in Fig. 15. Consecutive jumps in $\bar{e}$ occur with increasing number of atoms N_m at the transition points between the curves 1, 2, ... in Fig. 14. This explains the saw-tooth behavior first considered by Vincent (1969).

The strain curves in Fig. 15 also illustrate the fact that the two branches of the envelopes not only narrow down rapidly but also approach the N_m axis, characteristic of zero strain. Furthermore, the number $\Delta N_m = N_{m+1} - N_m$ of atoms between the introduction of consecutive dislocations approaches the normal dislocation spacing $P = a/(b-a)$ for undistorted springs. These results were predicted previously by Brooks (1952), Frank and van der Merwe (1949a, b), Vincent (1969), and Jesser and van der Merwe (1971, 1972).

One may surmise that the transition process from m to $m+1$ dislocations, in particular for small m, must be somewhat violent. It has been suggested (Jesser and van der Merwe, 1972) that this dynamic process characterizes a brief interval of time during which the resistance offered by the substrate for "rigid" translation or rotation of an island is appreciably smaller than otherwise. Favorable interactions with moving particles may accordingly set it in motion. The probability of such an event will presumably be proportional to the fraction of time spent in this state.

Matthews (1971) has also shown that the apparent sporadic "rigid" rotations of small islands, as have been concluded on the basis of moiré fringe observations, may be explained in terms of the discrete introduction of misfit dislocations without the need for a rotation of the crystallites. Another interesting speculation is that the disorientation of an island may decrease with size if the number of twist misfit dislocations defining its misorientation is conserved (Jesser and van der Merwe, 1972).

References

Ball, C. A. B. (1970). *Phys. Status Solidi* **42**, 357.
Ball, C. A. B., and van der Merwe, J. H. (1970). *Phys. Status Solidi* **38**, 335.
Cabrera, N. (1964). *Surface Sci.* **2**, 320.

Chopra, K. L. (1969). *Phys. Status Solidi* **32**, 489.
Drechsler, M., and Nicholas, J. F. (1969). *J. Chem. Phys. Solids* **28**, 2609.
du Plessis, J. C., and van der Merwe, J. H. (1965). *Phil. Mag.* **11**, 43.
Fletcher, N. H. (1964). *J. Appl. Phys.* **35**, 234.
Frank, F. C., and van der Merwe, J. H. (1949a). *Proc, Roy. Soc.* **A198**, 205.
Frank, F. C., and van der Merwe, J. H. (1949b). *Proc. Roy. Soc.* **A198**, 216.
Frank, F. C., and van der Merwe, J. H. (1949c). *Proc. Roy. Soc.* **A200**, 125.
Frenkel, J., and Kontorowa, T. (1938). *Phys. Z. Sowj.* **13**, 1.
Jesser, W. A., and Kuhlmann-Wilsdorf, D. (1967a). *Phys. Status Solidi* **19**, 95.
Jesser, W. A., and Kuhlmann-Wilsdorf, D. (1967b). *Phys. Status Solidi* **21**, 533.
Jesser, W. A., and van der Merwe, J. H. (1971). *Phil. Mag.* **24**, 295.
Jesser, W. A., and van der Merwe, J. H. (1972). *Surface Sci.* **31**, 229.
Matthews, J. W. (1972). *Surface Sci.* **31**, 241.
Nabarro, F. R. N. (1947). *Proc. Phys. Soc. (London)* **59**, 256.
Niedermayer, R. (1968). *Thin Films* **1**, 25.
Pashley, D. W. (1965). *Advan. Phys. (Phil. Mag. Suppl.)* **14**, 327.
Royer, L. (1928). *Bull. Soc. Fr. Min.* **51**, 7.
Schneider, H. G. (1969). *In* "Epitaxy and Endotaxy" (H. G. Schneider, ed.). VEB Deutscher Verlag für Grundstoffindustrie, Leipzig.
Szostak, R. F., and Molière, K. (1966). *In* "Basic Problems in Thin Films" (R. Niedermayer and H. Mayer, eds.), p. 10. Vandenhoeck and Ruprecht, Göttingen.
Timoshenko, S. (1934). "Theory of Elasticity," pp. 12–134. McGraw-Hill, London.
van der Merwe, J. H. (1950). *Proc. Phys. Soc. (London)* **A63**, 616.
van der Merwe, J. H. (1963a). *J. Appl. Phys.* **34**, 117.
van der Merwe, J. H. (1963b). *J. Appl. Phys.* **34**, 123.
van der Merwe, J. H. (1964). *In* "Single Crystal Films" (M. H. Francombe and H. Sato, eds.), p. 139. Pergamon, Oxford.
van der Merwe, J. H. (1966). *In* "Basic Problems in Thin Film Physics" (R. Niedermayer and H. Mayer, eds.), p. 122. Vandenhoeck and Ruprecht, Gottingen.
van der Merwe, J. H. (1970). *J. Appl. Phys.* **41**, 4725.
van der Merwe, J. H. (1971). *Surface Sci.* **31**, 198.
van der Merwe, J. H., and van der Berg, N. G. (1972). *Surface Sci.* **32**, 1.
Vermaak, J. S., and van der Merwe, J. H. (1964). *Phil. Mag.* **10**, 785.
Vermaak, J. S., and van der Merwe, J. H. (1965). *Phil. Mag.* **12**, 453.
Vincent, R. (1969). *Phil. Mag.* **19**, 1127.

CHAPTER 7

ENERGY OF INTERFACES BETWEEN CRYSTALS: AN AB INITIO APPROACH

N. H. Fletcher and K. W. Lodge

Department of Physics
University of New England
Armidale, Australia

I. Introduction

For many purposes it is helpful to distinguish between two different approaches to problems involving distortions of crystals: the atomic and the metric. An atomic view is generally a pictorial one that takes account of individual atoms or molecules, their arrangements, and their interactions with one another; it is useful in making first-order models but is difficult when even semiquantitative calculations are required. A metric approach concentrates attention on the crystal lattice and often regards it as embedded in a uniform elastic medium, with certain limitations being placed upon the types of lattice discontinuity that are permitted. The degree of abstraction is thus higher than for an atomic approach and leads to some sacrifice in visualizing the physical situation, but at the same time certain patterns become much more explicit and semiquantitative calculations are quite readily possible.

In the discussion of crystal interfaces these two approaches can be fairly readily recognized. The pioneering work of Burgers (1940) and Bragg (1940) began with an atomic picture in which each crystal was supposed to be continuous up to the interface. They then abstracted this to a metric model in which the connections between the two parts of the bicrystal were provided by an array of dislocations. This abstraction provides an excellent description of the interface for small mismatches between the two crystals—orientation differences up to about 15° or lattice-parameter differences up to about 20%—but for boundaries of greater mismatch than this the dislocations begin to overlap and the picture is no longer clear.

The classic work of Read and Shockley (1950) showed how this dislocation model could be made into a semiquantitative theory and yielded the well-known result

$$E(\theta) = \tau\theta(A - \ln\theta) \tag{1}$$

for the energy E per unit area of an interface with misorientation angle θ. The elastic coefficient τ was given explicitly in terms of the known elastic parameters of the crystal material, while the quantity A, which was related to the unknown "core energy" of a dislocation, was chosen to give best fit with experiment. Quite similarly, for a boundary between two crystals with fractional lattice-parameter difference δ, we find (Fletcher, 1971)

$$E(\delta) = \tau'\delta(A' - \ln\delta) \tag{2}$$

This simple model was extended to orientations for which a row of equally spaced dislocations appears inadequate by the introduction of hierarchies of dislocations (Read and Shockley, 1950; du Plessis and van der Merwe, 1965), but while this artifice gives a successful numerical result it constitutes something of a bar to the visualization of the model in atomic terms. Further important extensions of this approach to deal more particularly with epitaxial systems were introduced by Frank and van der Merwe (1949) and further developed by van der Merwe (1950) and his co-workers and by Fletcher (1964). These matters have been discussed in detail in earlier chapters and their considerable contribution to our understanding set out. The method suffers limitations because of the uncertainty of its application to interfaces of large mismatch, the necessity of making a further ad hoc introduction of crystal symmetry, and the somewhat artificial treatment of dislocation cores. We shall not, however, discuss these points here (see Fletcher, 1971).

Among treatments of interfaces that we might classify as primarily atomic are those based upon a simplified picture of atomic arrangements in the interface—"island" models (Mott, 1948), coincidence-lattice models (Brandon, 1966), and the like. A recent survey has been given by Gifkins (1969), and some of these matters are also discussed in the present volume. The 0-lattice concept introduced by Bollmann (1967) serves to formalize many of these ideas and to provide a basic means of describing and predicting atomic relationships across an interface.

Attractive and useful as these models are, they provide very little more than qualitative information about interface energy, even when they are extended to include some of the features of a dislocation description of the interface (Brandon, 1966; Bollmann, 1967). This is understandable, since the reasons for the development of the models were generally to provide a physical picture rather than to serve as an aid to computation.

The only basically atomic approach to a direct calculation of interface energy that has so far been developed appears to be that of Fletcher and his students (Fletcher, 1964, 1967; Fletcher and Adamson, 1966; Lodge 1970). The purpose of this chapter is to describe that method, to show its relationship to the various atomic and metric models mentioned above, and to show how it can be used to calculate interface energy in a real situation. Finally we shall make some remarks about interfacial entropy and interfacial free energy at a finite temperature.

II. The Interface Problem from First Principles

Before we begin to make any sort of calculation about an interface we must decide exactly what we are trying to calculate. Interfaces between real crystals will generally tend to modify their structure with time, so that the free energy of the system of which they are a part is minimized. This minimization will, however, generally be subject to one or more constraints governing the relative orientations of the two crystals (as when two grains grow together in a solidifying melt) or of the boundary relative to one of the crystals (as when an epitaxial film is deposited on a substrate of high melting point). We shall also generally assume that the two crystals are mutually insoluble.

We seek therefore, to be realistic, a minimum in the free energy of the system through variation of the interface configuration subject to the given constraints. Fortunately many of the systems in which we are interested are at temperatures well below the melting points of the crystalline components or of any possible grain-boundary eutectic, so that we may reasonably minimize the energy, thus calculating effectively the configuration and energy of the boundary at 0°K. This is the approach we shall take, returning in the last section of the chapter to consider the effect of finite temperature on the calculation.

Because we are concerned in this book with the epitaxial growth of one crystalline material upon another, we can usefully restrict the range of possible problems to those in which we are given one semiinfinite crystal possessing a plane surface of known orientation. We then seek to determine the structure and energy of the interface between this crystal and another, which may be of finite thickness, growing upon it with arbitrary orientation. The actual preferred growth orientations will then be those that yield minima in the interfacial energy.

III. Symmetry Considerations and Coincidence Lattices

Naive ideas about epitaxial growth place emphasis upon the principle that coherent overgrowth of crystal material B on crystal A is likely to occur if some undistorted crystal plane of B can be laid down on top of the exposed face of A in such a way that a large fraction of the B atoms can be made to coincide with the sites of A atoms. We might go further and say that the greater the number of coincidences per unit area, the lower will be the energy of the resulting interface. Both these principles are correct and must find expression in any complete theory of interface energy. They obviously rest heavily on consideration of the symmetry and structure of the two crystals involved, and the geometrical working out of these ideas constitutes the theory of coincidence lattices. We shall take a little time to consider this before going on to the main development of our ideas.

Consider two crystals of quite general structure and imagine that the lattice of each is extended to fill all space so that these two lattices interpenetrate. Let us then translate one lattice relative to the other so that they have one lattice point as a common origin. For two arbitrary crystals there will in general be no other lattice point in common, but there will always be pairs of lattice points between which the spacing is arbitrarily small, if we go far enough away from the origin. We can therefore, by infinitesimal adjustment of orientation angle and lattice parameter, bring three other pairs of lattice points into coincidence with negligible change in the physics of the situation. If these pairs of points are chosen so that they are not collinear, and not coplanar with the origin, then this action will have generated a three-dimensional grid of common lattice points that is called the "coincidence lattice." Its properties have been investigated by Friedel (1926) and more recently by Ranganathan (1966).

From our present point of view, possible low-energy boundaries between these two crystals in the given relative orientation are geometric planes containing a high density of coincidence-lattice points, and the reciprocal of the density of these points in the selected plane gives a measure of the grain-boundary energy. It is possible to extend these ideas to include dislocated coincidence boundaries (Brandon, 1966) but this need not concern us here.

Once we have established a coincidence lattice between two crystals we can proceed one step further, following the analysis of Bollmann (1967). Choosing one of the coincidence-lattice points as origin we can define a homogeneous linear transformation $\mathscr{A}$ performed about the point that establishes a one-to-one correspondence between the lattice points $\mathbf{r}$ of one crystal and the lattice points $\mathbf{r}'$ of the other by

$$\mathbf{r}' = \mathscr{A}\mathbf{r} \tag{3}$$

The transformation $\mathscr{A}$ will generally involve a rotation, a change in scale, and perhaps an angular distortion as well. In Fig. 1, for example, if we consider this to be an (001) section through two interpenetrating simple cubic lattices, $\mathscr{A}$ represents a relative rotation of $\tan^{-1}(1/2) = 26°34'$ about the [001] axis together with a uniform expansion by a factor $5^{1/2}$.

Looking again at Fig. 1 and remembering that the transformation $\mathscr{A}$ was applied about any one of the coincidence-lattice points, we see that we could have achieved exactly the same result by applying the same transformation $\mathscr{A}$

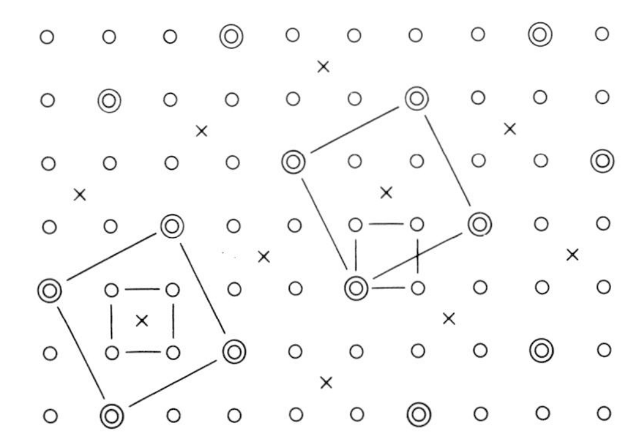

FIG. 1. A (001) section through two interpenetrating cubic lattices, represented by ○ and ◯, respectively, related by a transformation $\mathscr{A}$, showing points of the coincidence lattice and with points of the 0-lattice marked by ×.

about any one of the set of points marked ×. The set of all these points together with the coincidence-lattice points constitutes what Bollmann has called the "0-lattice" in the plane. In fact the 0-lattice, like the coincidence lattice, is a three-dimensional concept and consists of the complete set of points about which the transformation $\mathscr{A}$ could have been applied to achieve the same result. For the particular case shown in Fig. 1, the 0-lattice really consists of a set of lines running parallel to [001] and passing through the 0-lattice points in the (001) plane. For a more general transformation the 0-lattice is a discrete three-dimensional grid of points.

The 0-lattice has an analytical significance that we shall consider later. For the present we simply note the symmetry of both crystals about the points of the 0-lattice, which means that, no matter what elastic relaxation may take place when the physical interface is formed, the 0-lattice points will remain stationary. It can also be shown, though this is most useful for situations more complex than that shown in Fig. 1, that the 0-lattice gives directly the intensity maxima in the moiré pattern obtained by superposing the interface atoms of the two crystals (Bollmann, 1967).

IV. Interface Energy as a Variational Problem

Following this apparent digression, let us return to the program outlined in Section II: that we should write down an expression for the total energy of our bicrystal system and then minimize this with respect to the atomic configuration of the interface, subject to the constraints assumed for the physical problem. Formally we might proceed as follows.

Consider the potential energy $V_0(\mathbf{r})$ of a B atom at a point $\mathbf{r}$ just outside the plane surface of a crystal of A atoms located at points $\mathbf{R}_i$. If the interaction potential between individual A and B atoms is $v_{AB}(\mathbf{r})$, then

$$V_0(\mathbf{r}) = \sum_i{}^{-} v_{AB}(\mathbf{r}-\mathbf{R}_i) \tag{4}$$

where the minus on the summation indicates that it extends only over the lower half-space. If the B atom is actually part of a semiinfinite crystal of B with atom positions $\mathbf{r} = \mathbf{R}_i'$, then the energy of the entire system can be written

$$E_0{}^{\mathrm{T}} = \sum_i{}^{+} \sum_j{}^{-} v_{AB}(\mathbf{R}_i'-\mathbf{R}_j) + \tfrac{1}{2} \sum_{i\neq j}{}^{-} v_{AA}(\mathbf{R}_i-\mathbf{R}_j) + \tfrac{1}{2} \sum_{i\neq j}{}^{+} v_{BB}(\mathbf{R}_i'-\mathbf{R}_j') \tag{5}$$

the last two terms representing the self-energies of the two semiinfinite crystals.

So far this formulation contains no variational parameters, but in principle we could simply vary all the atomic positions $\mathbf{R}$ and $\mathbf{R}'$ within small limits so as to keep the interface position fixed and thus determine the configuration giving minimum energy. This is, of course, quite impracticable because of the number of atoms involved, so that we must devise some simplification.

To do this we note two things. In the first place, the interactions $v_{AB}(\mathbf{r})$ are generally of reasonably short range so that they should extend over only a few atomic layers on either side of the interface. In the second place, the variations of the self-energy terms describe essentially the elastic behavior of two semiinfinite crystals subject to stresses on their free surfaces. Provided these stresses are not too large, we should then be able to use continuum elasticity theory to approximate the variation of these last two terms.

To this end, let us suppose that all the atoms near the interface are varied in position so that $\mathbf{R}_i$ moves to $\mathbf{R}_i+\mathbf{F}_i-\mathbf{F}_0$, etc., where $\mathbf{F}_0$ is a small translation of the whole semicrystal and the $\mathbf{F}_i$ are small individual atomic displacements. The total energy then becomes

$$E^{\mathrm{T}} = \sum_i{}^{+} \sum_j{}^{-} v_{AB}(\mathbf{R}_i'+\mathbf{F}_i'-\mathbf{R}_j-\mathbf{F}_j+\mathbf{F}_0) + E_{A_0}^{\mathrm{T}} + E_{A}^{\mathrm{T}}(\mathbf{F}_1,\mathbf{F}_2,\ldots) + E_{B_0}^{\mathrm{T}} + E_{B}^{\mathrm{T}}(\mathbf{F}_1',\mathbf{F}_2',\ldots) \tag{6}$$

where $E_{A_0}^{\mathrm{T}}$ and $E_{B_0}^{\mathrm{T}}$ are the self-energies of the two undistorted semicrystals, $E_A{}^{\mathrm{T}}(\mathbf{F}_1,\mathbf{F}_2,\ldots)$ is the elastic strain energy in semicrystal A resulting from the displacement of the surface atom at $\mathbf{R}_i$ by an amount $\mathbf{F}_i$, etc., and similarly for B.

In this form the problem is now solvable in principle, although the number of independent parameters $\mathbf{F}$ and $\mathbf{F}'$ would still make a practical calculation prohibitively tedious. However, a great simplification is possible in special cases when there is a high density of coincidence-lattice points in the interface. When this happens, we note that symmetry requires the coincidence-lattice points to remain fixed during elastic relaxation and also requires the pattern of atomic displacements to repeat in each coincidence-lattice cell. The constancy of 0-lattice points during relaxation imposes further symmetry requirements on the distortions $\mathbf{F}$. In these special cases then, the variational formulation given in (6), together with explicit forms for $E_A{}^{\mathrm{T}}(\mathbf{F}_i)$ and $E_B{}^{\mathrm{T}}(\mathbf{F}_i)$ should make calculation relatively simple, since the number of independent variational parameters is small.

However, two problems with this approach have meant that it has not been seriously followed up. In the first place, the assumption of simple linear elastic behavior implied by the form of (6) is not justified for the individual atomic displacements implied in the variational problem. This necessarily introduces errors of unknown magnitude into the result. Second, there seems to be no simple way in which results of general utility can be derived from the variational function given in (6); all we can do is to calculate specific cases and then attempt to generalize.

In the next section we shall see that, by adhering to the general philosophy of the variational method but by carrying out all the operations in reciprocal space, we can overcome both these objections and derive valid and useful general results.

V. The Coincidence Boundary in Reciprocal Space

Returning to Eq. (4), we can write the potential energy of a B atom at position $\mathbf{r}$ outside a plane face of an A crystal with atomic positions $\mathbf{R}$ as

$$V_0(\mathbf{r}) = \sum_i{}^{-} v_{\mathrm{AB}}(|\mathbf{r}-\mathbf{R}_i|) \equiv \sum_{\mathbf{R}}{}^{-} v(\mathbf{r}-\mathbf{R}) \tag{7}$$

where we shall adopt now the simplified notation on the right. This potential can be written in terms of its Fourier components $V_0(\mathbf{k})$ as

$$V_0(\mathbf{r}) = (N/8\pi^3) \int V_0(\mathbf{k}) \exp(i\mathbf{k}\cdot\mathbf{r})\, d\mathbf{k} \tag{8}$$

where N is the (infinite) number of atoms in the crystal,

$$V_0(\mathbf{k}) = (1/N) \sum_{\mathbf{R}}{}^{-} v(\mathbf{k}) \exp(-i\mathbf{k}\cdot\mathbf{R}) \tag{9}$$

and $v(\mathbf{k})$, the Fourier transform of the atomic potential, is given by

$$v(\mathbf{k}) = \int v(\mathbf{r}) \exp(-i\mathbf{k}\cdot\mathbf{r})\, d\mathbf{r} \tag{10}$$

With this artifice we can now write down the total interaction energy between the two crystals by summing $V_0(\mathbf{r})$ over all the atomic positions $\mathbf{R}'$ of the B crystal, assumed undistorted, to give

$$E_0^{\mathrm{T}} = \sum_{\mathbf{R}'}{}^{+} V_0(\mathbf{R}') = (1/8\pi^3) \sum_{\mathbf{R}'}{}^{+} \sum_{\mathbf{R}}{}^{-} \int v(\mathbf{k}) \exp[i\mathbf{k}\cdot(\mathbf{R}'-\mathbf{R})]\, d\mathbf{k} \tag{11}$$

Now we know that a sum like $\sum_{\mathbf{R}} \exp(i\mathbf{k}\cdot\mathbf{R})$, taken over an infinite crystal, vanishes unless $\mathbf{k}$ is a vector of the reciprocal lattice of that crystal. The sums over $\mathbf{R}$ and $\mathbf{R}'$ in (11) are only over semiinfinite crystals, so we cannot make exactly this statement. We can, however, reexpress the crystal geometry if necessary so that, instead of using a normal primitive cell, we use a cell of the same volume having two of its primitive translations lying in the plane of the interface so as to define a surface lattice for each crystal. We denote the surface lattice vectors by $\mathbf{R}_s, \mathbf{R}_s'$, respectively, and the areas of the surface unit cells by A, A'. We can also define a surface reciprocal lattice for each crystal, following the usual rules, and denote the vectors of this lattice by $\mathbf{G}_s$, $\mathbf{G}_s'$, respectively. Similarly, $\mathbf{k}_s$ is the component of $\mathbf{k}$ parallel to the surface.

Since the sums in (11) over $\mathbf{R}_s$ and $\mathbf{R}_s'$ are infinite, we can perform these in one of two equivalent ways to write either

$$\sum_{\mathbf{R}_s} \exp(i\mathbf{k}_s \cdot \mathbf{R}_s) = (4\pi^2/A)\, \delta(\mathbf{k}_s - \mathbf{G}_s) \tag{12}$$

or

$$\sum_{\mathbf{R}_s'} \exp(i\mathbf{k}_s \cdot \mathbf{R}_s') = N_s'\, \delta_{\mathbf{k}_s, \mathbf{G}_s'} \tag{13}$$

where N_s' is the total number of B atoms in the interface. Substituting (12) and (13) into (11), integrating with respect to $\mathbf{k}_s$, and dividing by the interface area $N_s'A'$, we can write the interface energy per unit area as

$$E_0 = (1/2\pi AA')\, \delta_{\mathbf{k}_s, \mathbf{G}_s}\, \delta_{\mathbf{k}_s, \mathbf{G}_s'}\, \mathscr{V}_0(\mathbf{k}_s, B_3) \exp(i\mathbf{k}_s \cdot \mathbf{B}_s) \tag{14}$$

with a sum over all $\mathbf{G}_s$ and $\mathbf{G}_s'$ being understood and with

$$\mathscr{V}_0(\mathbf{k}_s, B_3) = \sum_{R_3'}^{+} \sum_{R_3}^{-} \int v(\mathbf{k}) \exp[ik_3(R_3' - R_3 + B_3)]\, dk_3 \tag{15}$$

where the subscript 3 represents the component of a vector normal to the interface, and we have redefined $\mathbf{R}$ and $\mathbf{R}'$ so that each is measured from an origin fixed on a lattice point of its respective crystal and $\mathbf{B}$ is the vector joining these two origins.

If we remember that so far we are dealing with two completely undistorted semicrystals, then (14) has a specially simple significance. The two Kronecker deltas require that the only nonvanishing terms have $\mathbf{G}_s = \mathbf{G}_s' = \mathbf{k}_s$, which means that they arise from the points of coincidence of the two reciprocal surface lattices $\mathbf{G}_s$ and $\mathbf{G}_s'$, each point of coincidence contributing an energy $\mathscr{V}_0$, which depends on the $\mathbf{G}_s$ involved and also on the relative displacement of the two crystals. Now the reciprocal surface lattices of two crystals will have points of coincidence only if the direct surface lattices have coincidences, so that what we have evaluated is the energy of a coincidence boundary before any elastic relaxation takes place. Notice that this energy is quite explicitly given and that it depends upon the interatomic interaction potential $v(\mathbf{r})$ through its Fourier components $v(\mathbf{k})$, upon the extent of the coincidence through the deltas, which select only the coincidence terms in reciprocal space, and upon the relative displacement of the two crystals through the vector $\mathbf{B}$.

Actual evaluation of the interface energy is relatively straightforward, since the sums in (15) need to be taken over only a few pairs of layers on either side of the interface, the vector $\mathbf{B}$ being appropriately chosen for each pair to agree with the specified physical situation. If only the orientations of the two crystals and of the boundary are fixed, then we have the three components of a single vector **B** that can serve as variational parameters. The minimum-energy configuration will possess certain symmetry, generally corresponding to $\mathbf{k}_s \cdot \mathbf{B}_s = n\pi$. There may, however, also be energy maxima or subsidiary minima corresponding to other symmetric configurations. This situation is illustrated in Fig. 2. The true minimum-energy configuration will depend upon the form of $v(\mathbf{r})$ and on the lattice geometry.

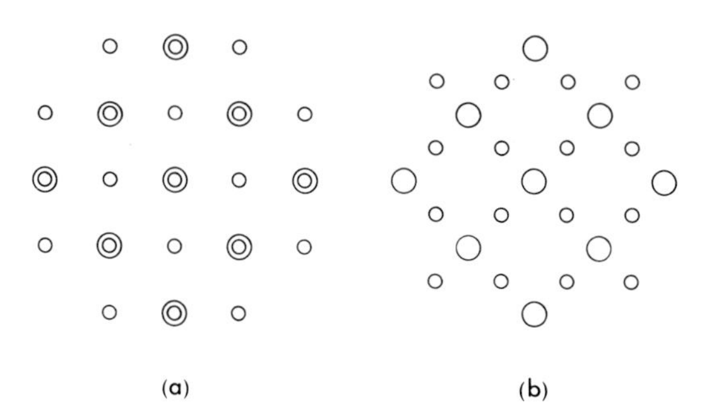

FIG. 2. An example of generalized matching between (100) faces of two fcc crystals with lattice parameters in the ratio $2^{1/2}:1$ and relative orientation 45°. Depending upon the form of the interaction potential either one of the configurations (a) or (b) might represent the minimum energy.

An important thing that we should point out about this approach is that not only is it quantitative and related to a priori atomic interaction potentials, but it also includes considerations of crystal symmetry automatically. In fact, the analysis of terms contributing to E_0 in (14) is formally very similar to the 0-lattice analysis discussed in Section III, the distinction being that here the whole treatment is carried out in reciprocal space so that there is a duality rather than a direct correspondence between the geometry in the two cases.

VI. The Interface Problem in Reciprocal Space

To treat a more general interface, and indeed to treat a coincidence boundary to a better approximation, it is necessary to allow for elastic displacement of atoms near the interface and for the effect of this upon more distant parts of the crystals. The method of approach is essentially that of Section IV, except that again the analysis is performed in reciprocal space.

The displacements that we will include in the theory are those in which a surface A atom, initially at $\mathbf{R}$, moves to $\mathbf{R}+\mathbf{F}(\mathbf{R})$ and the B atom at $\mathbf{R}'$ moves to $\mathbf{R}'+\mathbf{F}'(\mathbf{R}')$. Atoms deeper in the crystal will also move, but we may suppose this to be largely as a result of elastic interactions with their nearest-neighbor atoms of like kind rather than because of interactions across the interface.

While the vectors $\mathbf{F}$ are confined to describing the displacement of surface atoms, they are not necessarily parallel to the interface.

Now we have seen that for a general interface we can always define a coincidence lattice, although its primitive vectors may be extremely large. We can therefore construct the reciprocal lattice $\mathbf{K}$ of this superlattice with all vectors $\mathbf{K}$ lying parallel to the interface and, because of the periodicity in real space, it is necessary that all possible displacements $\mathbf{F}(\mathbf{R})$ can be written as a Fourier series

$$\mathbf{F}(\mathbf{R}) = -\sum_{\mathbf{K}} \mathbf{F}_{\mathbf{K}} \exp(i\mathbf{K}\cdot\mathbf{R}) \tag{16}$$

or, since $\mathbf{F}(\mathbf{R})$ is necessarily real,

$$\mathbf{F}(\mathbf{R}) = -\sum_{\mathbf{K}}{}^{+} 2[\mathbf{D}_{\mathbf{K}} \sin(\mathbf{K}\cdot\mathbf{R}) + \mathbf{C}_{\mathbf{K}} \cos(\mathbf{K}\cdot\mathbf{R})] - \mathbf{C}_0 \tag{17}$$

where $\mathbf{C}$ and $\mathbf{D}$ are real vectors. The corresponding expansion for the B crystal is

$$\mathbf{F}'(\mathbf{R}') = \sum_{\mathbf{K}}{}^{+} 2[\mathbf{D}_{\mathbf{K}}' \sin(\mathbf{K}\cdot\mathbf{R}') + \mathbf{C}_{\mathbf{K}}' \cos(\mathbf{K}\cdot\mathbf{R}')] + \mathbf{C}_0' \tag{18}$$

The change in sign relative to (17) is for reasons of symmetry and clearly

$$\mathbf{C}_0 + \mathbf{C}_0' = \mathbf{B} \tag{19}$$

If we substitute these distortions into (11), then the interaction part of the total interface energy becomes

$$E_{\mathrm{i}}^{\mathrm{T}} = \frac{1}{8\pi^3} \sum_{\mathbf{R}'}^{+} \sum_{\mathbf{R}}^{-} \int v(\mathbf{k}) \exp[i\mathbf{k}\cdot(\mathbf{R}'+\mathbf{F}'-\mathbf{R}-\mathbf{F})]\, d\mathbf{k} \tag{20}$$

To reduce this rather complicated expression we can make use of a Bessel function expansion (Watson, 1944; Fletcher and Adamson, 1966) to write

$$\exp[2i\mathbf{k}\cdot\mathbf{C}_{\mathbf{K}} \cos(\mathbf{K}\cdot\mathbf{R})] = \sum_{m=-\infty}^{\infty} i^m J_m(2\mathbf{k}\cdot\mathbf{C}_{\mathbf{K}}) \exp(im\mathbf{K}\cdot\mathbf{R}) \tag{21}$$

and

$$\exp[2i\mathbf{k}\cdot\mathbf{D}_{\mathbf{K}} \sin(\mathbf{K}\cdot\mathbf{R})] = \sum_{m=-\infty}^{\infty} J_m(2\mathbf{k}\cdot\mathbf{D}_{\mathbf{K}}) \exp(im\mathbf{K}\cdot\mathbf{R}) \tag{22}$$

Similar expressions arise in the theory of sidebands in frequency-modulated radio transmissions, although the analysis is usually carried out in a rather more limited way than has been done here. The significant thing about these two expansions is that, when they are substituted into (20), along with similar expressions for $\mathbf{F}'$, the sums over $\mathbf{R}_s$ and $\mathbf{R}_s'$ no longer have forms like (12) and (13) but rather like

$$\sum_{\mathbf{R}_s} \exp[i(\mathbf{k}_s - m\mathbf{K}_1 - n\mathbf{K}_2)\cdot\mathbf{R}_s] = N_s \delta_{\mathbf{k}_s, \mathbf{G}_s + m\mathbf{K}_1 + n\mathbf{K}_2} \tag{23}$$

so that when we come to calculate the energy per unit area, as in (14), we have contributions not just from those $\mathscr{V}_0(\mathbf{k}_s)$ for which $\mathbf{k}_s = \mathbf{G}_s = \mathbf{G}_s'$, but also from those with $\mathbf{k}_s = \mathbf{G}_s = \mathbf{G}_s' + m\mathbf{K}_1 + n\mathbf{K}_2$ and $\mathbf{k}_s = \mathbf{G}_s' = \mathbf{G}_s + m\mathbf{K}_1 + n\mathbf{K}_2$ for all possible vectors $\mathbf{K}_1$ and $\mathbf{K}_2$ corresponding to allowed distortion components and for all integers n and m. The Fourier components of the distortion have thus coupled together potential components $\mathbf{G}_s$ and $\mathbf{G}_s'$ that are not coincident.

If we follow the same sort of procedure that led from (11) to (14), then we find from (20) that the interaction energy per unit area of interface can be written

$$\begin{aligned} E_i = {} & (1/2\pi AA') \sum_{\mathbf{k}_s} \mathscr{V}_0(\mathbf{k}_s, B_3) \exp(i\mathbf{k}_s \cdot \mathbf{B}_s) \\ & \times \left\{ \prod_{\mathbf{K}}^{+} J_0(2\mathbf{k} \cdot \mathbf{D}_{\mathbf{K}}) J_0(2\mathbf{k} \cdot \mathbf{C}_{\mathbf{K}}) \right\} \left\{ \prod_{\mathbf{K}}^{+} J_0(2\mathbf{k} \cdot \mathbf{D}_{\mathbf{K}}') J_0(2\mathbf{k} \cdot \mathbf{C}_{\mathbf{K}}') \right\} \\ & \times \left\{ \delta_{\mathbf{k}_s, \mathbf{G}_s} \delta_{\mathbf{k}_s, \mathbf{G}_s'} + \sum_{n=1}^{\infty} \sum_{\mathbf{K}}^{+} [\,]_1 \, \delta_{\mathbf{G}_s + n\mathbf{K}, \mathbf{G}_s'} \delta_{\mathbf{k}_s, \mathbf{G}_s'} \right. \\ & \left. + \sum_{n=1}^{\infty} \sum_{\mathbf{K}}^{+} [\,]_{1'} \, \delta_{\mathbf{G}_s + n\mathbf{K}, \mathbf{G}_s'} \delta_{\mathbf{k}_s, \mathbf{G}_s} + \cdots \right\} \end{aligned} \tag{24}$$

The bracket $[\,]_1$ is defined by

$$[\,]_1 \equiv \left[\frac{J_n(2\mathbf{k}_s \cdot \mathbf{D}_{\mathbf{K}})}{J_0(2\mathbf{k}_s \cdot \mathbf{D}_{\mathbf{K}})} + i^n \frac{J_n(2\mathbf{k}_s \cdot \mathbf{C}_{\mathbf{K}})}{J_0(2\mathbf{k}_s \cdot \mathbf{C}_{\mathbf{K}})} \right] \tag{25}$$

and similarly for $[\,]_{1'}$ with $\mathbf{C}_{\mathbf{K}}$ and $\mathbf{D}_{\mathbf{K}}$ replaced by $\mathbf{C}_{\mathbf{K}}'$ and $\mathbf{D}_{\mathbf{K}}'$. There are further terms in expansion (24), involving the coupling of $\mathbf{G}_s$ and $\mathbf{G}_s'$ by two vectors $\mathbf{K}$, but we need not be concerned with these here. Explicit expressions for these second-order coupling terms are given by Fletcher and Adamson (1966) and a numerical error in these expressions is corrected by Lodge (1970).

These expressions (24) and (25) look forbiddingly complex, but in fact $\mathbf{k}_s \cdot \mathbf{C}_{\mathbf{K}}$ and $\mathbf{k}_s \cdot \mathbf{D}_{\mathbf{K}}$ are generally small quantities so that only terms with $n = 1$ or 2 need usually be considered and the second-order coupling terms are also small. We have therefore succeeded in expressing the interaction energy E_i across the interface in a relatively simple form involving a set of variational parameters $\mathbf{C}_{\mathbf{K}}$, $\mathbf{D}_{\mathbf{K}}$, $\mathbf{C}_{\mathbf{K}}'$, and $\mathbf{D}_{\mathbf{K}}'$. Their number is potentially the same as the number of variational parameters $\mathbf{F}$ shown in (6) for the same problem in direct space but we shall see that it is possible to reduce the number of significant $\mathbf{K}$-values to a quite small set so that the variational problem then becomes tractable. In addition, the elastic problem for each semicrystal is soluble in simple terms and does not do violence to our physical assumptions, as we shall see in the next section.

VII. The Elastic Problem

In our initial formulation we agreed to treat interactions across the interface on an atomic basis, as we have done, and to approximate the energy behavior of the two homogeneous semicrystals by continuum elasticity theory. Since the basis distortions envisaged in the theory are sinusoidal surface waves with wavevector $\mathbf{K}$ and amplitudes and polarizations specified by $\mathbf{C_K}$ and $\mathbf{D_K}$, this elastic problem can be fairly readily solved and, because of the continuous nature of the surface displacements, no reasonable physical assumptions are violated. In addition, since these elementary distortions for the elastic problem are all orthogonal, their contributions to the elastic energy can be individually evaluated and then simply summed.

A detailed discussion of the elastic problem has been given by van der Merwe (1950, 1963) both for the case of a semiinfinite crystal and for a crystal of finite thickness. A more recent investigation has been made by Lodge (1970). Without going into details of the analysis we can fairly easily see that, if the surface of a semiinfinite crystal is distorted by the imposition of a sinusoidal displacement

$$\mathbf{F_K}\exp(i\mathbf{K}\cdot\mathbf{r}) = 2[\mathbf{D_K}\sin(\mathbf{K}\cdot\mathbf{r})+\mathbf{C_K}\cos(\mathbf{K}\cdot\mathbf{r})] \tag{26}$$

where we have now replaced the discrete variable $\mathbf{R}$ by the continuous variable $\mathbf{r}$, then the amplitude of the displacement decays more or less expotentially away from the surface with a characteristic penetration depth of order $2\pi/K$. If we take the smallest allowed $\mathbf{K}$ for the problem, then this distance is equal to the largest of the primitive vectors of the surface coincidence lattice. Any crystal overgrowth that is much thicker than this is essentially infinite from an energy point of view, while for thinner overgrowths their finite thickness must be taken into account.

For the case of a semiinfinite A crystal subject to surface displacements $\mathbf{F_K}$, Lodge (1970) shows that the elastic energy per unit area is

$$E_e^A = \mu\sum_{\mathbf{K}}{}^{+} K\Bigg[C_{K_2}^2 + D_{K_2}^2 + \frac{4(1-\sigma)}{3-4\sigma}(C_{K_1}^2+D_{K_1}^2+C_{K_3}^2+D_{K_3}^2) + \frac{4(1-2\sigma)}{3-4\sigma}(D_{K_1}C_{K_3}-C_{K_1}D_{K_3})\Bigg] \tag{27}$$

where subscript K_1 refers to the component of a displacement in the direction of $\mathbf{K}$, the K_3-component is normal to the interface, and K_2 is chosen so that the 1, 2, 3 directions form a right-handed set. The quantities μ and σ are, respectively, the shear modulus and the Poisson's ratio for the crystal material A, assumed isotropic for simplicity. An exactly similar expression with μ, σ replaced by μ', σ' applies to crystal B.

From (27) we can see that elastic displacements with different wavevectors **K** are in fact independent but that for any given **K** the longitudinal displacements (subscript K_1) and the normal displacements (subscript K_3) are coupled, as is clearly necessary on physical grounds. The distortion components of the two crystals A and B could be taken as independent variational parameters, but a considerable simplification occurs if we instead apply the simple mechanical condition that elastic stresses must balance across the interface when it is in its equilibrium configuration. Again components with different **K**-vectors are orthogonal and a set of six relations between the twelve components of $\mathbf{C_K}$, $\mathbf{D_K}$, $\mathbf{C_K}'$, and $\mathbf{D_K}'$ can be derived. Because of the K_1, K_3 distortion coupling, these relations are also coupled, their explicit form being given by Lodge (1970) as

$$\mu C_{K_2} = \mu' C'_{K_2} \tag{28}$$

$$\mu D_{K_2} = \mu' D'_{K_2} \tag{29}$$

$$\mu(C_{K_3}+D_{K_1}) = [\mu'/(3-4\sigma')](C'_{K_3}+D'_{K_1}) \tag{30}$$

$$\mu'(C'_{K_3}-D'_{K_1}) = [\mu/(3-4\sigma)](C_{K_3}-D_{K_1}) \tag{31}$$

$$\mu(C_{K_1}-D_{K_3}) = [\mu'/(3-4\sigma')](C'_{K_1}-D'_{K_3}) \tag{32}$$

$$\mu'(C'_{K_1}+D'_{K_3}) = [\mu/(3-4\sigma)](C_{K_1}+D_{K_3}) \tag{33}$$

Rather similar treatment could be given to the case of an overgrowth crystal of finite thickness following the lead given by van der Merwe (1963) but this has not yet been done. Following his discussion and that given above we should expect the semiinfinite results to be approximately applicable to all but the very thinnest layers.

VIII. The Variational Problem in Reciprocal Space

We can now put together all the information contained in (24) and (27) to write for the total interface energy per unit area

$$E = E_i + E_e^A + E_e^B = E_i + E_e \tag{34}$$

The variational parameters in the problem are the $\mathbf{C_K}$, $\mathbf{D_K}$, $\mathbf{C_K}'$, and $\mathbf{D_K}'$ for all the allowed values of **K** but these can be reduced simply to the $\mathbf{C_K}$ and $\mathbf{D_K}$ by use of the relations (28)–(33). Let us now look at the totality of possible **K**-vectors and their associated distortions to see how they can be enumerated.

One of the great attractions of the variational approach is the opportunity it provides for simplification. An approximate result can always be obtained by considering just a few of the variational parameters that are judged on physical or mathematical grounds to be most important and, if one's intuition is good, this result may approach quite closely to the fuller treatment achieved with many more parameters.

To derive a simple criterion for this choice let us remember that distortions for different **K**-vectors are independent and then consider the form of E for a single $\mathbf{D_K}$ only, such that $\mathbf{G}_s = \mathbf{G}_s' + \mathbf{K}$. From (24), (25), (27), and (34) if we expand the Bessel functions and keep first-order terms only, we find

$$E \simeq E_0(\mathbf{B}) + Q(\mathbf{B})\, D_{\mathbf{K}} + PKD_{\mathbf{K}}^2 \tag{35}$$

where E_0 is the coincidence part (if any) of the boundary energy, $Q(\mathbf{B})$ contains the atomic interactions across the interface, $\mathscr{V}_0(\mathbf{G}_s)$ and $\mathscr{V}_0(\mathbf{G}_s')$, P represents the elastic coefficients, and $\mathbf{B}$ is, of course, the displacement of one crystal relative to the other. Clearly $P > 0$, and $\mathbf{B}$ can be chosen so that $Q < 0$.

If we minimize with respect to $D_{\mathbf{K}}$ and $\mathbf{B}$, we find

$$E \simeq E_0 - (Q^2/4PK) \tag{36}$$

with

$$D_{\mathbf{K}} = -Q/2PK \tag{37}$$

Clearly the approximation is valid only for $\mathbf{G}_s \cdot \mathbf{D_K} \lesssim 1$ but from it we can deduce two things: the contribution of a given $D_{\mathbf{K}}$ to the energy is greatest if **K** is very small, and the contribution is large if the potential components $\mathscr{V}_0(\mathbf{G}_s)$ and $\mathscr{V}_0(\mathbf{G}_s')$ for the two points are large. The energy (36) actually appears to have an infinite negative value as we approach $\mathbf{K} = 0$, the coincidence-boundary configuration, but the real situation is that the coincidence boundary of any order represents a cusped minimum in the energy, the depth of this minimum depending on the strength of the potential components $\mathscr{V}_0(\mathbf{G}_s)$ brought into coincidence. This already tells us a great deal about interface behavior and makes it plain that it is not only the geometry that is important but also the detailed form of the interaction potential $v(\mathbf{r})$.

To select the distortion wavevectors to be included in the calculation we therefore examine the pattern of the interface in reciprocal space and select a suitable number of short **K**-vectors joining reciprocal lattice points where the potential is large. There is an upper limit to the magnitude of any physically significant **K**-vector set by the criterion that it must lie within the first Brillouin zone of both the crystals concerned, but we may often select vectors that are limited to being much smaller than this. The choice will also be determined to some extent by the range of the interaction potential $v(\mathbf{k})$ in reciprocal space. An example of such an initial choice of **K**-vectors is shown in Fig. 3.

When it comes to second-order terms, the most important and therefore the only ones worth including are those made up from combinations of first-order **K**-vectors. Thus, for example, if **K** and **K**′ link pairs of reciprocal lattice

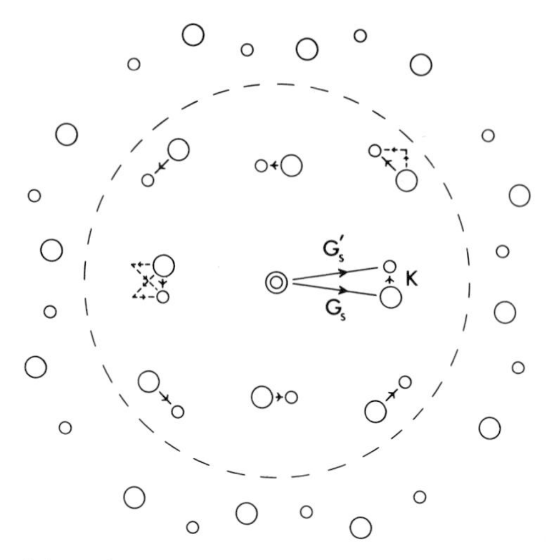

FIG. 3. Portion of the reciprocal lattice, near the origin, for two similar cubic crystals related by a twist boundary of small angle. The dashed circle represents either the range of the potential $v(\mathbf{k})$ or the arbitrary region inside which distortion vectors $\mathbf{K}$ are considered. The principal distortion $\mathbf{K}$-vectors are shown, together with the way they can combine to give second-order terms.

points in first-order, we can always find a pair of points linked by $\mathbf{K} \pm \mathbf{K}'$, and this term in E_i should be included since it requires no additional contribution to the elastic energy E_e.

The practical feasibility of this whole scheme depends on the possibility of selecting a sufficiently small set of important $\mathbf{K}$-vectors so that the computation becomes tractable. This in turn depends partly on the range of $v(\mathbf{k})$ in reciprocal space and partly on the geometry of the interface. The method becomes difficult for very small-angle boundaries because of the many small $\mathbf{K}$ involved, but this is not of much importance since it is just this region that is simply and validly covered by the dislocation theory.

One of the main contributions of the formalism itself is to place on a firm quantitative basis the energy considerations relating to coincidence and near-coincidence boundaries. The coincidence boundaries are shown to represent local cusped minima in the interface energy and their behavior is shown to be critically related to the actual form of the interatomic potential. Crystal symmetry is automatically taken into account and there is no necessity for any artificial picture involving hierarchies of dislocations. A simple dislocation is, in fact, represented by the distortions associated with a single set of vectors $n\mathbf{K}$, where $\mathbf{K}$ is the first-order distortion wavevector for a particular set of reciprocal lattice points. Inclusion of alternate couplings of the type $n(\mathbf{K}+\mathbf{K}')$ is equivalent to the introduction of another dislocation set of higher order.

IX. The Interaction Potential

In many problems in metal-crystal physics it is usual to assume a simple analytic form for the interaction potential $v(\mathbf{r})$, both the Lennard–Jones potential

$$v(\mathbf{r}) = v_0[(r/r_0)^{-12} - 2(r/r_0)^{-6}] \tag{38}$$

and the Morse potential

$$v(\mathbf{r}) = v_0\{\exp[-2a(r-r_0)] - 2\exp[-a(r-r_0)]\} \tag{39}$$

being common choices. In either case the parameters v_0 and r_0 are chosen to match as well as possible the lattice parameters, the cohesive energy, and the elastic properties derived from experiment. Similarly in the treatment of ionic crystals we might usefully assume a Coulombic or shielded Coulombic attraction or repulsion together with a hard-sphere or other appropriate repulsion at short distance.

These potentials are particularly useful in direct lattice calculations because of their simple analytical form. Apart from this and the r^{-6} term in (38) they have no particular physical significance. In the present situation what we need for a simple model calculation is a potential that has a simple analytical form for $v(\mathbf{k})$ in reciprocal space and that adequately represents the real interatomic potential. Since there is a dual relation between $v(\mathbf{k})$ and $v(\mathbf{r})$, in the sense that if $v(\mathbf{r})$ has a short range then the range of $v(\mathbf{k})$ is long and vice versa, it is rather difficult to find a suitable model potential for trial calculations. For real calculations, of course, we must try to use the real form of $v(\mathbf{r})$ derived as accurately as possible from calculations and experimental data, so that this difficulty cannot be avoided. Even for a real potential, however, we shall see that there is some freedom of choice associated with the exact form of the potential within the inaccessible repulsive core, and we may use this freedom to advantage in real calculations.

A suitable model potential for exploratory calculations introduced by Fletcher (1967) has the form

$$v(\mathbf{k}) = H[(a+b)/a]^3\{\exp[2.5k(a+b)-4]+1\}^{-1} \tag{40}$$

where H is a scale constant. This potential is designed for the case where the atoms of crystals A and B have radii proportional to a and b, respectively, so that the lattice parameters of the two crystals vary in the same way. The form of the potential in real and in reciprocal space for the case $a = b$ is shown in Fig. 4. It can be seen that the potential $v(\mathbf{k})$ has an appreciable value only for $\mathbf{k} \lesssim 3\pi/a$, while $v(\mathbf{r})$, its form in real space, has a physically reasonable shape and is small for $\mathbf{r} \gtrsim 1.5a$. With the particular form given by (40), $v(\mathbf{r})$ has a minimum at $r = \frac{1}{2}(a+b)$, the depth of which is independent of b/a. The potential is, of course, spherically symmetrical.

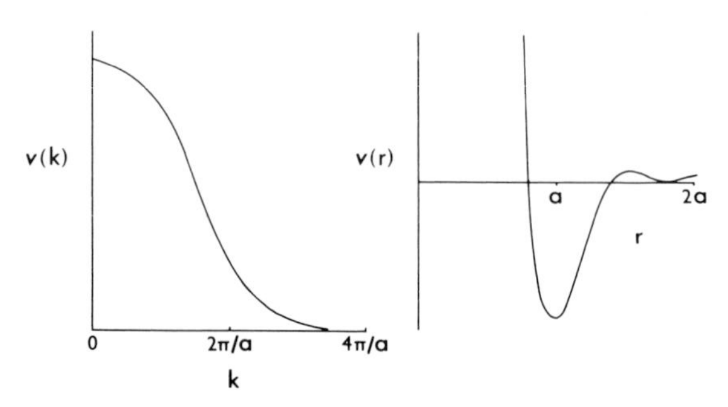

FIG. 4. (a) The Fourier transform $v(\mathbf{k})$ and (b) the direct potential $v(\mathbf{r})$ used in the model calculation, both drawn for the case $b = a$.

In the next section we shall detail the results of some calculations with this potential, which show the general features expected from the formal development. We must remember, however, that a model calculation with such a potential, instructive though it may be, has only a limited resemblance to reality. We can improve the resemblance by choosing the few available parameters to give agreement with important physical data but we cannot be sure how valid are the other predictions of the model.

It is more satisfying, from a physical point of view, to derive a more realistic potential from first principles and then to perform our best calculation without the aid of any adjustable parameters. This we shall try to do in Section XI.

X. Calculations with a Model Potential

A set of survey calculations with the model potential (40) was carried out by Fletcher (1967) to check the general feasibility of the method for interface-energy calculations and to examine quantitatively its predictions for misfit boundaries of various types. The system chosen for analysis was a twist boundary between (100) faces of two fcc crystals A and B having atomic potential parameters a and b, respectively. The cubic-lattice parameters thus turn out to be about $2.5a$ and $2.5b$ for the two crystals, and the surfaces are simple square lattices with parameters slightly less than a and b.

For the purposes of the calculation several additional simplifications were made: integral (15) for $\mathscr{V}_0$ was evaluated as a sum over a relatively small number of equally spaced k_3-values, only one atomic layer on either side of the interface was taken into account in evaluating E_i, and a simplified elastic analysis was used. This last involved neglect of the interaction term in (27) and replacement of the other terms by

$$E_e^A = (M/a^3) \sum_K{}^+ \sum_i (C_{K_i}^2 + D_{K_i}^2) \tag{41}$$

With the choice of 5.8 for the constant H in (40) to give a convenient arbitrary energy scale, a value of 500 for M made the various elastic moduli of the order expected physically. In retrospect this oversimplification was probably undesirable, but it is hardly worthwhile to recompute the results with less crude approximations.

Figure 5 shows the energy of the twist boundary as a function of angle θ for various values of the ratio b/a. When $b = a$ there is a deeply cusped energy minimum at the orientation of exact fit, $\theta = 0$, and the interface energy then rises smoothly in the same manner as predicted by the dislocation theory result (1). Instead of decreasing for $\theta > 20°$ and then becoming unphysically negative, however, the interface energy reaches a plateau for $\theta > 20°$. The predicted energy is also symmetrical about $\theta = 45°$ and shows a cusp for $\theta = 90°$ identical with that at $\theta = 0°$, as is physically necessary.

The behavior of interface energy as a function of misfit ratio $b/a = 1+\delta$ is shown more explicitly in Fig. 6. For $\theta = 0$ there is a logarithmic cusp of the

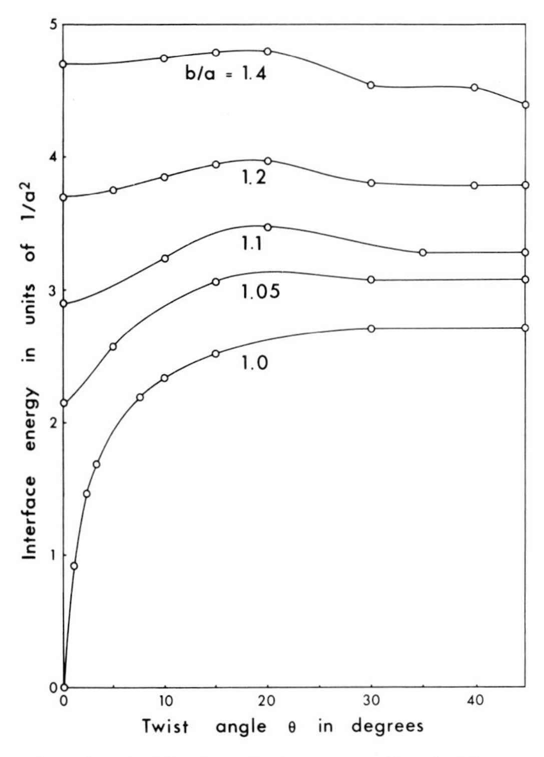

FIG. 5. Energy, in units of $1/a^2$, of a twist boundary of angle θ between two crystals with lattice-potential parameters a and b, respectively (Fletcher, 1967).

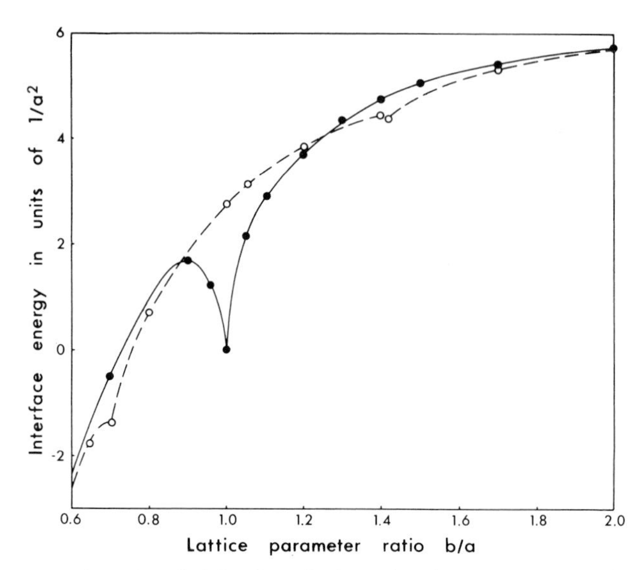

FIG. 6. Energy, in units of $1/a^2$, of a twist boundary between two crystals, with lattice-potential parameters a and b, respectively, for boundary twist angles of 0 (●) and 45° (○) (Fletcher, 1967).

type predicted by dislocation theory (2) for $b = a$, although this is superposed on a general rising trend of interface energy with parameter b which is a result of the particular form of the potential (40).

The theory also predicts the existence of subsidiary minima at points $\theta = 45°$, $b = 2^{1/2}a$, and $\theta = 0$, $b = 2a$, corresponding to more general coincidence boundaries. The first of these cusps is clearly visible in Fig. 6 but the second is not, presumably because the magnitude of the relevant Fourier components of the potential is too small.

From the results of the computation for the distortion vectors $\mathbf{C}_\mathbf{K}$ and $\mathbf{D}_\mathbf{K}$ it is also possible to determine atomic positions in the elastically relaxed interface. This is generally of little practical interest but there is a possibility of identifying the major **K**-vectors by electron-diffraction methods applied to thin films.

Using the theory with a simple model potential like this we are thus able to reproduce the results of the dislocation theory for small misfits and to make an estimate of interface energy for large misfits. For the particular potential chosen the number of variational parameters was only small and the calculation only involved quite a small amount of computer time. The more difficult cases of real potentials and of tilt boundaries, however, may no longer be simple computational problems.

As a second example of the use of the model potential (40), let us apply it to estimate the grain-boundary energy in aluminum. To this end we use the full treatment of the elastic problem set out in Section VII and choose the constant H in the potential (40) for $b = a$ to give best agreement with the experimental elastic moduli used in (27). To do this, the total energy of an fcc crystal with interaction potential (40) is evaluated and, upon minimization with respect to the lattice constant a_0, including interactions up to fourth neighbors, we find

$$a_0 = 1.3838a \tag{42}$$

The same minimization gives the bulk modulus in terms of H and thus allows H to be evaluated. This is considered to be a better procedure in the present case than to use the sublimation energy, because of the importance of relatively small atomic displacements in the theory. The value of H and the experimental values used for μ and σ are

$$H = 4.25 \times 10^{-35}\ \mathrm{erg\,cm^3}, \quad \mu = 2.62 \times 10^{11}\ \mathrm{dyne\,cm^{-2}}, \quad \sigma = 0.345 \tag{43}$$

The assumption of isotropic elastic properties for aluminum is a reasonable approximation but by no means exact.

Once more in this calculation only one atomic layer on either side of the interface was taken into account, since this is a reasonably good approximation when the interface plane is of low index. The results of the model calculation are shown in Fig. 7. The calculated energy of a high-angle twist boundary relative to the zero-angle (no grain-boundary) energy is 1014 erg cm^{-2}. The only available experimental value is 630 ± 100 erg cm^{-2}, determined by Astrom (1957) using a calorimetric method. The agreement is reasonable but not very close.

XI. An Ab Initio Potential for Aluminum

To illustrate the problems encountered in a calculation based upon a potential derived from ab initio physical argument, rather than on a criterion of calculational simplicity, we shall consider the case of aluminum. There are two reasons for this choice: aluminum is a metallurgically important material so that it is of value to understand its properties as fully as possible, and there is a reasonable body of theoretical work on which we can base our calculations. The argument we follow is essentially that given by Lodge (1970).

Aluminum is a metal, and the effective interaction between its atoms is therefore composed of two parts: the repulsive interaction between the aluminum ion cores and the shielding effect of the conduction electrons on this interaction. The shielding effect of the electrons is complicated by the

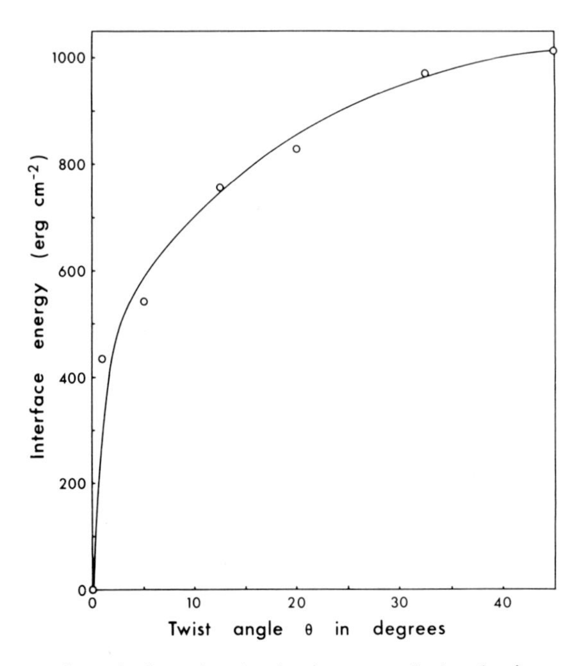

FIG. 7. Energy of a twist boundary in aluminum as calculated using a model potential.

electronic band structure of the metal, and the sharp decrease to zero of the electron density above the Fermi energy limits its shielding effectiveness at short wavelengths so that the total atomic-interaction potential resembles an exponentially shielded repulsion with, superposed upon it, an oscillation related to the shielding cutoff at the Fermi energy. It is the principal minimum produced by these oscillations that is responsible for the overall attractive potential between atoms. The matter has been discussed in detail by Ziman (1964) and by Harrison (1966).

It would be out of place to enter into a detailed discussion of this matter here and we content ouselves with reproducing, as the solid line in Fig. 8, the interaction potential calculated by Harrison (1966). At small distances the potential has the form of a simple Coulomb repulsion Z^2e^2/r, where Ze is the effective ionic charge.

One of the principal difficulties with using Harrison's potential in a calculation of grain-boundary energy arises from this highly repulsive core potential at short distances. The infinity in the potential at $r = 0$ leads to a large spread of $v(\mathbf{k})$ in reciprocal space, and any truncation of the expansion leads to large oscillations in $v(\mathbf{r})$ that completely mask the true form of the potential.

It is possible, however, to reduce this difficulty by a simple artifice. We note that, because of the limited energy available, ion cores are never forced to overlap appreciably, and it is therefore quite immaterial what is the form of $v(\mathbf{r})$ at very small $\mathbf{r}$, so long as it is large and positive. We may therefore use the modified core potential.

$$v'(\mathbf{r}) = \begin{cases} (Z^2e^2/R)[2-(r/R)], & r < R \\ Z^2e^2/r, & r > R \end{cases} \tag{44}$$

where R is chosen small enough (actually 3.5 a. u. or about 1.9 Å) that overlap to the altered region never occurs. These two potentials join smoothly as shown in Fig. 9 and the effect on the range of $v(\mathbf{k})$ is very great, as shown in Fig. 10. Using the modified potential it now becomes possible to truncate $v(\mathbf{k})$ in $\mathbf{k}$-space without doing violence to its representation of the important part of the potential, as shown in Fig. 8. We see that a cutoff at $k_M = 4.64$ a.u. (8.8 Å^{-1}) represents the potential quite well, although somewhat underestimating the depth of the primary minimum, while a cutoff at $k_M = 2.7$ a.u. (5.1 Å^{-1}) gives the correct overall behavior but overestimates the depth of this minimum by about a factor of two.

Here we arrive at the prime difficulty of the variational method when applied to real potentials, and this is simply computational. For the aluminum structure a cutoff in the potential $v(\mathbf{k})$ at $k_M = 2.7$ a.u. is found to include 20 pairs of reciprocal lattice points with $G_s < k_M$, which would coincide

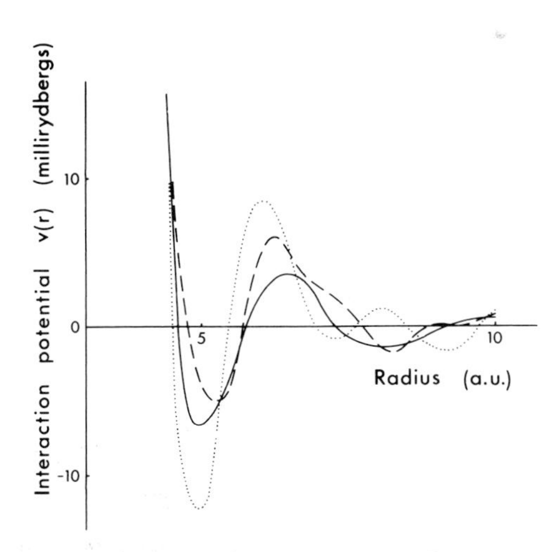

FIG. 8. Effective interaction potential between ions in aluminum as calculated by Harrison (solid line) and the modifications introduced by using $\mathbf{k}_M$ cutoffs of 4.6365 a.u. (dashed line) and 2.7 a.u. (dotted line) (Lodge, 1970).

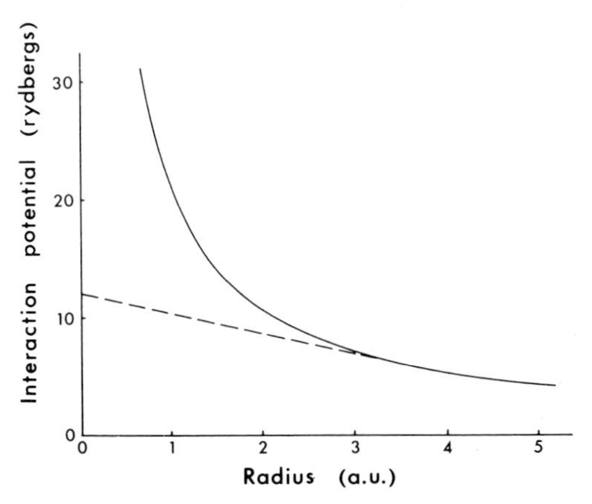

FIG. 9. The direct Coulomb interaction (solid line) compared with the altered direct interaction used in the calculation (dashed line) (Lodge, 1970).

at $\theta = 0$. This implies 20 primary **K**-vectors and 120 distortion components, although symmetry considerations reduce the number of independent components to 15. For a potential cutoff at $k_M = 4.64$ a.u., this number is increased by about a factor of 3. At small twist angles θ, the primary **K**-vectors

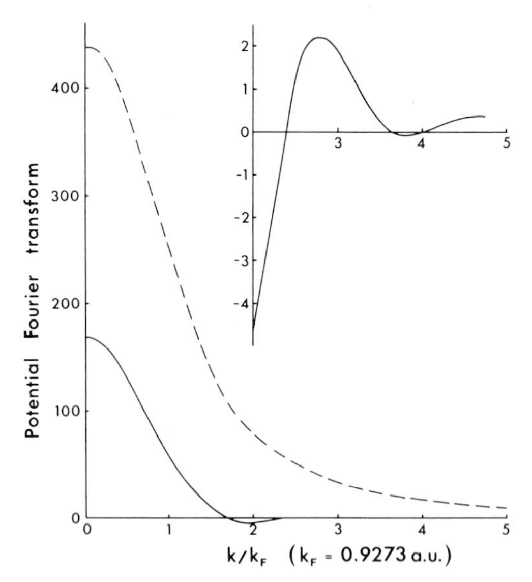

FIG. 10. Comparison between the Fourier transform of Harrison's potential for aluminum (dashed line) and the Fourier transform of the modified interaction potential. The units are rydbergs (Bohr radius)^3. The tail of the modified potential is shown on an expanded scale in the inset (Lodge, 1970).

are all small, convergence is slow, and second-order terms are important so that the calculation, while feasible, is computationally difficult. For larger twist angles, **K**-vectors joining reciprocal lattice points that are not equivalent at $\theta = 0$ become more important but, generally speaking, the second-order contributions are much smaller and the energy depends primarily on those few distortions for which **K** happens to be small.

Lodge (1970) has carried out calculations for a twist boundary in aluminum using a potential truncated at $k_M = 2.7$ a.u. and trying various simplifications and extensions. The convergence was poor for $\theta < 5°$, except for the identical-fit case $\theta = 0$, but the familiar cusped minimum was apparent. The boundary energy was found to have a plateau for θ between about 10 and 45° with an energy of about 1700 erg cm^{-2}, which does not accord very well with the experimental value of 630 ± 100 erg cm^{-2} found by Astrom (1957).

In retrospect the reasons for this descrepancy are clear. In his calculation, Lodge truncated the potential itself at $k_M = 2.7$ a.u., and this, as is plain from Fig. 8, has the effect of deepening the primary potential minimum by about a factor of two with a consequent increase by this factor in the binding energy across the interface at $\theta = 0$. In the spirit of our variational method it should not be that we truncate the potential at a particular value of k_M but rather that only distortion vectors **K** joining reciprocal lattice points G_s with $G_s < k_M$ should be included. All coincident reciprocal-lattice points should always be included in the energy summation.

It is actually quite easy to revise Lodge's calculation in this way since, in general, there are no coincidences besides $k = 0$ except for special coincidence-boundary configurations. Ideally the coincidence sum should be taken over all values of **k** but in fact, from Fig. 8, a sum to $k = 4.64$ a.u. is fairly adequate. When this revised calculation is performed and another error in the computer program corrected, the high-angle grain-boundary energy is found to be about 590 erg cm^{-2}. This is in excellent agreement with experiment, although this may be fortuitous.

XII. Surface Entropy and Free Energy

As we discussed in Section II, the quantity that we have considered and attempted to calculate has, so far, been the interfacial energy, whereas the quantity that is relevant in most physical problems is actually the interfacial free energy. For a liquid–vapor interface the surface free energy is identical with the surface tension, but for solid boundaries or interfaces these two quantities are physically and numerically quite distinct. The surface free energy is a scalar quantity (although its value depends in general on surface orientation) and it must always be positive, or the solid will spontaneously disintegrate to create extra surface. The surface tension, on the other hand, is generally a tensor describing surface stress and may have either positive or negative components, depending on the nature of the forces between atoms in the crystal. The same distinction applies to crystal interfaces.

If we denote the surface energy by ε, the surface free energy by γ, and the surface entropy by σ, then at temperature T we have as usual

$$\gamma = \varepsilon - T\sigma \tag{45}$$

so that γ and ε are equal at 0°K and

$$\sigma = -\partial\gamma/\partial T$$

For liquid–vapor interfaces we know that the surface entropy σ is positive so that γ decreases with rising temperature and vanishes at the critical point. For the interface between a crystal and its supercooled melt, however, there is indication from nucleation experiments that γ increases with rising temperature so that σ is negative. This behavior accords with the experimental fact that there does not appear to be any critical point associated with the solid–liquid transition.

If we consider a simple grain boundary, then the thermodynamic properties of the material on either side of the boundary are identical, while the boundary region itself is characterized by having a higher energy per atom than the bulk material and a less ordered configuration. We should thus expect both the surface energy and surface entropy to be positive in this case. It is more difficult to make the same assertion about an interface between two different crystals without detailed thermodynamic and statistical analysis, but it seems likely that it is still true. This conclusion implies that the free energy of a crystalline interface should decrease as the temperature is raised, so that all our calculated energies will be overestimates if they are taken to represent the free energy at a finite temperature.

It is difficult to estimate the numerical magnitude of the surface entropy without a very detailed analysis. In the first place there are vibration modes that differ in frequency from those of the bulk crystal and therefore contribute to surface entropy; there are also configurational terms that depend on the degree of disorder in the interface. It is tempting to assume simply, in the case of grain boundaries, that the free energy goes to zero at the melting point, but this is much too extreme a variation since melting is a phase change of first order rather than the higher order associated with a liquid–vapor critical point. It seems likely, rather, that the interface free energy near the melting point has probably at least half the magnitude it has at 0°K. Part of this change will be associated with configurational entropy terms, and part will be due to the effect of the decrease in shear modulus μ as the temperature is raised.

XIII. Relation to Epitaxial Growth

In our discussion we have seen that it is possible to calculate, quite generally, the energy of an interface between two materials provided that the interaction potential between pairs of atoms is known together with the elastic moduli of the two bulk crystals. In practice the calculation is feasible provided that the interface constitutes a low-index crystallographic face for each of the two crystals and that the interaction potential can be reduced to a form that does not have too large an extension in reciprocal space.

Extension of this treatment to the case where one of the crystals is of small thickness is relatively straightforward, since this only involves modification of the elastic energy component. If, however, we consider a further extension to the case where one of the crystals is of finite lateral extent, then we encounter more fundamental difficulties because the lattice sums parallel to the interface are no longer infinite. If the patch of overgrowth is finite but sufficiently large to contain several periods of the shortest distortion wavevector **K**, then it is legitimate to use the results for an overgrowth of infinite lateral extent as a good approximation. If this condition is not fulfilled, then the overgrowth island will be relatively small and an energy calculation in real space along the lines indicated in Section IV is probably the best way to proceed. Bearing these limitations in mind, we can now proceed to draw some general conclusions from our treatment, which will be of use in considering epitaxial growth problems.

Oriented overgrowth of one crystal upon another can occur when there is a well-defined minimum in the interfacial free energy for some particular orientational relationship of the two crystals relative to the specified boundary. Generally the boundary will be fixed relative to the substrate crystal and the orientational freedom will be that of the overgrowth.

For two arbitrary crystals of different materials there will not, in general, be any exact coincidence boundaries so that, instead of a cusped minimum as for the case $b/a = 1.0$ in Fig. 5, we are more likely to find a simple minimum as for the case $b/a = 1.05$ in the same figure. This means that, because of the finite curvature at the trough of the minimum, we should expect to find a small distribution in angle for different overgrowth islands rather than a precise orientation.

To estimate the expected spread we might reasonably evaluate the total interfacial free energy F for an island nucleus of critical size and we should then expect to find a distribution $\Delta\theta$ in orientation angle such that $\Delta F(\theta) \sim kT$. Applying this criterion to the case of $b/a = 1.05$ in Fig. 8 and assuming the reasonable value of 100 erg cm^{-2} for the plateau value of interface energy, a critical island embryo of 1000 atoms and a substrate temperature of 400°K, we find a spread in orientation of $\Delta\theta \simeq \pm 2°$. If deposition were carried out at a lower substrate temperature under the same deposition conditions, then the size of the critical embryo would decrease much more rapidly than kT so that the spread $\Delta\theta$ would increase. This calculation is, of course, simply for illustration and quite different values may be appropriate in physical situations of interest.

Another important possibility is that there may be more than one possible orientation of the overgrowth that approximates to a coincidence boundary of some order and therefore represents a local minimum in the interfacial free energy. Generally there will be one of these minima that is substantially deeper than the others, but occasionally we may find two minima of comparable depth. An example is the epitaxial growth from the vapor of ice onto an (001) face of iodine (Bryant *et al.*, 1959) where growth is observed with either the (0001) or the $(2\bar{1}\bar{1}0)$ planes of ice parallel to the interface. The growth habit of the ice islands either as hexagons or rectangles makes distinction between the two growth modes obvious.

Examination of the interface geometry shows that each of these configurations differs by about 6% from a simple coincidence boundary. It does not, of course, immediately follow that the interface energies are nearly equal in the two cases, since this may depend on details of the interaction potential, but it is not surprising that this is so. If we can calculate the interface energy in the two configurations, then it is possible to apply standard heterogeneous nucleation theory (e.g., Fletcher, 1963) to evaluate nucleation rates for the two crystal habits as a function of temperature and supersaturation. The critical embryos in this case are certainly large enough for our interface theory to be applied with good approximation.

This last remark has introduced the subject of nucleation theory, and it may be appropriate to conclude with a few comments on this topic. Broadly speaking, we are here concerned only with heterogeneous nucleation on a plane substrate, and it is useful to distinguish two extreme cases representing high and low supersaturations, respectively.

In the high-supersaturation regime we are concerned with the deposition of a relatively nonvolatile material, such as a metal, onto a substrate whose temperature is very much less than that of the vapor source. The effective supersaturation at the substrate surface is therefore very high and the critical embryo for epitaxial growth contains only a few atoms. It would be pointless to try to apply any of our techniques to this case since the assumptions from which they were derived are not even approximately valid. Even the application of classical nucleation theory represents an extrapolation that can hardly be justified, and the problem should properly be treated by straightforward cluster statistical techniques.

In the low-supersaturation regime, on the other hand, we are concerned with deposition of a relatively volatile material onto a substrate at a temperature nearly equal to that of the source so that the supersaturation is only of the order of a few percent. Alternatively, the situation may involve deposition from a solution with only a small supersaturation. In such cases the critical embryo typically contains thousands or even millions of atoms, the interface area may be of order 10^{-10} cm^2 or larger, and the embryo may be many molecular layers in thickness. Under these circumstances classical nucleation theory can be applied with a good deal of confidence and the techniques outlined above can be used to evaluate the interface energy, which plays such a critical role in determining the manner of the epitaxial growth.

References

Astrom, H. U. (1957). *Ark. Fys.* **13**, 69.

Bollmann, W. (1967). *Phil. Mag.* **16**, 363, 383.

Bragg, W. L. (1940). *Proc. Phys. Soc. London* **52**, 54.

Brandon, D. G. (1966). *Acta Met.* **14**, 1479.

Bryant, G. W., Hallett, J., and Mason, B. J. (1959). *J. Phys. Chem. Solids* **12**, 189.

Burgers, J. M. (1940). *Proc. Phys. Soc. London* **52**, 23.

du Plessis, J. C., and Van der Merwe, J. H. (1965). *Phil. Mag.* **11**, 43.

Fletcher, N. H. (1963). *J. Chem. Phys.* **38**, 237.

Fletcher, N. H. (1964). *J. Appl. Phys.* **35**, 234.

Fletcher, N. H. (1967). *Phil. Mag.* **16**, 159.

Fletcher, N. H. (1971). *Advan. Mater. Res.* **5**, 281–314.

Fletcher, N. H., and Adamson, P. L. (1966). *Phil. Mag.* **14**, 99.

Frank, F. C., and Van der Merwe, J. H. (1949). *Proc. Roy. Soc. London* **A198**, 205, 216

Friedel, G. (1926). "Leçons de Cristallographie." Berger Levrault, Paris.

Gifkins, R. C. (1969). *In* "Interfaces Conference Melbourne 1969" (R. C. Gifkins, ed.), pp. 29–51. Butterworths, London and Washington, D.C.

Harrison, W. A. (1966). "Pseudopotentials in the Theory of Metals," Chapters 1, 2, 8. Benjamin, New York.

Lodge, K. W. (1970). Structure and Energy of Crystal Interfaces. M.Sc. thesis, Univ. of New England.

Mott, N. F. (1948). *Proc. Phys. Soc. London* **60**, 391.

Ranganathan, S. (1966). *Acta Cryst.* **21**, 197.

Read, W. T., and Shockley, W. (1950). *Phys. Rev.* **78**, 275.

van der Merwe, J. H. (1950). *Proc. Phys. Soc. London* **A63**, 616.

van der Merwe, J. H. (1963). *J. Appl. Phys.* **34**, 117, 123, 2420.

Watson, G. N. (1944). "Theory of Bessel Functions," p. 14. Cambridge Univ. Press, London and New York.

Ziman, J. M. (1964). *Advan. Phys.* **13**, 89.

CHAPTER 8

COHERENT INTERFACES AND MISFIT DISLOCATIONS

J. W. Matthews

IBM Thomas J. Watson Research Center
Yorktown Heights, New York

I. Introduction

The energy of an interface between a pair of unstrained crystals in parallel orientation depends on the misfit between their lattices (Nabarro, 1940; Frank and van der Merwe, 1949; van der Merwe, 1950; Brooks, 1952; Fletcher, 1964, 1966; Fletcher and Adamson, 1966; Friedel, 1964). As a result of this, finite crystals placed in perfect contact strain elastically to reduce the energy of the interface between them (Mott and Nabarro, 1940; Frank and van der Merwe, 1949; van der Merwe, 1963, 1964). If the difference between the stress-free lattice parameters of the crystals is small (<10–15%) the elastic strain is such as to bring the lattice constants closer together. If the misfit is large, then the strain brings the interface toward or into a nearby low-energy configuration (du Plessis and van der Merwe, 1965; Fletcher and Adamson, 1966). The strains needed to do this may either increase or decrease the difference between the lattice parameters of the crystals. The studies of misfit strain made so far have been concerned with crystals with similar lattice parameters. As a result, this chapter describes bicrystals in which the misfit strains bring the lattice parameters of the crystals closer together.

The first theoretical treatment of the accommodation of misfit between crystals was made in 1940 by Mott and Nabarro, who considered the formation of precipitates in alloys by the simple interchange of atoms and showed that, in the early stages of precipitation in systems of this type, the precipitate and matrix strain elastically to bring the atomic planes of precipitate and matrix into register at the interface. This registry persists until a critical precipitate size is reached (Nabarro, 1940). Thereafter, the energy of the system is lowered if the precipitate breaks away and destroys the continuity, or *coherence*, of the atomic planes. Nabarro estimated the size that precipitates would attain before breakaway by determining the size at which the elastic energy released during breakaway became equal to the rise in interfacial energy. The value he used for the elastic strain energy was precise for isotropic materials but his estimate of the interfacial energy increase was approximate. He assumed that the rise in interfacial energy was equal to the energy needed to melt a monatomic layer coincident with the interface.

A more accurate calculation of interfacial energy and its dependence on the fit of atomic planes was made in 1949 by Frank and van der Merwe. One of the important predictions of their calculations is that the interface, after breakaway, is divided into regions of good register separated from one another by regions where the register is bad. The regions of bad fit resemble crystal dislocations and because of this are called "interfacial" or "misfit" dislocations. The calculations of Frank and van der Merwe have been followed by many others (van der Merwe, 1950, 1963, 1964, 1972; Brooks, 1952; Fletcher, 1964, 1966; Friedel, 1964; Ball and van der Merwe, 1970; Ball, 1970; van der Merwe and van der Berg, 1972). They are described in Chapters 6 and 7. Although some tautology results from it this chapter also contains a description of the interfacial structure expected in epitaxial bicrystals. The description is divided into two parts. The first is concerned with epitaxial deposits that exhibit monolayer or monolayerlike growth, and the second with deposits formed by the nucleation, growth, and coalescence of three-dimensional islands (Pashley, 1965). Much of the second part is devoted to effects that the abrupt relaxation of misfit strain—which accompanies the creation of individual misfit dislocations (Vincent, 1969)—has on the orientation of moiré fringe patterns visible in electron images of islands. Two explanations for the presence of misaligned moiré fringes are given. One of these involves a true misalignment of the islands and is expected in deposits weakly bound to their substrate. The other involves a distortion of the unit cell. It is expected in all epitaxial islands but particularly noticeable in strongly bound ones.

The description of the expected interface structures is followed by a review of the experimental observations and by a comparison of the observations with predictions. The behavior and role of misfit dislocations during interdiffusion is also described.

Before we begin the main body of this chapter, it should be emphasized that there is a great deal of theoretical and experimental work concerned with coherency strain and misfit dislocations that is omitted. This is the work that has been done on precipitates and inclusions in alloys. It is reviewed by Hardy and Heal (1954), Kelly and Nicholson (1963), Aaronson *et al.* (1970), Brown and Ham (1971), and Matthews (1973).

II. Predictions

A. Accommodation of Misfit between a Thin Film and Its Substrate

One form of the calculation of the optimum elastic strain in a thin film on a thick substrate consists of determining the value of the strain ε that minimizes the sum of E_δ, the energy of the misfit dislocations, and E_ε, the energy of the elastic misfit strain (van der Merwe, 1964). An alternative but equivalent form described in Section V considers the forces on threading dislocations (Jesser and Matthews, 1967, 1968a, b; Matthews and Crawford, 1970).

To simplify the calculations in this section the film and substrate are assumed to be prepared from cubic crystals, and deviations from Hooke's law are neglected. In addition, the dislocations are assumed to be in edge orientation, to be arranged in a square grid, and to have Burgers vectors in the interface. If the plane of the interface between the crystals is {001}, {111}, or {110}, the energy associated with elastic strain ε parallel to the film plane is (Cahn, 1962)

$$E_\varepsilon = \varepsilon^2 Bh \tag{1}$$

where h is film thickness and

$$B = \frac{1}{2}(C_{11}+2C_{12}) \times \left[3 - \frac{C_{11}+2C_{12}}{C_{11}+2(2C_{44}-C_{11}+C_{12})(l^2m^2+m^2n^2+n^2l^2)}\right]$$

where C_{ij} are elastic constants, and l, m, and n the direction cosines that relate the direction normal to the interface to the cube axes. If the normal to the interface is not ⟨001⟩, ⟨111⟩, or ⟨110⟩, Eq. (1) is a good approximation but is not exact. An expression that is precise for all interfaces is given by Hilliard (1970). If the film is elastically isotropic, then

$$2C_{44} - C_{11} + C_{12} = 0 \qquad \text{and} \qquad B = 2G(1+\nu)/(1-\nu) \tag{2}$$

where G is the shear modulus and ν the Poisson ratio. The energy of an edge dislocation in the interface between a pair of crystals with shear moduli G_o and G_s is approximately

$$\tfrac{1}{2}Db[\ln(R/b)+1] \qquad \text{where} \quad D = G_o G_s b/\pi(G_o+G_s)(1-\nu) \tag{3}$$

with b the magnitude of the Burgers vector of the dislocation, and R the distance to the outermost boundary of the dislocation's strain field. The separation of misfit dislocations depends on the fraction of total misfit that is accommodated by dislocation lines. If the stress-free lattice parameters of the overgrowth and substrate are a_o and a_s, and the thickness of the film is very much less than the thickness of its substrate, then a convenient definition of the total misfit is

$$f = (a_s - a_o)/a_o \tag{4}$$

The convenient features of this definition are as follows. If a film is strained so that the lattices of film and substrate are in register at the interface, then $\varepsilon = f$. If the misfit is shared between dislocations and strain, then

$$f = \varepsilon + \delta \tag{5}$$

where δ is the misfit accommodated by dislocations. A positive value for f implies that the misfit strain is tensile and that misfit dislocations are positive Taylor dislocations (i.e., the extra atomic planes lie in the overgrowth). The separation of parallel misfit dislocations is

$$S = b/\delta \tag{6}$$

The energy of two perpendicular and noninteracting arrays of edge dislocations with separation S is approximately

$$E_\delta = D(f-\varepsilon)[\ln(R/b)+1] \tag{7}$$

The most appropriate value for R is difficult to determine. However, if $2S$ is less than the film thickness h, then $R \approx S$. If $2S > h$, then $R \approx h$ and the value of ε that minimizes $E_\delta + E_\varepsilon$ is

$$\varepsilon^* = (D/2Bh)[\ln(h/b)+1] \tag{8}$$

The largest possible value for ε^* is the misfit f. If the value of ε^* predicted by Eq. (8) is equal to or larger than f, the film will be strained to match the substrate precisely. If ε^* is smaller than f, then a portion of f, equal to $\delta = f - \varepsilon^*$, will be accommodated by dislocations. The thickness at which it becomes energetically favorable for the first misfit dislocations to be made is obtained by setting $\varepsilon^* = f$ in Eq. (8):

$$h_c = (D/2Bf)[\ln(h_c/b)+1] \tag{9}$$

If the film thickness is such that $\delta \gtrsim 2b/h$, then $R \approx S = b/(f-\varepsilon)$ and

$$\varepsilon^* = (-D/2Bh)\ln 2(f-\varepsilon^*) \tag{10}$$

The misfit strains predicted by this equation are slightly larger than those predicted by the more sophisticated calculations of van der Merwe and his colleagues (van der Merwe, 1964, 1972; Ball and van der Merwe, 1970; Ball, 1970; van der Merwe and van der Berg, 1972).

B. Accommodation of Misfit between Three-Dimensional Islands and Their Substrate

1. *Islands Strongly Bonded to the Substrate*

Figure 1 shows a tile-shaped island of width X, length Y, and height Z.

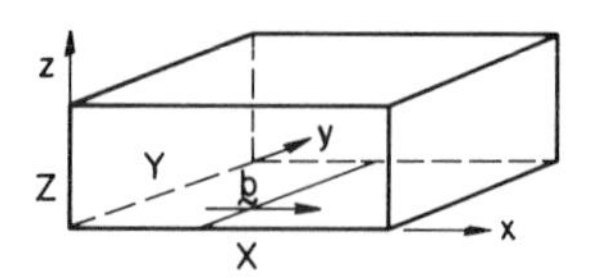

FIG. 1. A tile-shaped island on a substrate.

Suppose that $X = Y$, that misfit along x and y is f, and that misfit is accommodated by edge dislocations with Burgers vectors along x and y. The formation of misfit dislocations during the growth of such an island is illustrated in Chapter 3.6, Fig. 2 (see p. 370). The heavy continuous lines represent the borders of the island seen from above. The fine continuous lines are old misfit dislocations. The fine dashed lines are new ones. The figure shows that first one dislocation is made, which may lie along either x or y. After further growth a second dislocation with line perpendicular to the first is formed. Further growth results in the creation of a third dislocation, which again may lie along x or y. Later, a fourth dislocation perpendicular to the third is made, and so on. If the island in Fig. 1 is strongly bound to its substrate, its thickness will be small compared with its dimensions in the film plane. The elastic energy associated with the misfit strain in such an island is approximately

$$E_\varepsilon = [G_o V/(1-\nu)]\,(\varepsilon_x{}^2 + \varepsilon_y{}^2 + 2\nu\varepsilon_x\,\varepsilon_y) \tag{11}$$

where ε_x and ε_y are the strains along x and y, V is the island volume, and ε_x and ε_y are given by

$$\varepsilon_x = f - (mb/x) \qquad \text{and} \qquad \varepsilon_y = f - (nb/y) \tag{12}$$

with m and n the numbers of misfit dislocations with Burgers vectors along x and y. If $X = Y$ and $m = n$, the elastic energy released when the $(2n+1)$th dislocation is made is

$$E_{2n+1} = \frac{G_o V}{1-\nu}\left\{\frac{2bf(1+\nu)}{X} - \frac{b^2}{X^2}(2n+1+2\nu n)\right\} \tag{13}$$

If there are n dislocations with **b** along x and $n-1$ with **b** along y, or $n-1$ dislocations with **b** along x and n with **b** along y, then the energy released when the $2n$th dislocation is made is

$$E_{2n} = \frac{G_o V}{(1-\nu)}\left\{\frac{2bf(1+\nu)}{X} - \frac{b^2}{X^2}(2n-1+2\nu n)\right\} \tag{14}$$

The line energy of a single misfit dislocation is

$$\frac{G_o G_s}{(G_o+G_s)}\,\frac{Xb^2}{2\pi(1-\nu)}\left(\ln\frac{Z}{b}+1\right) \tag{15}$$

If we assume that a new misfit dislocation is made when the elastic energy released by it is equal to its own line energy, and that $Z = pX = h$, where p is a constant, then

$$X_{2n+1} = \frac{b}{2fp(1+v)}\left[\frac{G_s}{2\pi(G_o+G_s)}\left(\ln\frac{h}{b}+1\right)+2n+1+2vn\right] \qquad (16)$$

and

$$X_{2n} = \frac{b}{2fp(1+v)}\left[\frac{G_s}{2\pi(G_o+G_2)}\left(\ln\frac{h}{b}+1\right)+2n-1+2vn\right] \qquad (17)$$

The abrupt changes in elastic misfit strain that occur at X_{2n+1} and X_{2n} give rise to sudden and readily observable changes in moiré fringe patterns (Vincent, 1969, Matthews, 1972a). For the remainder of this section we consider the effects that new misfit dislocations have on the orientation and separation of moiré fringes.

The separation of moiré fringes formed as a result of interference between the indeviated electron beam and a beam that has experienced hkl and $\overline{hkl}$ reflections in the overgrowth and substrate is (Menter, 1958)

$$1/\Delta\mathbf{g} \qquad (18)$$

where

$$\Delta\mathbf{g} = \mathbf{g}_{hkl} + \mathbf{g}_{\overline{hkl}} \qquad (19)$$

$\mathbf{g}_{hkl}$ and $\mathbf{g}_{\overline{hkl}}$ are reciprocal lattice vectors that define the (hkl) and $(\overline{hkl})$ planes. The direction of fringes is perpendicular to $\Delta\mathbf{g}$.

The reflecting planes involved in fringe formation are usually perpendicular to the specimen plane. If we restrict ourselves to reflecting planes of this type, then the planes that give rise to fringes in an island like the one in Fig. 1 are $(kh0)$ and $(hk0)$. The separation of fringes formed by $kh0$ reflections is

$$[(hm/X)^2+(kn/Y)^2]^{-1/2} \qquad (20)$$

The angle between the fringes and the x-axis is

$$\text{arc tan}\,(hmY/knX) \qquad (21)$$

If the fringes were formed in an unstrained and perfectly aligned island, then the angle between the fringes and the x-axis would be

$$\text{arc tan}\,(h/k) \qquad (22)$$

The misalignment of moiré fringes associated with the discrete nature of the relaxation of misfit strain is therefore

$$\text{arc tan}\,(hmY/knX) - \text{arc tan}\,(h/k) \qquad (23)$$

The orientation changes expected when $h = k = 1$ and $X = Y$ are given in Chapter 3.6, Fig. 3 (see p. 371). This figure shows that island growth will be accompanied by abrupt rotations away from and into perfect alignment. The rotations coincide with the creation of new misfit dislocations. Their magnitude decreases as island size increases.

If X and Y are allowed to vary independently of one another, then gradual changes in the orientation of fringes are expected. These accompany gradual changes in the ratio of X to Y. Also, abrupt changes in fringe orientation will not be restricted to rotation into and away from perfect alignment. Sudden rotations from one imperfect orientation into another will occur.

Moiré fringes misaligned for the reason described in this section have been observed in deposits of copper on nickel (Matthews, 1972a). They are described in Section VI, A. Other misaligned moiré fringes have been found in silver on mica (Matthews, 1960) and silver on molybdenite (Bassett 1960). However, these misaligned fringes cannot be satisfactorily accounted for using the ideas presented in this section: their misalignment is too large. Possible explanations for them are discussed below.

2. *Islands Weakly Bonded to the Substrate*

A large fraction of the study of epitaxy has been devoted to systems in which the bonding between deposit and substrate is weak. Examples of systems of this type are fcc metals on alkali halides, on molybdenite, and on mica. In these, and in other weakly bonded systems, growth begins with the generation of isolated three-dimensional islands (Pashley, 1965). The height of the islands is comparable with their dimensions in the film plane. One of the puzzling features of these weakly bound deposits is the misalignment—by rotation about the normal to the deposit plane—of a substantial fraction of the islands present. The first evidence for this misalignment was obtained by electron diffraction (Pashley, 1959). However, in recent years it has been studied by transmission electron microscopy. Microscopy has the advantage that the alignment of individual islands can be determined, and such things as the dependence of misalignment on island size or island shape can be found. Electron microscopy has revealed that the average misalignment of large islands is less than that of small ones (Matthews, 1960), and that growing islands sometimes rotate from one orientation into another (Bassett, 1960). A surprising feature of the rotation is that it does not always improve the alignment of an island. It sometimes brings an island nearer to epitaxy and sometimes brings it further away.

Several attempts to explain the presence of misaligned islands and the rotation of islands from one orientation into another have been made. Cabrera (1965) and Jesser and Kuhlmann-Wilsdorf (1968) have considered the nucleation of well- and badly aligned islands. They conclude that islands misaligned by the observed angles will be created early in film growth. However, their calculations do not provide an explanation for the fact that most misaligned islands are able to retain their misalignment for long periods of time. Explanations for the stability of misaligned islands have been suggested by Cabrera (1959), Tucker (1964, 1966), du Plessis and van der Merwe (1965), Fletcher and Adamson (1966), Bollmann (1967), and Reiss (1968). All these explanations are based on the assumption that there are minima in interfacial energy at particular values of misalignment and misfit. Cabrera, Tucker, du Plessis and van der Merwe, and Bollmann consider that the minima are found where some of the interfacial atoms belonging to one crystal lie in surface sites belonging to the other. Their model is thus an extension of the coincidence-lattice model for high-angle grain boundaries (Gleiter and Chalmers, 1972). Evidence that the coincidence model does correctly predict some of the low-energy boundaries between crystals with different lattice parameters has been obtained (Tucker, 1964, 1966; Simmons *et al.*, 1967; Edmonds *et al.*, 1971).

Reiss suggests that there are low-energy interfaces in addition to those predicted by the coincidence model. He has calculated the energy of the interface between a rigid square array of atoms and a substrate and the variation of interfacial energy with misalignment of the array. He shows that there are minima in the interfacial energy at angles given by

$$\gamma = (2l+1)/2N^{1/2} \tag{24}$$

where l is an integer and N the number of atoms in the array. The minima are deep when l is small and shallow when it is large. A badly aligned island therefore rotates into another orientation more readily than a well-aligned one. Indeed, Reiss concludes that a nucleus is only able to rotate if it is both badly misaligned and small.

It may be possible to account for the badly misaligned fringes in deposits of silver on mica, silver on molybdenite, or gold on molybdenite, using the coincidence model and the model of Reiss. However, the results of an attempt to do so have not been encouraging. Most islands are misaligned by angles too small to be accounted for using either model. Also, most observations suggest that islands are not misaligned by the discrete amounts predicted by the two models. A hypothesis that seems to provide a more satisfactory explanation for the results is as follows: The generation of misfit dislocations in systems where the interfacial bonding is weak is accompanied by a reversal in the sign of the misfit strain. This reversal is followed by a rotation of the islands away from the epitaxial orientation. Rotation is about normal to the interface plane and may be either clockwise or anticlockwise. It occurs in order to reduce the efficiency of misfit dislocations. The reduction in efficiency is such as to bring the misfit strain to approximately zero.

The remainder of this section is a determination of the misalignments that are expected to result from this process. A comparison of the predictions with experiment is made in Section VI.

Consider a circular interface between a three-dimensional island and substrate, and suppose that misfit dislocations—in edge orientation and with Burgers vectors in the interface—are made in pairs (Fig. 2). The assumption

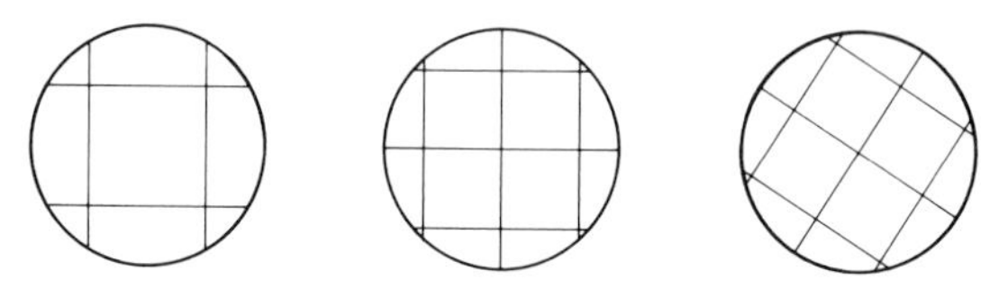

FIG. 2. A circular interface between a three-dimensional island and its substrate. The lines inside the island are misfit dislocations that are assumed, for convenience, to be formed in pairs.

that dislocations are made in pairs is artificial but convenient. It is convenient because it enables one to distinguish between the real rotation discussed in this section and the rotation of moiré fringes discussed in Section II, B, 1. Let the elastic strain in an island be ε_i before the introduction of a pair of dislocations and ε_f afterwards. If we begin with $n-1$ pairs of misfit dislocations under an island, then the initial elastic strain of the island is

$$\varepsilon_i = f - [(n-1)\,b/2r] \tag{25}$$

The elastic strain energy of the island is

$$E_{\varepsilon_i} = Ar^3\{f - [(n-1)\,b/2r]\}^2 \tag{26}$$

where A depends on the size of the island, the shape of the island, and the elastic constants of island and substrate. For a hemispherical island Cabrera (1965) finds it to be

$$A = [3/8(1-\nu)]\,[G_o\,G_s/(4G_s + G_o)] \tag{27}$$

The energy of the island after the introduction of the nth pair of misfit dislocations is

$$E_{\varepsilon_f} = Ar^3\,[f - (nb/2r)]^2 \tag{28}$$

Most of the experimental data available at present has been obtained from islands in which $n \gtrsim 5$. The length of misfit dislocation line, per unit area of interface, under islands of this type is approximately

$$2S/S^2 \tag{29}$$

S before the creation of the nth pair of dislocations is

$$S_i = 2r/(n-1)$$

Afterwards it is

$$S_f = 2r/n$$

The length of dislocation line in a circular interface is

$$L_i = \pi r(n-1)$$

before the introduction of the nth pair of dislocations and

$$L_f = \pi rn$$

afterwards. The energy of a network composed of $2(n-1)$ dislocations is

$$E_{\delta_i} = \pi r(n-1)\alpha\frac{Db}{2}\cdot\left(\ln\frac{r}{(n-1)b}+1\right) \tag{30}$$

where α is a measure of the strength of the bonds between overgrowth and substrate. It is close to unity when the bonds across the interface are comparable with those found in the overgrowth or substrate crystals, and close to zero when the interfacial bonding is weak. The change in interfacial energy that accompanies the creation of the nth pair of misfit dislocations is

$$\Delta E_\delta = \frac{\pi\alpha Dbr}{2}\left[n\ln\frac{(n-1)}{n}+\ln\frac{r}{(n-1)b}+1\right] \tag{31}$$

If we assume that the nth pair of misfit dislocations is made when it reduces the elastic energy by an amount equal to its own energy, then the radius at which the nth pair is formed is

$$r_{2n} = \frac{b}{rf}(2n-1)-\frac{\pi\alpha D}{2Af}\left[n\ln\frac{(n-1)}{n}+\ln\frac{r}{(n-1)b}+1\right] \tag{32}$$

If $\alpha = 1$, the second term is comparable with the first. However, if the bonding across the interface is very weak and $\alpha \ll 1$, the second term is negligible and

$$r_{2n} = (b/rf)(2n-1) \tag{33}$$

So far we have assumed that the creation of misfit dislocations is an abrupt and discontinuous process. At r infinitesimally less than r_{2n} there are $n-1$ pairs of misfit dislocations, and at r infinitesimally greater than r_{2n} there are n pairs. This assumption is unrealistic. However, if we continue to make it for a moment, then, in systems where $\alpha \ll 1$, the misfit strain at r_{2n} changes abruptly from

$$f-[(n-1)b/2r] \quad \text{to} \quad f-(nb/2r)$$

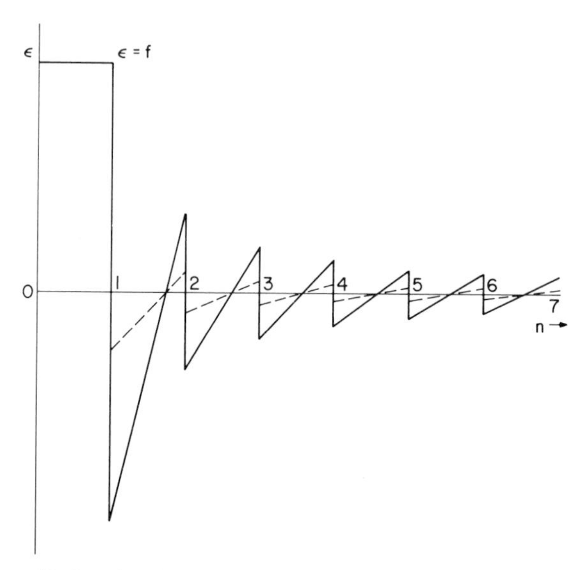

FIG. 3. The oscillations in misfit strain expected in islands weakly bound to their substrate.

and the elastic strain of the island oscillates between

$$+f/(2n-1) \qquad \text{and} \qquad -f/(2n-1)$$

as illustrated in Fig. 3.

We now discuss the assumption that the creation of misfit dislocations is an abrupt and discontinuous process. In fact, the formation of misfit dislocations is partly abrupt and partly gradual. The atoms at the edge of an island occupy sites in the substrate surface only if the misfit strain in the island is zero. If the strain in an island deviates from zero, the peripheral atoms become slightly displaced from sites in the substrate surface. This displacement is equivalent to the introduction of a fraction of a misfit dislocation. The strength of the fractional dislocation increases as island growth proceeds. Eventually the peripheral atoms pop over into, or, more precisely, almost into, the next site in the substrate surface. The pop-over changes the Burgers vector of the fractional dislocation from less than $\frac{1}{2}b$ to between $\frac{1}{2}b$ and b. Further growth of the island gradually increases the Burgers vector of the misfit dislocation from more than $\frac{1}{2}b$ to b. The effect of the gradual–abrupt–gradual introduction of misfit dislocations is illustrated qualitatively by the dashed line in Fig. 3.

Although the gradual introduction of misfit dislocations reduces the peak values of the misfit strain in islands it is unlikely to change significantly the rotation of islands discussed below. This is because the rotation needed to minimize the interfacial and misfit strain energy in weakly bound islands is the same for the solid and dashed lines in Fig. 3.

The rotation of an island after the introduction of a misfit dislocation takes place about the normal to the interface plane. It is accompanied by a much larger rotation of the misfit dislocations. The relation between island rotation ϕ and the rotation of misfit dislocations θ is

$$\phi = f\theta \tag{34}$$

If f [as defined by Eq. (4)] is positive, then ϕ and θ have the same sense. If f is negative, their senses are opposite. Rotation of misfit dislocations changes the character of the dislocations from pure edge to a mixture of edge and screw. This reduces the misfit strain from a negative value (using the sign convention implied by Fig. 3) to near zero. The magnitude of the misalignment can be determined by finding the rotation of the dislocations that minimizes the sum of E_ε and E_δ. The variation of E_ε with θ is given by

$$E_\varepsilon = Ar^3[f-(nb/2r)\cos\theta]^2 \tag{35}$$

and the variation of E_δ with θ by

$$E_\delta = \frac{\pi r n \alpha D}{2}\cdot(1-\nu\cos^2\theta)\left(\ln\frac{r}{nb}+1\right) \tag{36}$$

The value of θ that minimizes the sum of these energies is

$$\frac{1}{\cos\theta} = \frac{nb}{2rf} - \frac{\pi\alpha\nu D}{Ar}\cdot\left(\ln\frac{r}{nb}+1\right) \tag{37}$$

so long as the predicted value for $\cos\theta$ does not exceed unity. If the predicted value does exceed unity, then the actual value is unity and the island is perfectly aligned.

In weakly bonded systems $\alpha \ll 1$ and the second term on the right of Eq. (37) is small compared with the first. In systems of this type

$$\cos\theta = 2rf/nb \tag{38}$$

The maximum value of θ occurs at $r = r_{2n}$. In weakly bonded systems $r_{2n} = (b/4f)(2n-1)$. Thus the maximum value of θ in weakly bonded systems is

$$\cos\theta_{\max} = 1 - (1/2n) \tag{39}$$

If θ is small, then

$$\theta_{\max} = 1/n^{1/2} \tag{40}$$

and

$$\phi_{\max} = f/n^{1/2} \tag{41}$$

The predictions of Eq. (38) are shown graphically in Fig. 4.

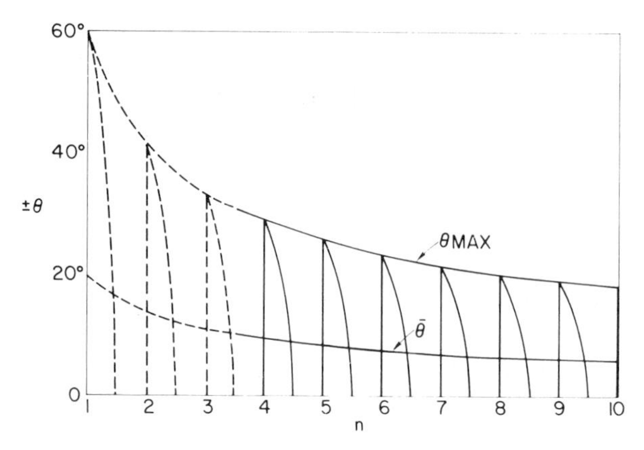

FIG. 4. The variation in alignment with island size expected in islands weakly bonded to their substrate.

In order to compare the predictions of this section with measurements made on deposits of silver on mica we need an experssion for the average value of $|\theta|$. In weakly bonded systems this average is

$$\bar{\theta} = \left(\frac{4n-1}{4}\right)^{1/2} - \left(\frac{2n-1}{2}\right) \operatorname{arc\,cos}\left(\frac{2n-1}{2n}\right) \tag{42}$$

The predictions of this expression are compared with experimental results for silver on mica in Section VI, B.

III. Observations of Coherency Strains

A. FILMS ON SUBSTRATES

The prediction that films would sometimes strain to match their substrate was made in 1949 but is was not until the mid-1960s that clear experimental confirmation of it was obtained. In the past seven or eight years many systems that exhibit coherency strains have been found. Examples are nickel on copper (Gradmann, 1964, 1966; Thompson and Lawless, 1966; Kuntze *et al.*, 1969; Matthews and Crawford, 1970), gold on silver (Matthews, 1966), palladium on gold (Matthews, 1966; Cherns and Stowell, 1973), platinum on gold (Matthews and Jesser, 1967), β-cobalt on copper (Jesser and Matthews, 1968a; Matthews, 1970; Fedorenko and Vincent, 1971), γ-iron on copper (Jesser and Matthews, 1967, 1968b), α-iron on gold (Wassermann and Jablonski, 1970), germanium on gallium arsenide (Matthews *et al.*, 1970), lead sulfide on lead selenide (Yagi *et al.*, 1971; Matthews, 1971), and garnet films on garnet substrates (Besser *et al.*, 1972; Matthews and Klokholm, 1972). In metals the predicted and experimental values for h_c agree fairly well with one another. The variation of elastic strain with thickness after coherence has been lost is also in fairly good agreement with predictions. However, the observed strains do seem to be significantly larger than predicted. These discrepancies have been ascribed to processes that impede the generation of misfit dislocations. The existence of these processes is revealed by Frank and van der Merwe's calculations. The first to invoke them in order to explain experimental observations was Cabrera (1965).

Clear evidence for discrepancies between the predictions of the equilibrium theories, such as those of van der Merwe (1972), and experiment is provided by observations made on copper oxide on copper (Borie *et al.*, 1962), germanium on gallium arsenide (Matthews *et al.*, 1970), and garnet films on garnet substrates (Besser *et al.*, 1972; Matthews and Klokholm, 1972). Copper oxide films 450 Å in thickness are strained by about 2% near the oxide–metal interface. This strain accommodates only 12% of the misfit between the two crystals but is about 300 times the expected strain. Films of germanium grown on gallium arsenide at 350°C are strained to fit their substrate until their thickness reaches 2 μm. The predicted critical thickness is about one-tenth of this value. Magnetic garnet films that are grown on nonmagnetic garnets, and used for bubble-domain devices, are almost invariably strained to fit their substrates exactly (Besser *et al.*, 1972; Matthews and Klokholm, 1972). This is so in spite of the fact that most of them have thicknesses many times h_c. Possible explanations for these results are discussed in Sections V, C and V, D.

B. Three-Dimensional Islands on Substrates

Although a very large number of deposits composed of isolated three-dimensional islands have been examined by electron microscopy, only a few examples of islands strained to fit their substrate have been found. Coherent islands have been seen in γ-iron on copper (Jesser and Matthews, 1968b), β-cobalt on copper (Jesser and Matthews, 1968a), and copper on nickel (Matthews, 1972a). A micrograph of coherently strained islands of β-cobalt on copper is shown in Fig. 5. The image contrast at the islands is typical of that found where there is one strongly diffracted beam. One side of the island is dark and the other one light. The light and dark sides meet along a line that is parallel to the diffracting planes.

The size that islands attain before the first misfit dislocation is made has been measured in γ-iron on copper and β-cobalt on copper. The results are in good agreement with the critical radii predicted by Jesser and Kuhlmann-Wilsdorf (1967a, b). The critical radii observed in cobalt–copper and iron–copper were 375 and 750 Å, respectively (Jesser and Matthews, 1968a, b).

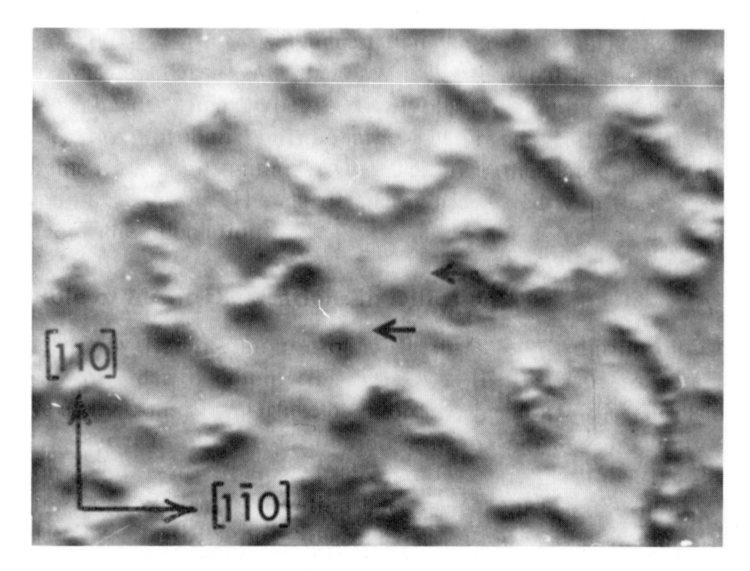

FIG. 5. An electron micrograph of islands of β-cobalt on copper. The arrows indicate islands strained to fit the substrate (reprinted from Jesser and Matthews, 1968a, by courtesy of the Philosophical Magazine).

The radii predicted (Jesser and Kuhlmann-Wilsdorf, 1967a, b) for these systems are 360 and 870 Å. The probable errors in the calculations and measurements are such that the agreement between these predicted and observed radii must be partly fortuitous.

IV. Observations of Misfit Dislocations

Misfit dislocations have been observed between thin films and substrates, between islands and substrates, and around precipitates in alloys. Some of these dislocations have resembled those envisaged by Frank and van der Merwe (1949) in that they have been in edge orientation and have had Burgers vectors parallel to the interface plane. Others have been more complicated than this. Some have had Burgers vectors inclined to the interface, some have been in mixed orientation, and some have been imperfect. The misfit dislocations described below have been selected to illustrate this diversity.

A. Edge Dislocations with Burgers Vectors in the Interface

Misfit dislocations in edge orientation and with Burgers vectors parallel to the interface have been observed in (001) films of lead sulfide on lead selenide (Matthews, 1961, 1971; Yagi *et al.*, 1971), in silicon doped by diffusion

(Washburn *et al.*, 1964), in copper on nickel (Matthews and Crawford, 1970), in gold on palladium (Matthews, 1964, 1966), and palladium on gold (Cherns and Stowell, 1973). An electron micrograph of lead sulfide on lead selenide in which misfit dislocations can be seen under a variety of diffracting conditions is present in Fig. 6.

The features labeled A are moiré fringes formed by 200 reflections in the lead sulfide and lead selenide lattices. Those labeled C are moiré fringes formed by 220 reflections. An approximately square network of misfit dislocations is discernible near B. Parallel dislocations are visible near D. The fact that only one set of dislocations is visible near D, and that the line direction of the invisible set is perpendicular to the reflecting planes, shows that the invisible dislocations are in edge orientation. The indices of the fringes at C show that the invisible dislocations at D have Burgers vectors whose component in the interface plane is parallel to $[1\bar{1}0]$. The symmetry of the (001) interface is such that the visible set of dislocations near D must be in edge orientation and have Burgers vectors along [110]. The magnitude of the edge component parallel to the interface can be determined very easily. Comparison of the fringes at C with the dislocations at D shows that the separation of the fringes is half the dislocation spacing. From this one can conclude that the dislocations have an edge component in the film plane that is twice the spacing of {220} planes. The spacing of {220} planes is $\frac{1}{4}a\langle 110\rangle$. Thus the component of Burgers vector in the film plane is $\frac{1}{2}a\langle 110\rangle$. This result, taken together with Frank and Nicholas' (1953) predictions of the complete dislocations that are stable in fcc crystals, shows that the misfit dislocations lead sulfide–lead selenide bicrystals have Burgers vectors of type $\frac{1}{2}a\langle 110\rangle$ that are parallel to the interface plane.

The generation of misfit dislocations with Burgers vectors parallel to the interface has been discussed by Washburn *et al.* (1964), Matthews (1966, 1971, 1972a, 1973), Vincent (1969), Yagi *et al.* (1971), and Cherns and Stowell (1973). The mechanisms suggested by them are described in Section V.

B. Complete Dislocations with Burgers Vectors Inclined to the Interface

At least half the misfit dislocations observed so far have had Burgers vectors inclined to the interface. Some of the systems in which inclined Burgers vectors have been found are doped silicon on silicon (Washburn *et al.*, 1964), gold on silver (Matthews, 1966), and gallium phosphide on gallium arsenide (Abrahams *et al.*, 1969, Matthews and Blakeslee, 1972). Inclined misfit dislocations have been present in these systems in spite of the fact that in every one of them the orientations of the interfaces have been

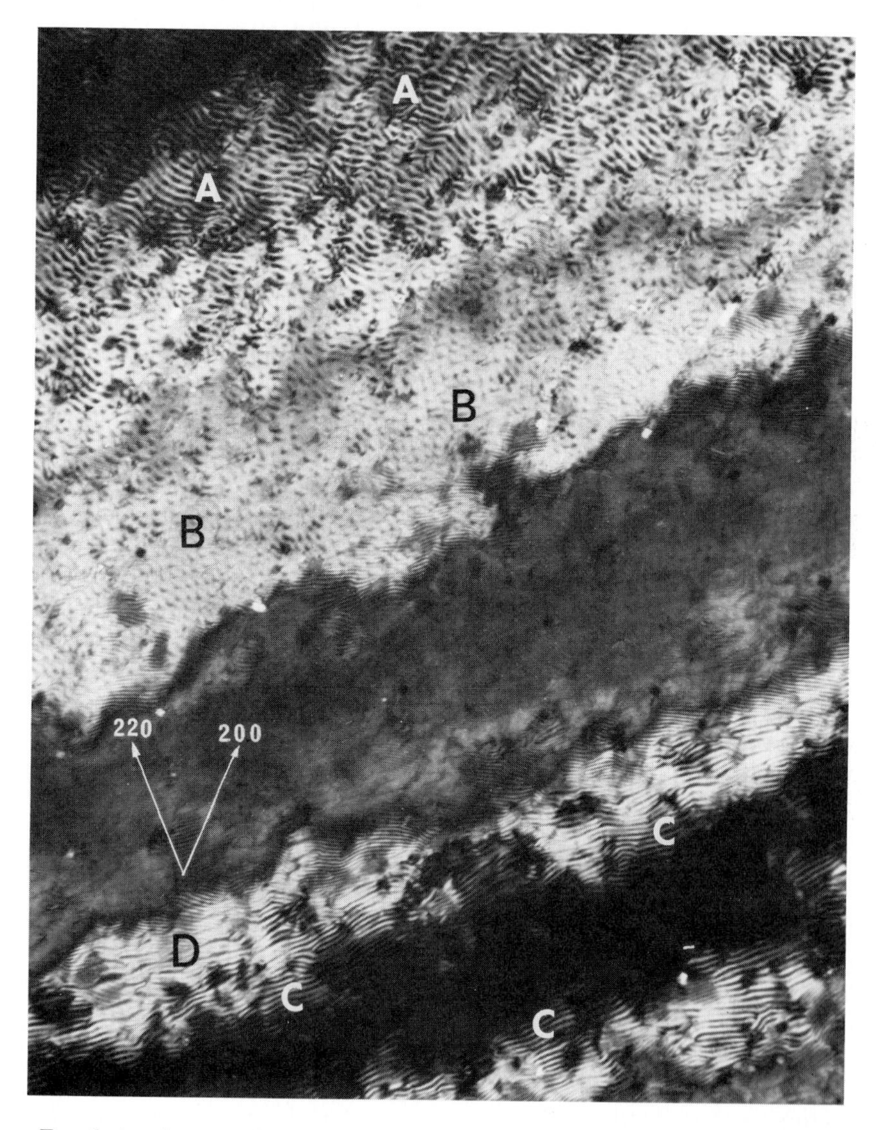

FIG. 6. An electron micrograph of misfit dislocations and moiré fringes in an (001) deposit of lead sulfide on lead selenide ($\times$ 185,000). See text for discussion of labeling.

such that stable, complete, edge dislocations with Burgers vectors in the interface could have been constructed in them. The explanation for the presence of dislocations with inclined Burgers vectors, rather than edge dislocations with Burgers vectors in the interface, is simply that dislocations with inclined Burgers vectors can move to the interface by glide on planes inclined to the interface but edge dislocations with Burgers vectors in the interface cannot. The generation of misfit dislocations by glide is discussed in Section V.

Electron micrographs of complete misfit dislocations with Burgers vectors inclined to the interface between (001) films of gold on silver are seen in Fig. 7. The Bragg reflection responsible for the image contrast was 220 in

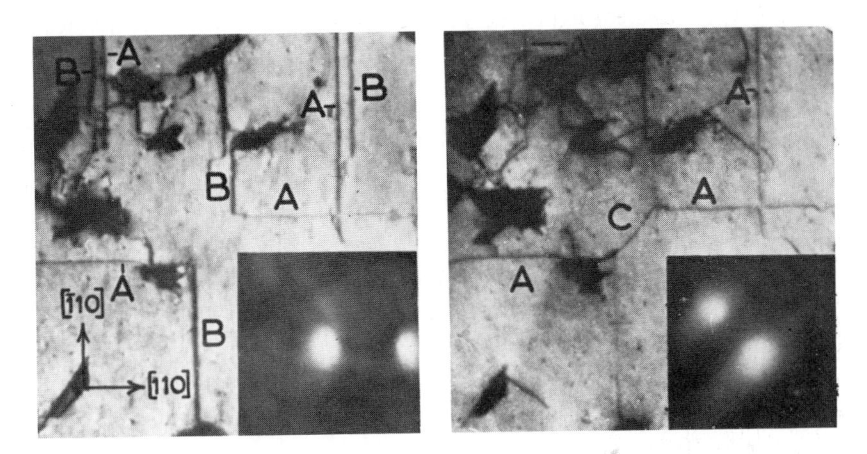

FIG. 7. Electron micrographs of misfit dislocations in a deposit of gold on silver. The Bragg reflection was 220 (a) and 200 (b).

Fig. 7a and 200 in Fig. 7b. The straight dislocations labeled A are visible in both images. Those labeled B are visible in (a) but are invisible or almost invisible in (b). Pairs of dislocations labeled A and B combine to form a dislocation C, which is visible in (b) but not in (a). These image features taken together with Frank and Nicholas' (1953) predictions of the stable dislocations in fcc metals, enables one to assign Burgers vectors to the dislocations. All the dislocations have Burgers vectors of type $\frac{1}{2}a\langle 110\rangle$. The possible Burgers vectors of the dislocations labeled A are $\pm\frac{1}{2}a[10\bar{1}]$ and $\pm\frac{1}{2}a[\bar{1}0\bar{1}]$. The possible Burgers vectors of the dislocations labeled B are $\frac{1}{2}a[0\bar{1}1]$ and $\pm\frac{1}{2}a[011]$. The possible Burgers vectors of the dislocation labeled C are $\pm\frac{1}{2}a[1\bar{1}0]$. Thus A and B are in mixed orientation and have Burgers vectors inclined at 45° to the interface plane. The misfit they accommodate is proportional to their edge component projected into the interface. This means that

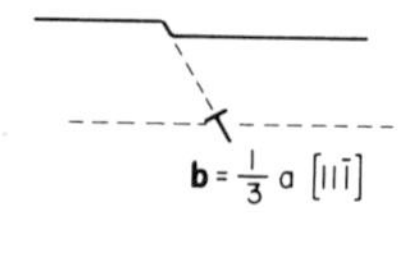

FIG. 8. An imperfect misfit dislocation with Burgers vector of type $\frac{1}{3}a\langle 111\rangle$.

they accommodate only half as much misfit as edge dislocations with Burgers vectors of type $\frac{1}{2}a\langle 110\rangle$ that lie in the interface plane. They are however the most efficient complete misfit dislocations that can be made by glide to the interface on $\{111\}$ slip planes. It is worth noting that the line directions of A and B are consistent with their generation by glide. Their lines are accurately parallel to the [110] and [1$\bar{1}$0] directions in the film plane. The $\{111\}$ glide planes intersect the interface along $\langle 110\rangle$ directions.

In fcc crystals a dislocation with Burgers vector $\frac{1}{2}a\langle 110\rangle$ has two possible $\{111\}$ glide planes. In certain bicrystal films, of which (001) and (111) are examples, the misfit accommodated by a complete dislocation does not depend on the glide plane it uses. It is possible for a misfit dislocation to be made by glide on one $\{111\}$ plane over part of its length, and by glide on the other $\{111\}$ plane over the remainder. The change from one glide plane to the other is called "cross slip." Examples of cross slip in fcc bicrystals have been observed by Matthews (1966) and Matthews and Jesser (1967).

C. PARTIAL DISLOCATIONS WITH BURGERS VECTORS INCLINED TO THE INTERFACE

If the stacking-fault energy of a film is low or negative, or if the misfit strain is larger than $\sim 1\%$, misfit dislocations may be imperfect (Jesser and Matthews, 1968a, 1968c; Matthews, 1968; Cherns and Stowell, 1973). They may be generated by climb (Cherns and Stowell, 1973) or by glide. Their geometries in films oriented with (001) parallel to their plane are illustrated in Figs. 8 and 9. Figure 8 shows a Frank partial with Burgers vector $\frac{1}{3}a[11\bar{1}]$; it has moved to the interface by climb on (11$\bar{1}$). Examples of dislocations with this geometry have been observed by Cherns and Stowell (1973) in films of palladium on gold.

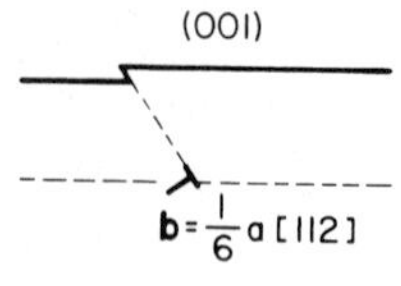

FIG. 9. An imperfect misfit dislocation with Burgers vector of type $\frac{1}{6}a\langle 112\rangle$.

Figure 9 shows a Shockley partial made by glide to the interface on $(11\bar{1})$. Examples of dislocations of this type have been seen in deposits of cobalt on nickel and cobalt on nickel–palladium alloys by Jesser and Matthews (1968a) and by Matthews (1968).†

The nature of stacking faults formed by the processes illustrated in Figs. 8 and 9 depends on the orientation of the interface and the sign of the misfit strain: If the misfit strain is compressive, and the interface (001), then the faults that terminate on Frank partials are intrinsic (Fig. 8). Those that terminate on Shockley partials are extrinsic (Fig. 9).

D. Supersessile Misfit Dislocations

Some of the imperfect misfit dislocations on (001) deposits of cobalt on nickel are supersessile. They lie at the junction of a pair of stacking faults as illustrated in Fig. 10. They are in edge orientation and their Burgers vectors

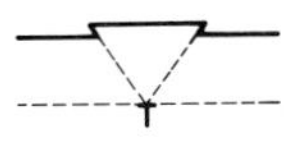

Fig. 10. A supersessile misfit dislocation in a fcc bicrystal oriented with (001) parallel to its plane.

are parallel to one of the ⟨110⟩ directions in the film plane. A micrograph of a cobalt–nickel bicrystal that contained a large number of them is seen in Fig. 11. The evidence that the defects in Fig. 11 have the geometry shown in Fig. 10 is provided by features, such as the one labeled A, that are produced when the pair of stacking faults associated with one supersessile terminate against one of the stacking faults associated with another. The geometry that one expects to find is shown in Fig. 12. The resemblance of this geometry to that visible at A in Fig. 11 is obvious.

Direct evidence that dislocations like those in Fig. 11 are supersessile is provided by experiments in which specimens were alloyed while they were under observation inside an electron microscope. These experiments showed that at no time in the alloying process do the supersessile dislocations move at all. They are eliminated (when the alloying process is almost complete) by migration of the partial dislocations at their ends.

† Recently Cherns (1974) has shown that it is very difficult to distinguish between the images of Frank and Shockley partial dislocations in electron micrographs of thin (001) bicrystals. One of the consequences of this is that the nature of imperfect misfit dislocations in (001) bicrystals of palladium on gold, or cobalt on nickel or nickel–palladium alloys, is still uncertain.

FIG. 11. An electron micrograph of supersessile misfit dislocations in cobalt on nickel.

The formation of supersessile dislocations has not been observed directly, nor has the magnitude of their Burgers vector been measured. The predictions of Frank and Nicholas (1953) suggest that only two Burgers vectors are possible. They are $\frac{1}{3}a\langle 110\rangle$ and $\frac{1}{6}a\langle 110\rangle$. A reaction that may be involved in

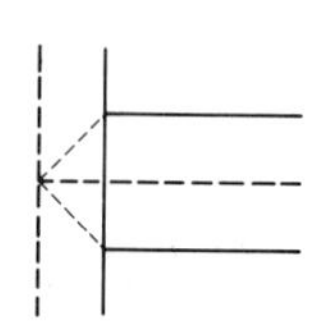

FIG. 12. The geometry of the defect formed when the pair of stacking faults associated with one supersessile misfit dislocation is impacted against one of the faults associated with another.

the generation of the first of these dislocations from a Shockley partial is:

$$\tfrac{1}{6}a[112] \rightarrow \tfrac{1}{3}a[110] + \tfrac{1}{6}a[\bar{1}\bar{1}2]$$

The first product dislocation is the supersessile. The second is a glissile partial that moves to the surface and escapes.

V. The Generation of Misfit Dislocations

A. Nucleation of Misfit Dislocations at the Edge of Islands

The misfit dislocations in deposits that exhibit island growth are usually generated at the junction of the island and substrate surfaces. The process has been observed directly by Yagi *et al.* (1971). The abrupt drop in misfit strain that occurs when a misfit dislocation is generated at the edge of an island has been detected by Vincent (1969). He measured the elastic strain in islands of tin on tin telluride as a function of island size and found that island growth was accompanied by a saw-tooth variation in misfit strain (see Chapter 3.6, Fig. 1, p. 370). Vincent also showed that a very satisfactory explanation for his observations is obtained if one postulates that misfit dislocations are created whenever their presence reduces the misfit strain to zero. The success of this postulate indicates that nucleation of misfit dislocations at the edges of islands of tin on tin telluride is a rather easy process. That this should be so is not surprising. If misfit dislocations are made when their presence cancels the misfit strain, then the strain responsible for their formation is

$$\varepsilon = b/X \tag{43}$$

where X is the dimension of the island in the film plane (Fig. 1). The shear stress at the edge of the island as a result of this strain is of order

$$G(Z/b)^{1/2}(b/X) \tag{44}$$

The stress predicted by this expression for tin on tin telluride (where $b = 3$ Å and $X = 4Z \simeq 200$ Å) is $G/15$. This stress is large enough to result in dislocation nucleation.

Although many misfit dislocations under three-dimensional islands are made at the island edge, this is not true of all of them. The misfit dislocations under islands of γ-iron on copper, for example, are made by glide to the interface on planes inclined to the interface plane. This process is discussed in Section V, C.

B. Nucleation of Misfit Dislocations at the Base of Cracks

A mechanism closely related to the one described in Section A has been found to operate in (001) films of lead sulfide grown at room temperature

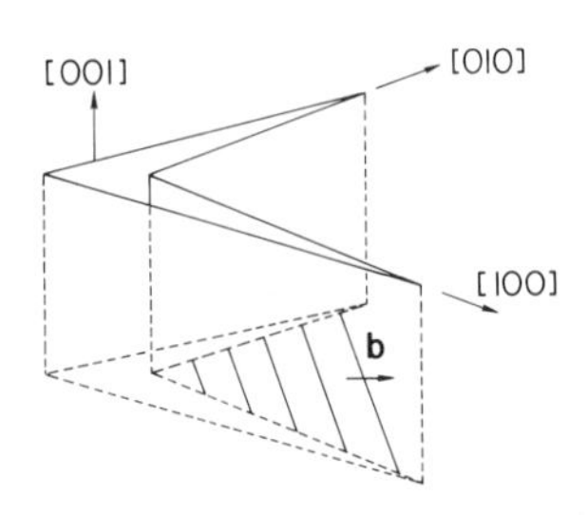

FIG. 13. An L-shaped crack in a thin (001) deposit of lead sulfide on lead selenide (reprinted from Matthews, 1971, by courtesy of the Philosophical Magazine).

on lead selenide. The mechanism is illustrated in Fig. 13. It begins with the formation of an L-shaped crack in the lead sulfide layer. The crack is formed as a result of the tensile misfit stress and lies on the (100) and (010) planes perpendicular to the film plane. Crack formation is followed by a parting of the crack faces. This is accompanied by the nucleation of edge dislocations at the point where the (100) and (010) crack faces meet the lead sulfide–lead selenide interface. The Burgers vectors of the dislocations are of type $\frac{1}{2}a\langle 110\rangle$ and lie in the film plane. Dislocation nucleation is followed by glide of the dislocations on the (001) interface. Glide is accompanied by an increase in line length. Thus, if a succession of dislocations is formed at an L-shaped crack, the longest in the series is the oldest, and the shortest one the youngest. The series taken as a whole is a triangular patch of parallel dislocation lines. Examples of these patches are present in Fig. 14.

An important feature of the dislocations in the triangular patches is that they accommodate misfit along one $\langle 110\rangle$ direction in the interface but not along the other. Thus, the lead sulfide above a triangular patch is strained to match lead selenide along one of the $\langle 110\rangle$ directions in the interface plane but not along the other.

The separation of the dislocations in the patches in Fig. 14 is 115 ± 15 Å. The separation expected if the dislocations accommodate all the misfit between lead sulfide and lead selenide is 140 Å. Thus the dislocations in the patches are so close to one another that they reverse the sign of the misfit strain. This sign reversal is believed to result from the Poisson contraction associated with the coherency strain parallel to the dislocation lines. Support for this belief is provided by calculations showing that it is energetically favorable for the elastic strain parallel to the Burgers vectors of the dislocations in a patch to change sign when the lead sulfide layer is 100 Å or so in thickness.

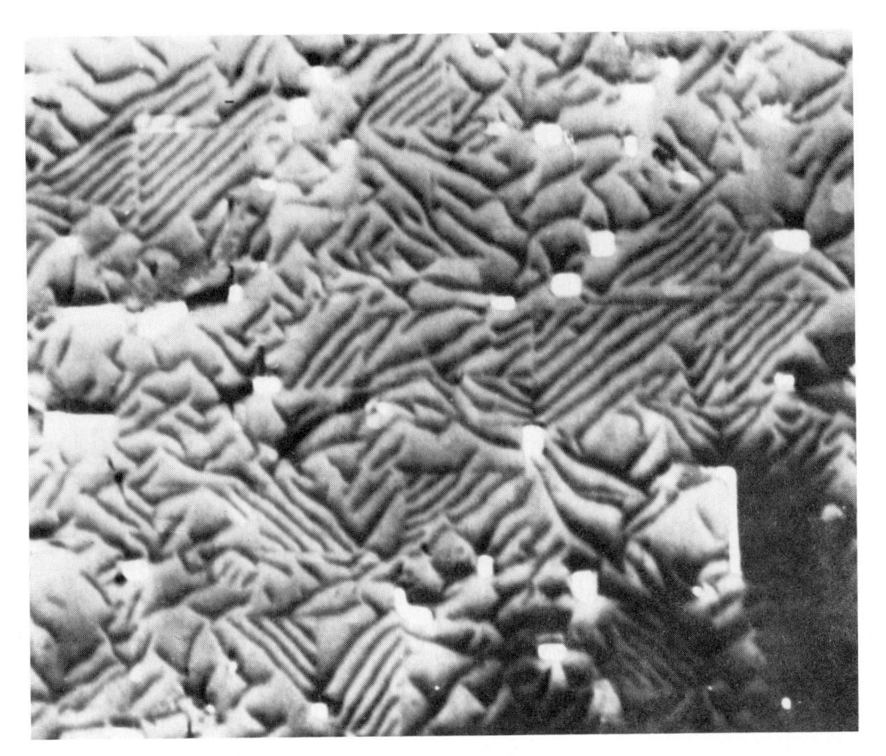

FIG. 14. A micrograph of misfit dislocations in a lead sulfide–lead selenide bicrystal oriented with (001) parallel to its plane.

C. THE GLIDE OF THREADING DISLOCATIONS

The misfit dislocations present in semiconductors doped by diffusion, and in deposits of one fcc metal on another, are often made by the glide of threading dislocations with Burgers vectors inclined to the interface plane. A form of the mechanism that gives rise to complete misfit dislocations is illustrated in Fig. 15. Figure 15a shows a dislocation line that extends from the substrate, across the interface, and through the overgrowth film. As a result of the misfit between the stress-free lattice parameters of film and substrate there are glide forces acting on this dislocation. These forces increase as deposit thickness increases and cause the dislocation to bow as shown in Fig. 15b. Eventually, the glide forces acting on the dislocation are able to overcome the forces that oppose glide. When this happens the length of threading dislocation in film and substrate move apart, as shown in Fig. 15c, and leave a length of misfit-dislocation line in the interface.

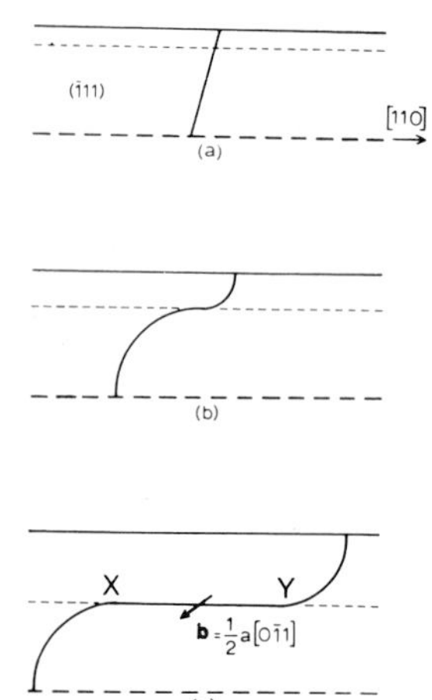

FIG. 15. The generation of a length XY of complete misfit dislocation by the glide of a threading dislocation. See text for discussion of parts.

A second form of the process is illustrated in Fig. 16. Figure 16a shows a threading dislocation similar to that seen in Fig. 15a. As a result of the misfit stress in the overgrowth the portion of the threading dislocation that lies in the overgrowth dissociates into two partial dislocations separated by a stacking fault. When a critical thickness is reached, the front partial moves away as shown in Fig. 16b, leaving an imperfect misfit dislocation in the interface and a stacking fault that extends from the interface to the surface of the film. The Burgers vectors of the dislocations in Fig. 16 are shown in Thompson's (1953) notation. The positive dislocation direction is from substrate to overgrowth if the overgrowth is strained in tension by the substrate. It is from the overgrowth to substrate if the overgrowth is compressed.

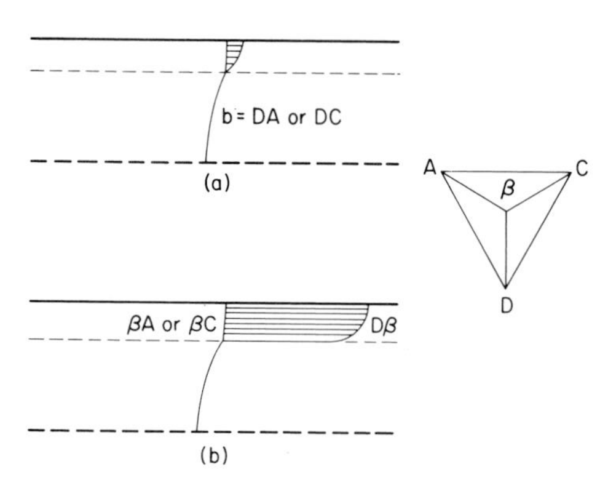

FIG. 16. Generation of a partial misfit dislocation by the dissociation and glide of a threading dislocation. See text for discussion.

The film thickness at which the complete dislocation in Fig. 15, or the partial in Fig. 16, moves to generate a length of misfit-dislocation line can be calculated from the forces on the dislocations. For simplicity, and because this situation is common in practice, we will assume that the film is much thicker than its substrate, and that the misfit between the stress-free lattice parameters of film and substrate does not depend on direction. Under these conditions the force tending to move the dislocation in the overgrowth to the right is

$$F_\varepsilon = Bbhf \cos\lambda \tag{45}$$

where λ is the angle between the slip direction and that direction in the interface that is perpendicular to the line of intersection of the slip plane and the interface. F_ε is opposed by the line tension of the dislocation formed in the interface, by the tension in the surface step created by the dislocation, and, if the misfit dislocation is imperfect, by the stacking fault. In addition, as will be discussed later, it is opposed by the Peierls–Nabarro friction stress. The line tension in the misfit dislocation is approximately

$$F_1 = Db \cdot (1 - \nu\cos^2\alpha)\,[\ln(h/b) + 1] \tag{46}$$

where α is the angle between the misfit dislocation and its Burgers vector. The force due to the tension in the surface step created by the migration of the dislocation is

$$F_s = \sigma_o b \sin\alpha \tag{47}$$

where σ_o is the overgrowth's surface energy. The force due to the stacking fault is

$$F_\gamma = \gamma_o h/\cos\phi \tag{48}$$

where γ_o is the stacking-fault energy and ϕ the angle between the specimen surface and the normal to the slip plane. If the driving force is equated to the sum of the retarding forces, then the following equation for h_c, the thickness at which the generation of misfit dislocations will begin, is obtained

$$h_c = \frac{Db(1 - \nu\cos^2\alpha)\,[\ln(h/b) + 1] + \sigma_o b \sin\alpha}{Bbf\cos\lambda - \gamma_o/\cos\phi} \tag{49}$$

If the film thickness exceeds h_c, then the elastic strain in the film will decrease in order to keep F_ε equal to $F_1 + F_s + F_\gamma$. This can take place by the elongation of lines like XY in Fig. 15. The variation of elastic strain with thickness once h_c has been passed is

$$\varepsilon^* = \frac{1}{Bh\cos\lambda}\left\{D(1 - \nu\cos^2\alpha)\left(\ln\frac{h}{b} + 1\right) + \sigma_o \sin\alpha + \frac{\gamma_o h}{b\cos\phi}\right\} \tag{50}$$

This equation is expected to be fairly accurate in systems where F_ε, F_1, F_s, and F_γ are the only significant forces, and where h is less than half the separation of misfit dislocations. In systems where h exceeds half the dislocation separation $[\ln(h/b)+1]$ should be replaced by $[1-\ln 2(f-\varepsilon^*)]$. If forces in addition to F_ε, F_1, F_s, and F_γ are significant, they should be included in the estimates of h_c and ε^*. Examples of forces that might need to be included are the forces between gliding dislocations, the forces between gliding and interfacial dislocations, the force between gliding dislocations and obstacles such as stacking faults, twins, and grain boundaries, and the force needed to move dislocations by glide. The last one is particularly important in semiconducting crystals with the diamond or sphalerite structure. The force needed to move dislocations by glide through semiconductors with these structures has been discussed by Haasen (1957), Celli *et al.* (1963), Gilman (1965, 1968), and many others (Haasen and Alexander, 1968). Haasen assumes that there is a microcrack coincident with the dislocation core and that dislocation motion proceeds by diffusion of the core. On this model the friction force opposing the motion of a threading dislocation is approximately

$$F_f = \frac{h}{\cos\phi}\frac{vkT}{bD_o}\exp\frac{U}{kT} \tag{51}$$

where v is the dislocation velocity and $D_o \exp(-U/kT)$ the diffusion coefficient of the core. The relation between v and $\dot{\delta}$, the rate of change of the misfit accommodated by dislocations, is

$$v = 2\dot{\delta}/\rho b \cos\lambda \tag{52}$$

where ρ is the density of threading dislocations that migrate to generate the misfit-dislocation line. If Eq. (52) is used to eliminate v from Eq. (51) and F_ε is equated to the sum of the foreces that oppose the glide of threading dislocations, a differential equation is obtained. Its solution describes the conversion of ε to δ as a function of time t:

$$\delta_t = \delta_\infty(1-\exp-\beta t) \tag{53}$$

where

$$\beta = \frac{Bb^3\rho\cos\phi\cos^2\lambda}{2}\frac{D_o}{kT}\exp\left(\frac{-U}{kT}\right)$$

with δ_∞ the misfit accommodated by dislocations at $t=\infty$. δ_∞ is equal to $f-\varepsilon^*$, and ε^* given by Eq. (50). The predictions of Eq. (53) are as follows. If $\beta t \ll 1$, then $\delta_t \approx 0$, and all misfit is accommodated by elastic strain. If $\beta t \gg 1$, then $\delta_t = \delta_\infty$, and the misfit strain is given by Eq. (50).

A comparison between the predictions of Eq. (53) and some experimental results obtained for deposits of germanium on gallium arsenide is made in

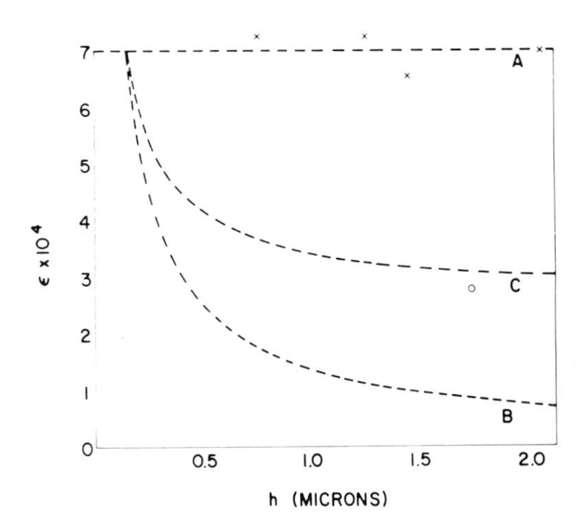

FIG. 17. A summary of predictions and experimental results for deposits of germanium on gallium arsenide (reprinted from Matthews *et al.*, 1970, by courtesy of the Journal of Applied Physics). Curve A: $\beta t \ll 1$; curve B: $\beta t \gg 1$; curve C: $\beta t = 1$.

Fig. 17. The germanium films were grown at 350°C and cooled to room temperature shortly thereafter. Experimental measurements of the strain in these specimens are shown by crosses. One sample was annealed at 600°C for 30 min after the growth of the germanium film. A measurement made on this sample is shown by the circle in Fig. 17.

The predictions of Eq. (53) for (110) deposits of germanium on gallium arsenide are shown in Fig. 17. Curve A is for films in which $\beta t \ll 1$, curve B for films in which $\beta t \gg 1$, and curve C for films in which $\beta t = 1$. The calculated value of βt for the as-grown germanium films is 10^{-3}. One would therefore expect experimental measurements made on the as-grown films to be near A. Figure 17 shows that this is indeed so. The value of βt for the specimen annealed at 600°C is ~ 4. An experimental measurement made on this sample would be expected to lie close to but above B. The position of the circle in Fig. 17 confirms this.

So far in this section the specimens have been assumed to consist of uniformly thick films on planar substrates. These are, however, not the only specimens in which misfit dislocations are made by glide on planes inclined to the interface. Dislocations formed by glide on inclined slip planes have been observed in thin deposits of iron on copper. These deposits consisted of isolated three-dimensional islands of fcc or γ-iron. A micrograph of some of the islands and the misfit dislocations beneath them is present in Fig. 18. Figure 19 shows how the dislocations are believed to have been formed.

FIG. 18. Micrograph of three-dimensional islands of γ-iron on a (001) copper substrate. The straight lines visible inside the images of the islands are misfit dislocations formed by glide on {111} planes inclined to the interface ($\times$ 125,000).

Figure 19a shows a dislocation with Burgers vector inclined to the interface and a line that extends from one surface of the specimen to the other. Glide forces act on this dislocation as a result of the misfit stresses present in the island and substrate. The forces increase as island size increases and eventually overcome the line tension in the threading dislocation. At this stage the

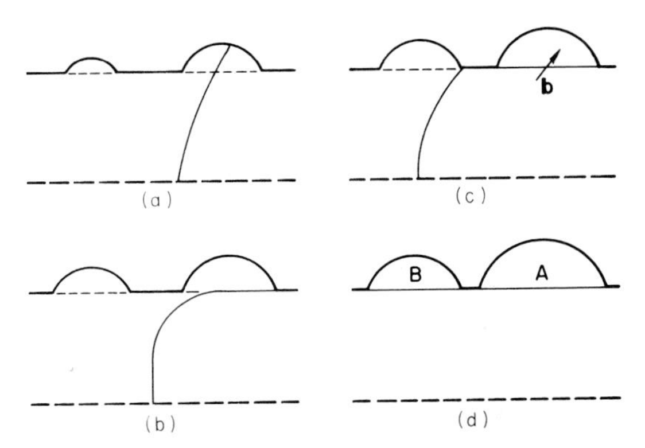

FIG. 19. Stages in the generation of misfit dislocations under islands of γ-iron on copper (see Fig. 18 and text discussion) (reprinted from Jesser and Matthews, 1968, by courtesy of the Philosophical Magazine).

lengths of dislocation in island and substrate move apart to make a length of misfit dislocation. The motion of the portion in the island ceases when it reaches the edge of the island. However, the dislocation in the substrate moves to the island edge and then rapidly crosses the gap to the next island. Passage across the gap is rapid because the glide force due to the misfit stress is not opposed by the tension in the misfit dislocation. If the next island is large enough, the dislocation in the substrate moves under it. This process accounts for the fact that misfit dislocations under neighboring islands frequently lie along the same straight line (Jesser and Matthews 1968b).

Dislocations formed by glide sometimes react with one another to form dislocation lines that accommodate misfit more efficiently. These reactions seem to be particularly prevalent in systems where the interface between the crystals is not sharp. They have been studied in semiconductors doped by diffusion by Washburn *et al.* (1964) and by Abrahams *et al.* (1969).

D. CLIMB OF THREADING DISLOCATIONS

The observations of Yagi *et al.* (1971) and Cherns and Stowell (1973) have demonstrated that misfit dislocations may be made by climb to the interface as well as by glide. The thickness at which climb will begin can be estimated in the same way as the critical thickness for glide was determined in Section C. If we assume that misfit dislocations made by climb are in edge orientation, and that the normal to the climb plane makes an angle ϕ_c with the specimen plane, then the climb force exerted by the elastic strain is

$$F_\varepsilon = Bbhf \cos \phi_c \tag{54}$$

The line tension in the misfit dislocation is

$$F_l = Db[\ln(h/b)+1] \tag{55}$$

If ϕ_c is fairly small, then the tension in the strip of surface created or eliminated by the climbing dislocation is

$$F_s = \pm\sigma_o b \tag{56}$$

The positive sign is to be used if the misfit strain is tensile, and the negative sign if it is compressive. F_γ is given by Eq. (48).

If the driving force is equated to the sum of the forces that oppose climb, then

$$h_c = \frac{D[\ln(h/b)+1] \pm \sigma_o}{Bf - \gamma_o/b \cos\phi_c} \tag{57}$$

and the optimum elastic strain after coherence is lost is

$$\varepsilon^* = \frac{1}{Bh\cos\phi_c}\left[D\left(\ln\frac{h}{b}+1\right) + \frac{\gamma_o h}{b\cos\phi} \pm \sigma_o\right] \tag{58}$$

E. Nucleation of Misfit Dislocations in Thin Films

If there are insufficient threading dislocations to relieve elastic misfit strain by climb or glide, or if dislocation motion is prevented by obstacles such as grain boundaries and stacking faults, it is possible for new dislocations to be created. These new dislocations elongate by climb or glide to make lengths of misfit-dislocation line. Two forms of this process are illustrated in Fig. 20. Figure 20a shows the nucleation of a complete dislocation and Fig. 20b the nucleation of a partial. The nucleation of dislocations has been considered

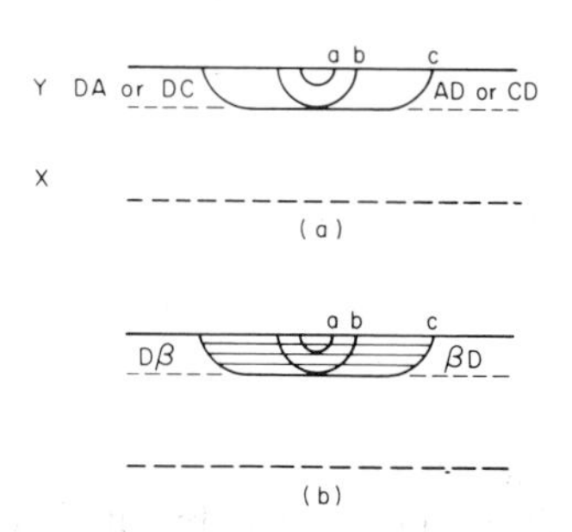

Fig. 20. Generation of misfit dislocations by the nucleation and growth of climb or glide loops. The loops in (a) are complete and those in (b) are partial. If we assume that the film thickness is h_c, then the loops labeled a are below critical size, the loops labeled b are critical, and those labeled c are stable.

by Frank (1950) and Hirth (1963). The calculation given here resembles Hirth's. The energy of a half-loop made by glide is approximately (Nabarro, 1967)

$$\pi R \cdot \frac{Gb^2}{4\pi}\left\{\frac{1}{2}\left(1+\frac{1}{1-\nu}\right)\right\}\left(\ln\frac{8R}{e^2 b}+1\right) \tag{59}$$

where R is the loop radius. The elastic energy released by the loop is

$$\frac{\pi R^2}{2}\cdot b \cdot \frac{2G(1+\nu)\,\varepsilon}{(1-\nu)}\cos\lambda\,\cos\phi \tag{60}$$

If the loop is imperfect, there is a stacking fault associated with it whose energy is

$$\pi R^2 \gamma/2 \tag{61}$$

The energy of the surface step created by the loop is

$$\pm 2R\sigma_o b \sin\alpha \tag{62}$$

The positive sign is to be used when new surface atoms are exposed by the loop and the negative one when surface atoms are removed or covered up. The sum of (59)–(62) is zero when $R = 0$, rises to a maximum value, the activation energy needed for loop nucleation, and then decreases. The radius at which the energy is a maximum is

$$R_c = \left[\frac{G_o b^2}{8\pi}(2-\nu)\left(\ln\frac{8R}{e^2 b}+2\right) \pm 2\sigma_o(1-\nu)\, b \sin\alpha\right] \times \left[2G_o(1+\nu)\,\varepsilon b \cos\lambda\,\cos\phi - \gamma(1-\nu)\right]^{-1} \tag{63}$$

A similar expression is obtained for loops formed by climb. However, it is worth emphasizing that the expression [analogous to Eq. (63)] for climb loops applies to loops generated in an equilibrium point defect population. Nonequilibrium point defect concentrations may raise or lower the stresses needed for the nucleation of climb loops.

If the film thickness is such that a half-loop of radius R_c cannot fit into the film, then loop nucleation will not occur, and the film will remain coherent. This size barrier to nucleation provides an explanation for the fact that very thin films are able to sustain larger elastic stresses than dislocation-free whiskers. A half-loop can fit into a film if $R \leqslant h/\cos\phi$. Thus, the critical thickness for coherence loss by dislocation nucleation is

$$h_c = R_c \cos\phi \tag{64}$$

This equation predicts critical thicknesses fairly close to those given by Eqs. (49) or (57).

The condition that film thickness exceed h_c is a necessary condition for loop nucleation but not a sufficient one. If film thickness exceeds h_c, and the activation energy needed to create a loop of critical size is not available, then loop nucleation will not take place. If we assume that the activation energy available does not exceed 50 kT, then the value of f needed for nucleation to proceed at temperatures close to room temperature is greater than 1 or 2%. Evidence that misfit strains greater than 1 or 2% do result in the nucleation of climb loops has been obtained by Cherns and Stowell (1973).

F. Generation of Misfit Dislocations by a Climblike Process

Dislocations may elongate during the growth of one crystal on another by a process that bears a superficial resemblance to climb but does not involve the diffusion of point defects to or from a dislocation line. The mechanism is illustrated in Fig. 21. Figure 21a shows a portion of a dislocation loop with Burgers

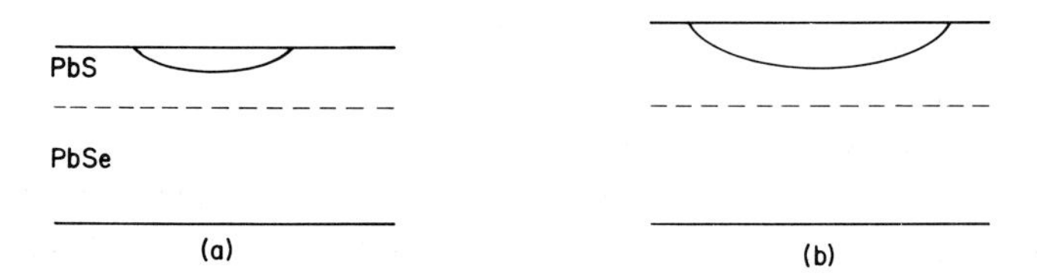

Fig. 21. Stages in the elongation of a misfit dislocation during the growth of one crystal on another. The Burgers vector of the dislocation is parallel to the interface plane (reprinted from Matthews, 1971, by courtesy of the Philosophical Magazine). (a) Dislocation loop; (b) same loop after more deposit material was added.

vector parallel to the interface plane. Figure 21b shows this loop after the addition of more deposit material: the length of dislocation line in the deposit has increased. If misfit dislocations are made by the process in Fig. 21, and observations made on deposits of lead sulfide on lead selenide suggest that they may be (Matthews, 1971), then parts of the misfit dislocation lines do not lie in the interface.

G. Dissociation of Frank Partials

Cherns and Stowell (1973) have found that most of the misfit dislocations in (001) deposits of palladium on gold are Frank partial dislocations made by climb (see Section IV, C). These partial dislocations can dissociate to form a Shockley partial and a complete edge dislocation. The edge dislocation has a Burgers vector parallel to the (001) interface plane and is thus an ideal misfit dislocation. The reaction is

$$\tfrac{1}{3}a[111] \rightarrow \tfrac{1}{6}a[\bar{1}\bar{1}2] + \tfrac{1}{2}a[110]$$

The Shockley partial moves to the surface of the film over the original climb plane and removes the stacking fault while it does so. The reaction is promoted by the stacking fault and by the misfit strain.

VI. Misfit Dislocations and the Alignment of Deposits

A deposit elastically strained to match its substrate is perfectly aligned. However, when misfit dislocations are made, this perfection is usually destroyed. Two mechanisms for this were considered in Sections II, B, 1 and II, B, 2. In this section experimental support for the mechanisms in Section II, B will be presented. In addition, two other processes that lead to misalignment will be described. All the mechanisms taken together suggest a simple rule of thumb (Matthews, 1972b): *the relaxation of elastic misfit strain leads to a variation in the orientation of crystal planes (in radians) that is approximately equal to the misfit* f.

A. Islands Strongly Bonded to Their Substrate

If copper is deposited onto a clean nickel surface a specimen with geometry similar to that considered in Section II, B, 1 is obtained (Matthews 1972a). Small copper islands are generated early in deposit growth and are strained to match the nickel substrate. Larger islands are separated from the substrate by misfit dislocations. The dislocations are in edge orientation, have Burgers vectors in the interface, and are arranged in a square net. An electron micrograph of these dislocations is seen in Fig. 22.

To observe moiré fringe patterns misaligned by the angles predicted in Section II, B, 1, it is convenient to choose reflecting planes that are perpendicular to the film plane and inclined at 45° to the Burgers vectors of the misfit dislocations. Reflecting planes that satisfy this requirement in deposits of copper on nickel are (200) and (020). A micrograph in which type 200 moiré fringes are present is shown in Fig. 23. In areas where the copper is perfectly aligned and strained equally in all directions in the interface the fringes lie along [010]. This is approximately true in the lower portion of the figure, where the copper completely covered the nickel surface. Large deviations from the ideal [010] direction are noticeable near the center of the figure. The fringes in the islands labeled A are inclined at 45° to [010]. Fringes misaligned by this angle are expected if the islands are strained to match the substrate along one of the [110] directions in the interface but not along the other. The suggestion that the islands labeled A are strained in this way is consistent with their size. It is also consistent with the absence of features that could be interpreted as misfit dislocations with lines along $[1\bar{1}0]$.

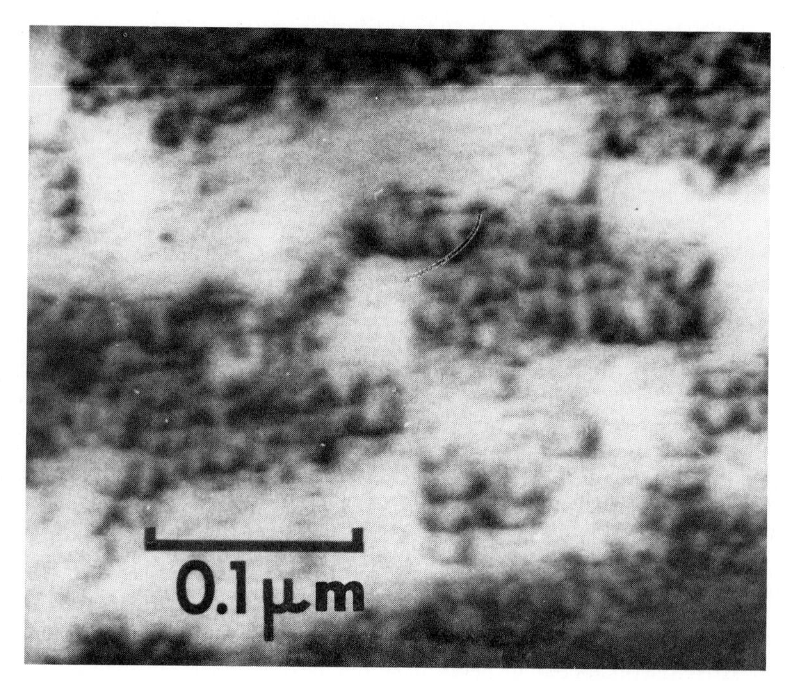

FIG. 22. An electron micrograph of misfit dislocations under islands of copper on nickel (reprinted from Matthews, 1971, by courtesy of Surface Science). The dislocations are arranged in an approximately square array and have Burgers vectors parallel to the $\langle 110 \rangle$ directions in the (001) specimen plane.

The misalignment of moiré fringes predicted in Section II, B, 1 and revealed by Fig. 23 is not the result of a true rotation. The unit cell of the copper in the island is distorted, and a rotation of the reflecting planes involved in fringe formation is associated with this distortion. These planes are indicated by dashed lines in Fig. 24. Their rotation during the creation of misfit dislocations is

$$\phi = \pm b/2X \tag{65}$$

If we assume (Vincent, 1969) that the first misfit dislocation is made when it cancels the misfit strain along its Burgers vector, then

$$X = b/f \tag{66}$$

and the variation in the orientation of the dotted planes is

$$2\phi = f \tag{67}$$

The variation in the alignment predicted by Eq. (67) exists so long as there is a single dislocation under each island. A second dislocation with line perpendicular to the first removes the rotation. A third introduces a new rotation

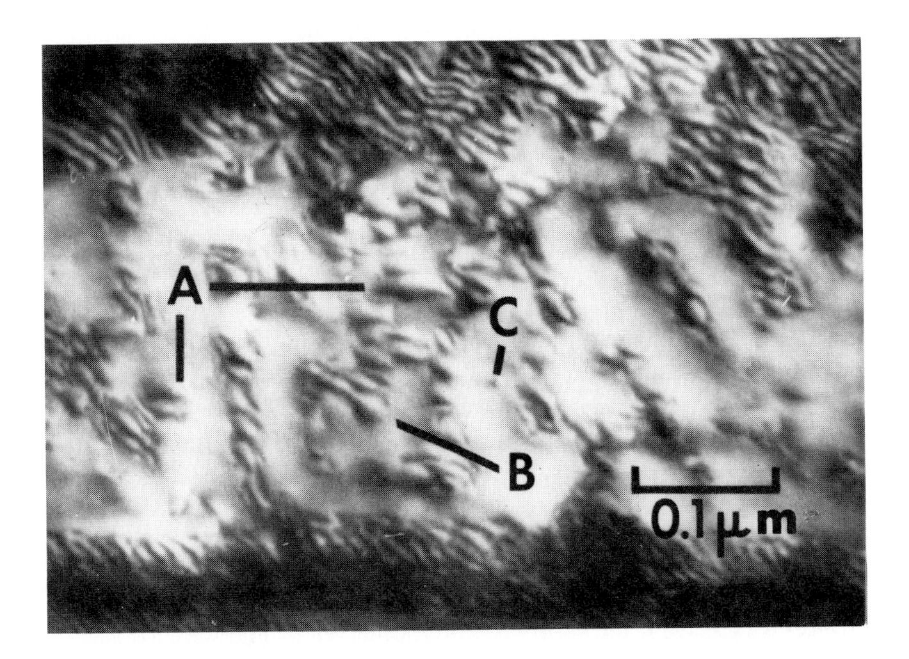

FIG. 23. Micrograph of moiré fringe patterns formed by 200 reflections in a deposit of copper on nickel (reprinted from Matthews, 1972, by courtesy of Surface Science).

half as large as that associated with the first. A fourth removes the rotation once more. Thus the misalignment of the dotted planes is zero when the island fits its substrate or when the number of misfit dislocations under the island is even. If the number of dislocations is $m+n$ and $m=n+1$, then the misalignment of the dotted planes is

$$\phi = \pm f/2(n+1) \tag{68}$$

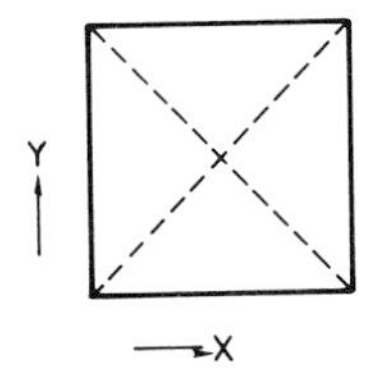

FIG. 24. A small island of copper on nickel. The dashed lines within the island represent 200 and 020 reflecting planes.

B. ISLANDS WEAKLY BOUND TO THEIR SUBSTRATE

The predictions of Section II, B, 2 suggest that the upper limit of $\phi_{\max}$ is $\sim \pm f$ and so agree with the rule of thumb given at the start of this section.

The agreement between the experimental results (Bassett, 1960; Matthews, 1960) and Section II, B, 2 is also good. A comparison of the experimental results for silver on mica, with predictions of Section II, B, 2 is made in Fig. 25. The hatched areas in Fig. 25 show experimental values for the average misalignment of islands that were obtained many years ago (Matthews, 1960). The values of $\phi_{\max}$ and $\bar{\phi}$ predicted by Eqs. (34), (39), and (42) are shown by continuous lines. The agreement between the predicted and experimented values of $\bar{\phi}$ is clearly very good. The predictions were made assuming that Burgers vectors of the misfit dislocations—measured in the silver lattice—were of type $\frac{1}{6}a\langle 112\rangle$ and lay in the interface plane.

The rotations observed by Bassett (1960) in deposits of silver on mica are a little difficult to compare with predictions. This is because the radii of the rotating islands were not listed. However, measurements of rotation and radius made on the micrographs that Bassett has published agree with the predicted values for $\phi_{\max}$.

There is one rather interesting feature of Bassett's micrographs that has received little attention so far. It is that each of the abrupt rotations revealed by his pictures was accompanied by an increase in the number of moiré fringes present in the island. It is easy to show that an increase in the number of fringes is expected if the abrupt rotations observed by him were associated with the creation of new misfit dislocations.

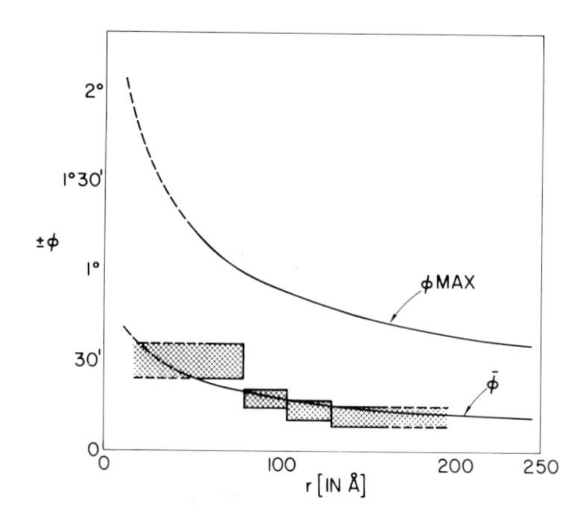

FIG. 25. A comparison of predictions of the misalignment of weakly bound islands with experimental results obtained on deposits of silver on mica. The shaded areas are experimental measurements of the average misalignment of the silver islands (Matthews, 1960). The solid line labeled $\bar{\phi}$ gives the average misalignment predicted on the assumption that the misfit dislocations in the silver–mica system have Burgers vectors of type $\frac{1}{6}a\langle 112\rangle$ (measured in the silver lattice).

C. Misalignment of {001} Planes in Deposits of Lead Sulfide on Lead Selenide

Most of the misfit dislocations in (001) films of lead sulfide on lead selenide are made at the base of L-shaped cracks as described in Section V, B. A particularly common combination of L-shaped cracks is illustrated in Fig. 26.

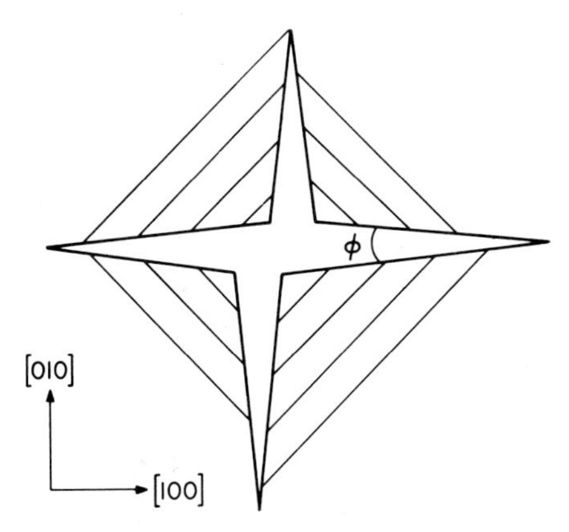

Fig. 26. A common configuration of cracks in (001) deposits of lead sulfide on lead selenide.

If the misfit dislocations in this figure have spacings slightly smaller than those expected when all misfit is accommodated by dislocations (see Section V, B), then the angle between the faces of the cracks is slightly larger than f (Matthews, 1971, 1972b).

D. Misfit Dislocations Generated by Glide on Planes Inclined to the Interface

Misfit dislocations are often generated by glide on slip planes inclined to the interface. This process results in lattice rotation for the same reason that the plastic deformation of large single crystals is accompanied by lattice rotation (Schmid and Boas, 1950). The rotation depends in a rather complicated way on the orientation of the sample, the slip systems that are activated, and the magnitude of the deformation. However, in thin films that are coherently strained at the start of their growth, the rule given earlier is often a good guide. To illustrate this we consider the misalignment associated with the glide of complete dislocations into an (001) interface between a pair of fcc crystals. The misfit accommodated by these dislocations is half that accommodated by ideal misfit dislocations with Burgers vectors in the interface. The separation of misfit dislocations made by glide is therefore

$$S = b/2(f-\varepsilon) \tag{69}$$

The screw component is

$$b_s = b/2 \tag{70}$$

This component causes a rotation about the normal to the interface. The maximum value of the rotation is

$$\phi_{max} = \pm b/s = \pm(f - \varepsilon) \tag{71}$$

If ε is only a small fraction of f, which is true when the film thickness exceeds b/f, then

$$\phi_{max} \approx \pm f \tag{72}$$

Thus, in the example we have chosen, the rule of thumb given at the beginning of this section is obeyed to within a factor of two. However, it is worth noting that there are instances where the rule is not obeyed. If misfit dislocations made by glide are in edge orientation, then $b_s = 0$ and the rotation about the normal to the interface is zero.

VII. Behavior of Misfit Dislocations during Diffusion

If a pair of miscible crystals, separated by a grid of misfit dislocation, is allowed to interdiffuse, the dislocations move to become distributed in the alloyed volume (Matthews, 1963, 1964). Some of the dislocations move away from the interface in one direction and some move away in the other. If the dislocations are in edge orientation and have Burgers vectors parallel to the interface plane, then the motion is by climb. If their Burgers vectors are inclined to this plane, then the motion is a mixture of climb and glide. The forces responsible for dislocation motion can be divided into four parts: the force exerted by the diffusion-induced stress field, the force due to the misfit strain, the force due to a nonequilibrium vacancy concentration (associated with the Kirkendall effect), and the force exerted by the dislocations on one another.

A. Force due to the Diffusion-Induced Stress

This force has been considered by Hirth (1964), Vermaak and van der Merwe (1964, 1965), and Geguzin *et al.* (1971). Consider two crystals A and B that are so thick that all the misfit between their stress-free lattice parameters is accommodated by dislocations. Suppose that the misfit is f along the x- and y-directions in the interface and that misfit along x and y is accommodated by edge dislocations with Burgers vectors parallel to x and y. If the crystals are completely miscible and the diffusion of A into B proceeds at the same rate as the diffusion of B into A, then the diffusion profile is given by

$$C_{a,b} = \tfrac{1}{2}\{1 \mp \operatorname{erf}[z/2(Dt)^{1/2}]\} \tag{73}$$

where C_a and C_b are the concentrations of A and B, and z is zero at the initial AB interface and positive in B. If the lattice constant of the alloy varies linearly with composition, then the variation of lattice constant along z is

$$\zeta = \tfrac{1}{2}(a_a + a_b) + \tfrac{1}{2}(a_b - a_a) \operatorname{erf}[z/2(Dt)^{1/2}] \tag{74}$$

To simplify the calculations of the force acting on the misfit dislocations we will follow Vermaak and van der Merwe (1964, 1965) and assume that the dislocations behave in a very orderly fashion. Adjacent parallel dislocations are considered to move away from the AB interface in opposite directions, as shown in Fig. 27, to form two planar, square arrays with mesh size $2b/f$.

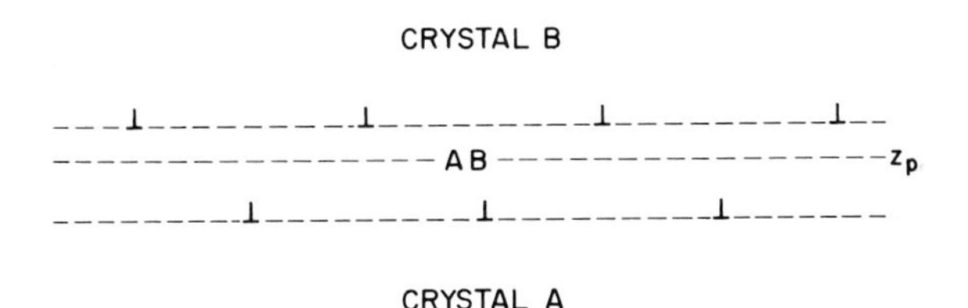

FIG. 27. Misfit dislocations between two miscible crystals A and B after some interdiffusion has occurred.

The assumption that dislocations divide into a pair of arrays that move in opposite directions is realistic in the early stages of interdiffusion (i.e., when $2(Dt)^{1/2} \lesssim b/f$). Later, however, the two arrays split up into planar subarrays (Vermaak and van der Merwe, 1964, 1965).

Experimental evidence for planar arrays of misfit dislocations in diffusion zones has been obtained by Washburn *et al.* (1964) and Abrahams *et al.* (1969). Particularly striking evidence for planarity is provided by some stereomicrographs that Mader (1972) has obtained.

The lattice parameter of the region that lies between the two dislocation arrays is

$$d = (a_a + a_b)/2 \tag{75}$$

The elastic strain along z in the region AB between the two arrays is therefore

$$\varepsilon_{ab} = (\zeta - d)/\zeta \tag{76}$$

The elastic strains in the regions A and B outside the arrays are

$$\varepsilon_a = (\zeta - a_a)/\zeta \qquad \text{and} \qquad \varepsilon_b = (\zeta - a_b)/\zeta \tag{77}$$

If the array between AB and B moves a distance dz, the change in elastic energy is

$$B(\varepsilon_{ab}^2 - \varepsilon_b{}^2)\, dz \tag{78}$$

There is thus a force per unit area of the array between AB and B, which is

$$F_{AB,B} = -B(\varepsilon_{ab}^2 - \varepsilon_b{}^2) \tag{79}$$

A similar expression holds for the force $F_{AB,A}$ on the array between AB and A. $F_{AB,B}$ and $F_{AB,A}$ are zero when

$$\varepsilon_{ab} = -\varepsilon_b = -\varepsilon_a \tag{80}$$

Thus the equilibrium positions of the arrays are defined by

$$\zeta = (a_a + 3a_b)/4 \qquad \text{and} \qquad \zeta = (3a_a + a_b)/4 \tag{81}$$

Substitution of these values into Eq. (74) shows that the equilibrium positions of the arrays are defined by

$$\pm\tfrac{1}{2} = \operatorname{erf}[z/2(Dt)^{1/2}] \tag{82}$$

Tabulated values of the error function show that the equilibrium positions of the arrays are

$$z_{eq} = \pm 0.944(Dt)^{1/2} \tag{83}$$

Thus the equilibrium separation of the arrays is approximately

$$2(Dt)^{1/2} \tag{84}$$

B. Force Associated with Misfit Strain

Coherency or misfit strains in thin epitaxial deposits can have rather striking and, at first sight, surprising effects on the behavior of misfit dislocations during diffusion. Consider a thin, but uniformly thick, deposit on a substrate of infinite thickness and suppose to begin with that the film is elastically strained to match its substrate exactly. In this circumstance the strain in film is f and the strain in the substrate is zero. If the specimen is heat treated and interdiffusion takes place, the concentration of overgrowth atoms as a function of z is

$$C_b = \frac{1}{2}\left[\operatorname{erf}\frac{z}{2(Dt)^{1/2}} + \operatorname{erf}\frac{2h-z}{2(Dt)^{1/2}}\right] \tag{85}$$

If the lattice parameters of AB alloys vary linearly with compostion, and interdiffusion is not accompanied by the generation of misfit dislocations, the misfit strain as a function of z is

$$\varepsilon = \frac{f}{2}\left[\operatorname{erf}\frac{z}{2(Dt)^{1/2}} + \operatorname{erf}\frac{2h-z}{2(Dt)^{1/2}}\right] \tag{86}$$

Now suppose a planar network of misfit dislocations is constructed in the crystal. If the dislocations are in edge orientation, have Burgers vector in the interface, and form a square network with mesh size S, then the misfit accommodated by dislocations is

$$\delta = b/S \tag{87}$$

The elastic misfit strain above the grid of dislocations is $\varepsilon - \delta$, and below it it is ε. If the grid moves upward a distance dz, then the energy change during this motion is

$$B[\varepsilon^2 - (\varepsilon - \delta)^2]\, dz \tag{88}$$

The force pushing the network toward the free surface of the film is therefore

$$B(\delta^2 - 2\varepsilon\delta) \tag{89}$$

This force is zero when $\delta = 2\varepsilon$. Thus the equilibrium position of the network is defined by

$$\delta = f\left[\operatorname{erf}\frac{+z}{2(Dt)^{1/2}} + \operatorname{erf}\frac{2h - z}{2(Dt)^{1/2}}\right] \tag{90}$$

The predictions of this equation are a little unexpected. If a thin film is grown on a thick substrate and $\delta < f$, then the force exerted on the network of misfit dislocations is such as to drive the dislocation from the interface into the *substrate*. If diffusion continues further, the force declines to zero, changes sign, and thereafter pushes the network back toward the free surface of the film. One of the results of this in-and-out motion is that misfit dislocations will sometimes remain inside a sample for an unexpectedly long time. A planar array of dislocations escapes from the surface when

$$2\varepsilon_s = \delta \tag{91}$$

where ε_s is the elastic strain that would exist at the surface of the film if no misfit dislocations were present in the sample. It is given by

$$\varepsilon_s = f \operatorname{erf}[h/2(Dt)^{1/2}] \tag{92}$$

A graph of $2\varepsilon_s$, or δ, against $2(Dt)^{1/2}$ (in units of h) is shown in Fig. 28. It predicts that an array that accommodates one-eighth of the misfit remains inside the sample until $2(Dt)^{1/2}$ is about $18h$.

The largest value of δ shown in Fig. 28 is $\sim f/2$. This limit has been set because the assumption that a grid of misfit dislocation remains planar is unrealistic when δ is much larger than $f/2$. If the misfit accommodated by dislocations is larger than $f/2$, the network divides itself into subnetworks resembling those described in Section VII, A.

It is worth noting that the calculations described in this section contain at least one serious omission. An array of misfit dislocations between a film and substrate is attracted to the free surface of the film. This attractive force has been neglected here. If it were included, it would reduce the distance that dislocations move into the substrate and the time needed for them to escape at the free surface.

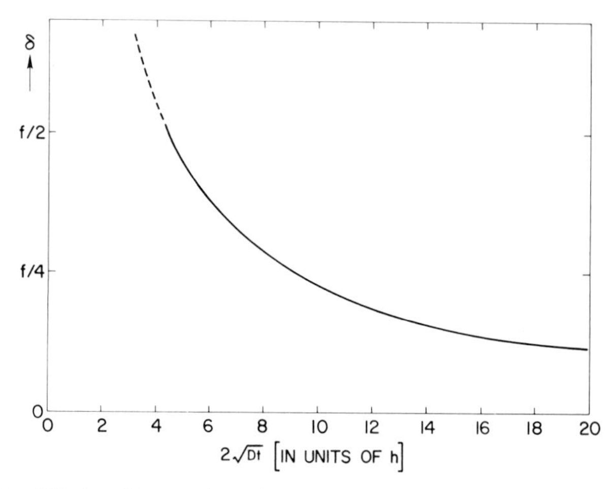

FIG. 28. The diffusion distance (in units of h) that is needed before a misfit dislocation network of strength δ will escape at the surface of a thin film grown on a substrate of infinite thickness. The film and substrate are miscible in all proportions, and the network of misfit dislocations remains planar throughout the diffusion process.

C. Force Associated with the Kirkendall Effect

In many bicrystals diffusion takes place by vacancy migration. If a pair of crystals A and B interdiffuse by this mechanism, then the diffusion of A into B usually proceeds at a rate significantly different from the diffusion of B into A. This leads to the Kirkendall effect, and a vacancy flux is associated with it in the direction opposite to that taken by the faster diffusing species. This vacancy flux gives rise to a paucity of vacancies on one side of the interface and to an excess on the other. These nonequilibrium vacancy concentrations exert a climb force on the misfit dislocations. If the large atoms diffuse more rapidly than the small, then this force opposes climb away from the original interface. If the small atoms diffuse more rapidly than the large, the force promotes climb. The influence of the vacancy flux associated with the Kirkendall effect on the climb of misfit dislocations was first discussed by Hirth (1964).

D. Force between Dislocations

The force between misfit dislocations influences the geometry of the arrays of climbing dislocations. Its effects have been discussed by Matthews and

Crawford (1965), Vermaak and van der Merwe (1964, 1965), Winchell *et al.* (1971), and Ayres and Winchell (1972). Their conclusions resemble one another fairly closely.

In the early stages of climb the forces between arrays like those in Fig. 28 are such as to keep them staggered. The dislocations in the upper array lie over the gaps between the dislocations in the lower. This configuration remains the stable one until the line joining a dislocation in one array to its nearest parallel neighbor in the other makes an angle of more than 45° to the interface plane. Thereafter, any small displacement from the staggered configuration brings the arrays into a region where they experience a glide force that tends to increase the displacement. This causes the arrays to move laterally with respect to one another until the dislocations in them are arranged vertically above one another.

As diffusion proceeds the two arrays break up to form subarrays (Vermaak and van der Merwe, 1964, 1965). This is accompanied by the arrangement of more and more dislocations into walls perpendicular to the plane of the original interface. These walls give rise, in specimens composed of thin films, to small grains like the one in Fig. 29. The lattice within the grain in this figure is curved to accommodate the misfit between the upper and lower surface of the specimen.

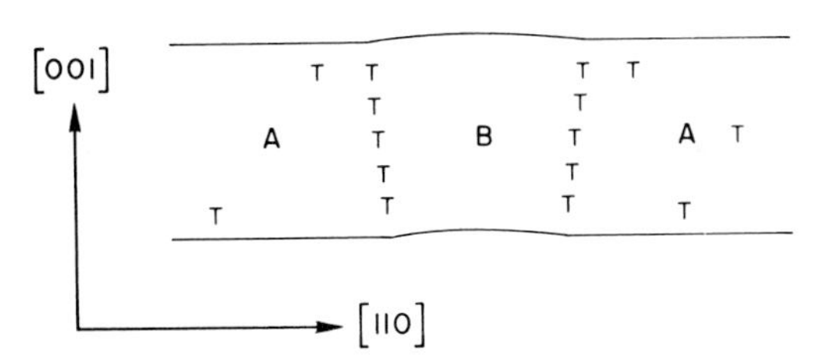

FIG. 29. A small curved grain made by the polygonization of misfit dislocations.

If a grain like the one in Fig. 29 is formed, and the lattice parameter of the bicrystal varies almost linearly from one surface of the specimen to the other, then the grain grows laterally to form a large curved grain like the one seen in Fig. 30.

The growth of a curved grain does three things. It removes misfit dislocations from the region occupied by the curved grain, it makes a tilt boundary, and it increases the area of the exposed surface of the film.

The energy change brought about by the removal of misfit dislocations from a grain whose radius (measured in the film plane) is r is

$$-\pi r^2 h_t \rho_m (Db/2) \ln(\rho_m^{-1/2}/b) \tag{93}$$

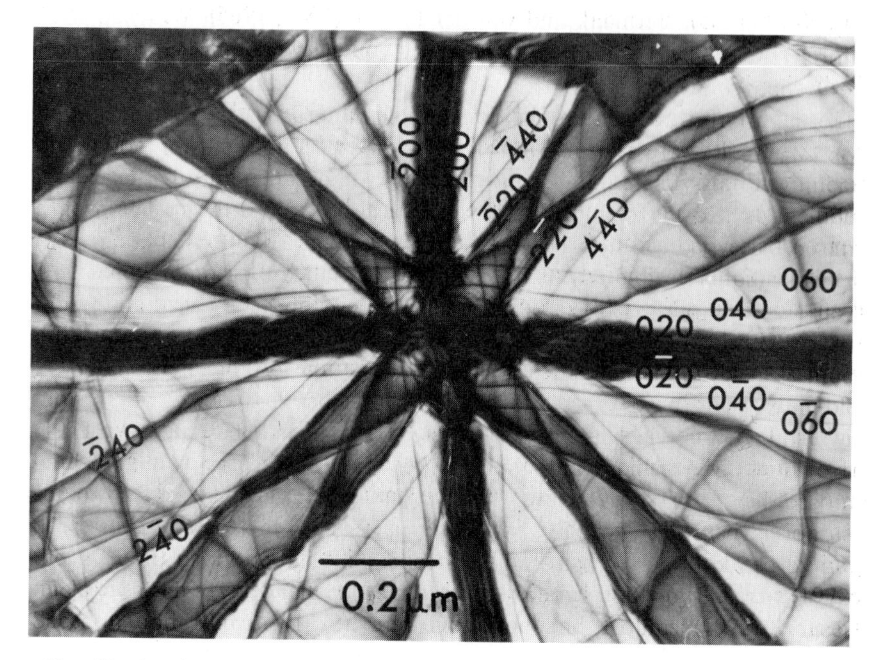

FIG. 30. An electron micrograph of a curved grain formed during diffusion between nickel and palladium films oriented with (001) parallel to the plane.

where ρ_m is the length of misfit dislocation line per unit volume and h_t the total thickness of the sample. If misfit is accommodated by two perpendicular sets of edge dislocations with Burgers vectors in the interface, then

$$\rho_m = 2f_g/bh_t \tag{94}$$

where f_g is the misfit between the lattice parameters of the upper and lower surfaces of the specimen. The energy of the grain boundary formed during the growth of the grain is

$$2\pi rh \cdot \tfrac{1}{2}D \cdot \alpha_g(\tfrac{1}{2} - \ln \alpha_g) \tag{95}$$

where α_g is the tilt across the boundary and depends on r:

$$\alpha_g = rf_g/h_t \tag{96}$$

The increase in the surface energy is

$$\pi(\sigma_o + \sigma_s)\,r\left(\frac{1-\cos\alpha_g}{\sin\alpha_g}\right) \tag{97}$$

The sum of (93), (95), and (97) indicates that the energy of the system decreases during the early life of a curved grain. This decrease proceeds until r reaches an optimum value r_{opt} and then increases, with r_{opt} given approximately by

$$r_{opt} = h_t[2D/(\sigma_o + \sigma_s) f_g]^{1/2} \tag{98}$$

At r_{opt} the generation of new surface proceeds rapidly enough to overcome the energy gained by placing misfit dislocations in a tilt boundary.

Experimental observations made so far are consistent with Eq. (98). The prediction that curved grains will grow out to an optimum radius agrees with observations made by Matthews and Jesser (1969), who examined nickel–palladium bicrystals in the hot stage of an electron microscope. Curved grains were found to grow rapidly out to a particular size and then stop growing. New grains were then nucleated immediately outside the old ones and grew to r_{opt}. This was followed by the nucleation of more grains outside the old ones. Eventually clusters of curved grains met one another, and the entire specimen was occupied by curved grains of uniform size.

A test of the prediction that r_{opt} is proportional to $h_t/(f_g)^{1/2}$ has been made in the following way. Four nickel–palladium samples identical in all respects except thickness were grown side by side in an ultrahigh vacuum system. They were then heat treated until about half their area was occupied by curved grains. r_{opt} and W_{hkl}, the separation of hkl and hkl bend contours, were then measured on each sample. f_g was obtained from W_{hkl}, using the following relationship:

$$f_g = \lambda_e h_g / d_{hkl} W_{hkl} \tag{99}$$

where λ_e is the electron wavelength and d_{hkl} the separation of the (hkl) planes. The results are shown graphically in Fig. 31 and confirm that r_{opt} is propor-

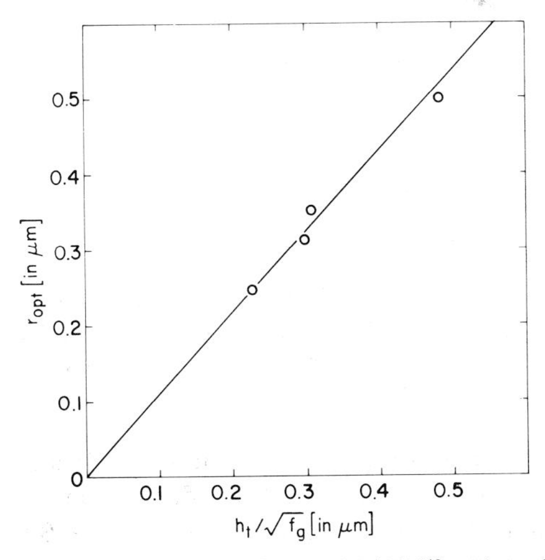

FIG. 31. Experimental measurements of r_{opt} and $h_t/(f_g)^{1/2}$ made on four nickel–palladium bicrystals.

tional to $h_t/(f_g)^{1/2}$. They also suggest that the average surface energy of the samples was $\sim$4000 erg/cm^2. This result is somewhat larger than expected. However, the discrepancy between the observed and expected values for surface energy is probably not too large. The probable errors in the thickness measurements are substantial. Also, there are rather large elastic stresses in and around the grains. The elastic energy associated with these stresses was neglected in the calculation of r_{opt}.

VIII. Role of Misfit Dislocations in Diffusion

In Section VII,A the motion of misfit dislocations during diffusion was discussed. In order to simplify this discussion the dislocations were assumed to move away from the interface in an orderly fashion. Adjacent dislocations were considered to move from the interface in opposite directions, and the direction taken by a particular dislocation was assumed to be the same over the entire length of the dislocation line. In fact, dislocation motion will be much less orderly than this. Adjacent dislocations will sometimes move in the same direction. Also, portions of particular dislocations will move away from the interface in one direction and other portions will move away in the other. This phenomenon gives rise to threading dislocations (Matthews, 1964; Washburn *et al.*, 1964; Abrahams *et al.*, 1969). A diagram showing the formation of threading dislocations is seen in Fig. 32. The hatched area in this

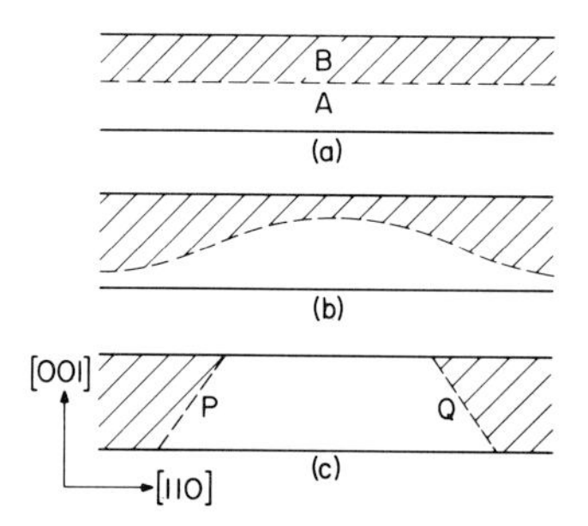

FIG. 32. Formation of threading dislocations by climb of a misfit dislocation (reprinted from Matthews and Crawford, 1965, by courtesy of the Philosophical Magazine). The misfit dislocation is represented by the dashed line. The hatched area is the additional sheet of atoms associated with the dislocation.

figure represents the additional plane of atoms associated with a misfit dislocation. It is clear that the threading dislocations retain their edge character throughout the climb process, and that the Burgers vectors of threading dislocations formed along a particular misfit dislocation alternate in sign.

The number of threading dislocations formed during diffusion is difficult to estimate precisely. However, it is easy to show that the number may be very large. The maximum number of threadings can be found by equating the force driving the arrays of misfit dislocations from the interface [Eq. (79)] to the line tension in the threading dislocations that tends to pull them back. The force driving the arrays away from the interface has a maximum value at $z = 0$ given by

$$\frac{2G(1+\nu)}{(1-\nu)} \cdot \frac{(a_a - a_b)^2}{(a_a + a_b)^2} \approx Gf^2 \tag{100}$$

If we assume that the line tension of the threading dislocations is $Gb^2/2$, then the density of threading dislocations has an upper limit equal to

$$\rho_{max} = 2f^2/b^2 \tag{101}$$

If $f = 0.02$ and $b = 2 \times 10^{-8}$ Å, then $\rho_{max} = 2 \times 10^{12}$ cm^{-2}.

Acknowledments

The author would like to thank S. Mader for helpful discussions, D. Cherns and M. J. Stowell for showing him some of their results prior to publication, and the Air Force Office of Scientific Research (AFSC) for financial support [Contract Number F44620-72-C-0061].

References

Aaronson, H. I., Laird, C., and Kinsman, K. R. (1970). Phase Transformations, p. 313. Amer. Soc. for Metals, Cleveland, Ohio.
Abrahams, M. S., Weisberg, L. R., Buiocchi, C. J., and Blank, J. (1969). *J. Mater. Sci.* **4**, 223.
Ayres, P. S., and Winchell, P. G. (1972) *J. Appl. Phys.* **43**, 816.
Ball, C. A. B. (1970). *Phys. Status Solidi* **42**, 357.
Ball, C. A. B., and van der Merwe, J. H. (1970). *Phys. Status Solidi* **38**, 335.
Bassett, G. A. (1960). *Proc. Eur. Reg. Conf. Electron Micros.* p. 270. de Nederlandse Vereniging voor Electronenmikroscopie, Delft.
Besser, P. J., Mee, J. E., Elkins, P. E., and Heinz, D. M. (1971). *Mater. Res. Bull.* **6**, 1111.
Bollmann, W. (1967). *Phil. Mag.* **17**, 363, 383.
Borie, B., Sparks, C. J., and Cathcart, J. V. (1962). *Acta Met.* **10**, 691.
Brooks, H. (1952). Metal Interfaces, p. 20. Amer. Soc. for Metals, Cleveland, Ohio.
Brown, L. M., and Ham, R. K. (1971). "Strengthening Mechanisms in Crystals" (A. Kelly and R. B. Nicholson, eds.), p. 12. Wiley, New York.
Cabrera, N. (1959). "Structure and Properties of Thin Films" (C. A. Nengebauer, J. B. Newkirk, and D. A. Vermilyea, eds.), p. 529. Wiley, New York.
Cabrera, N. (1965). *Mem. Sci. Rev. Met.* **62**, 205.
Cahn, J. W. (1962). *Acta Met.* **10**, 179.
Celli, V., Kabler, M., Ninomiya, T., and Thompson, R. (1963). *Phys. Rev.* **131**, 58.
Cherns, D. (1974). Ph.D. Dissertation, University of Cambridge, England.
Cherns, D., and Stowell, M. J. (1973). *Scripta Met.* **7**, 489.
du Plessis, J. C., and van der Merwe, J. H. (1965). *Phil. Mag.* **11**, 43.
Edmonds, T., McCarroll, J. J., and Pitkethly, R. C. (1971). *J. Vac. Sci. Technol.* **8**, 68.

Fedorenko, A. I., and Vincent, R. (1971). *Phil. Mag.* **24**, 55.
Fletcher, N. H. (1964). *J. Appl. Phys.* **35**, 234.
Fletcher, N. H. (1966). *Phil. Mag.* **16**, 159.
Fletcher, N. H., and Adamson, P. L. (1966). *Phil. Mag.* **14**, 99.
Frank, F. C. (1950). *Symp. Plast. Deformat. Cryst. Solids* p. 89. Carnegie Inst. of Technol. Pittsburgh, Pennsylvania.
Frank, F. C., and Nicholas, J. F. (1953). *Phil. Mag.* **44**, 1213.
Frank, F. C., and van der Merwe, J. H. (1949). *Proc. Roy. Soc.* **A198**, 205.
Friedel, J. (1964). "Dislocations," p. 184. Addison-Wesley, Reading, Massachusetts.
Geguzin, Ya. E., Boiko, Yu. I., and Strelkov, V. P. (1971). *Sov. Phys.-Solid State*, **12**, 1804
Gilman, J. J. (1965). *J. Appl. Phys.* **36**, 3195.
Gilman, J. J. (1968). *J. Appl. Phys.* **39**, 6086.
Gleiter, O., and Chalmers, B. (1972). *Progr. Mater. Sci.* **16**.
Gradmann, U. (1964). *Ann. Phys. Leipzig.* **13**, 213. 1966, ibid., **17**, 91.
Gradmann, U. (1966). *Ann. Phys. Leipzig* **17**, 91.
Haasen, P. (1957). *Acta Met.* **5**, 598.
Haasen, P., and Alexander, H. (1968). *Solid State Phys.* **22**, 27.
Hardy, H. K., and Heal, T. J. (1954). *Progr. Metal Phys.* **5**, 143.
Hilliard, J. E. (1970). Phase Transformations, p. 497. Amer. Soc. for Metals, Cleveland, Ohio.
Hirth, J. P. (1963). "Relation Between Structure and Strength in Metals and Alloys," p. 218. H.M. Stationery Office, London.
Hirth, J. P. (1964). "Single Crystal Films" (M. H. Francombe and H. Sato eds.), p. 173. Pergamon, Oxford.
Jesser, W. A., and Kuhlmann-Wilsdorf, D. (1967a). *Phys. Status Solidi* **19**, 95.
Jesser, W. A., and Kuhlmann-Wilsdorf, D. (1967b). *Phys. Status Solidi* **21**, 533.
Jesser, W. A., and Kuhlmann-Wilsdorf, D. (1968). *Acta Met.* **16**, 1325.
Jesser, W. A., and Matthews, J. W. (1967). *Phil. Mag.* **15**, 1097.
Jesser, W. A., and Matthews, J. W. (1968a). *Phil. Mag.* **17**, 461.
Jesser, W. A., and Matthews, J. W. (1968b). *Phil. Mag.* **17**, 595.
Jesser, W. A., and Matthews, J. W. (1968c). *Acta Met.* **16**, 1307.
Kelly, A. and Nicholson, R. B. (1963). *Progr. Mater. Sci.* **10**, 238.
Kuntze, R., Chambers, A., and Prutton, M. (1969). *Thin Solid Films* **4**, 47.
Mader, S. (1972). Private communication.
Matthews, J. W. (1960). *Proc. Eur. Reg. Conf. Electron Microsc.* p. 270. de Nederlandse Vereniging voor Electronenmikroscopie, Delft.
Matthews, J. W. (1961). *Phil. Mag.* **6**, 1347.
Matthews, J. W. (1963). *Phil. Mag.* **8**, 711.
Matthews, J. W. (1964). *In* "Single Crystal Films" (M. H. Francombe and H. Sato, eds.), p. 165. Pergamon, Oxford.
Matthews, J. W. (1966). *Phil. Mag.* **13**, 1207.
Matthews, J. W. (1968). *Phil. Mag.* **18**, 1149.
Matthews, J. W. (1970). *Thin Solid Films* **5**, 369.
Matthews, J. W. (1971). *Phil. Mag.* **23**, 1405.
Matthews, J. W. (1972a). *Surface Sci.* **31**, 241.
Matthews, J. W. (1972b). *Thin Solid Films* **12**, 243.
Matthews, J. W. (1973). *In* "A Treatise on Dislocations" (F. R. N. Nabarro, ed.). Dekker, New York (in press).
Matthews, J. W., and Blakeslee, A. E. (1972). *IBM Res. Rep.* RC 3854.
Matthews, J. W., and Crawford, J. L. (1965). *Phil. Mag.* **11**, 977.

Matthews, J. W., and Crawford, J. L. (1970). *Thin Solid Films* **5**, 187.
Matthews, J. W., and Jesser, W. A. (1967). *Acta Met.* **15**, 595.
Matthews, J. W., and Jesser, W. A. (1969). *J. Vac. Sci. Technol.* **6**, 641.
Matthews, J. W., and Klokholm, E. (1972). *Mat. Res. Bull.* **7**, 213.
Matthews, J. W., Mader, S., and Light, T. B. (1970). *J. Appl. Phys.* **41**, 3800.
Menter, J. W. (1958). *Advan. Phys.* **7**, 299.
Mott, N. F., and Nabarro, F. R. N. (1940). *Proc. Phys. Soc.* **52**, 86.
Nabarro, F. R. N. (1940). *Proc. Roy. Soc.* **A175**, 519.
Nabarro, F. R. N. (1967). "Theory of Crystal Dislocations," p. 75. Clarendon Press, Oxford.
Pashley, D. W. (1959). *Phil. Mag.* **4**, 319.
Pashley, D. W. (1965). *Advan. Phys.* **14**, 327.
Reiss, H. (1968). *J. Appl. Phys.* **39**, 5045.
Schmid, E., and Boas, W. (1950). "Plasticity of Crystals," p. 196. Chapman and Hall, London.
Simmons, G. W., Mitchell, D. F., and Lawless, K. R. (1967). *Surface Sci.* **8**, 130.
Thompson, E. R., and Lawless, K. R. (1966). *Appl. Phys. Lett.* **9**, 138.
Thompson, N. (1953). *Proc. Phys. Soc.* **B66**, 481.
Tucker, C. W. (1964). *J. Appl. Phys.* **35**, 1897.
Tucker, C. W. (1966). *J. Appl. Phys.* **37**, 3013.
van der Merwe, J. H. (1950). *Proc. Phys. Soc.* **A63**, 616.
van der Merwe, J. H. (1963). *J. Appl. Phys.* **34**, 117, 127.
van der Merwe, J. H. (1964). *In* "Single Crystal Films" (M. H. Francombe and H. Sato, eds.), p. 139. Pergamon, Oxford.
van der Merwe, J. H. (1972). *Surface Sci.* **31**, 198.
van der Merwe, J. H., and van der Berg, N. G. (1972). *Surface Sci.* **32**, 1.
Vermaak, J. S., and van der Merwe, J. H. (1964). *Phil. Mag.* **10**, 785.
Vermaak, J. S., and van der Merwe, J. H. (1965). *Phil. Mag.* **12**, 463.
Vincent, R. (1969). *Phil. Mag.* **19**, 1127.
Washburn, J., Thomas, G., and Queisser, J. H. (1964). *J. Appl. Phys.* **35**, 1909.
Wassermann, E. F., and Jablonski, H. P. (1970). *Surface Sci.* **22**, 69.
Winchell, P. G., Boah, J., and Ayres, P. S. (1971). *J. Appl. Phys.* **42**, 2612.
Yagi, K., Takayanagi, K., Kobayashi, K., and Honjo, G. (1970). *Int. Electron Microsc. Conf., 7th, Grenoble* p. 439. Soc. Fr. Microsc. Electron., Paris.
Yagi, K., Takayanagi, K., Kobayashi, K., and Honjo, G. (1971). *J. Crystal Growth* **9**, 84.

CHAPTER 9

LIST OF EPITAXIAL SYSTEMS

Enrique Grünbaum

Department of Physics and Astronomy
and School of Engineering
Tel-Aviv University
Ramat-Aviv, Tel-Aviv, Israel

I. Introduction

Since the publication of the review article by Pashley in 1956 (*Advan. Phys.* **5**, 173), in which about 150 epitaxial systems were listed, the number of publications has increased by an order of magnitude, as also evidenced by the number of specialized journals that have appeared since then. This enormous boom was a result, on the one hand, of the development of important new techniques for the preparation and observation of films and, on the other, of the large interest in single-crystal films for studies of their physical properties and for possible applications.

The present list refers to films in which a two-dimensional orientation is induced by a suitable crystalline substrate; systems with only a one-dimensional (fibrous) orientation have been omitted. All studies, ranging from those devoted to the basic mechanism of oriented growth either in the very initial or successive growth stages, to those seeking conditions for the growth of single-crystal films of high perfection, are cited. Exceptions are oriented layers grown chemically in studies of corrosion and oxidation, or during decomposition, to which a considerable number of publications are devoted, but they have been excluded here. Reference to publications on physical properties of single-crystal films has been made if they contain new information on the preparation of these films by means of oriented growth on single-crystal substrates.

The epitaxial systems have been classified according to the scheme described in Section II. This list refers to publications in all current journals as well as many proceedings of congresses and conferences. Some reviews with details on specific epitaxial systems have also been included. References to abstracts of conferences, internal reports, and doctoral theses have been omitted. Due to the scope of this endeavor, omissions and mistakes have been unavoidable, and the author would appreciate receiving additions and corrections to this list as well as reprints of new publications to be included in future supplements.

One of the main difficulties encountered in the preparation of the present survey has been the lack of details, in an appreciable number of publications and abstracts, on the substrates, methods of preparation, and methods of study of the epitaxial systems. Some publications on epitaxial films even fail to give any proof of the fact that they are single-crystal films. It would be highly desirable in the future that all studies on epitaxial films contain an abstract including information on the film material, the substrate, the methods of preparation, the methods of observation, and the main subjects dealt with.

II. Classification and Nomenclature

A. Column 1: Overgrowth or Film Material

Each material constituting the film or overgrowth (the element, and its corresponding alloys and compounds) is listed in order of increasing atomic number Z. For each material, the following *subdivisions* are given according to the *type of substrate*:

1. metal substrates;
2. semiconductor substrates (mainly graphite, Si, Ge; III–V compounds; PbS, ZnS, MoS_2, and other chalcogenides);
3. alkali halide substrates;

4. insulating substrates in general (mainly mica, sapphire, magnesium aluminate spinel and other oxides, calcium fluoride, and some other halides).

A further subdivision is made between materials with a large number of cases of autoepitaxy (homoepitaxy), and of epitaxy (heteroepitaxy), indicated by A and B, respectively, after the corresponding subdivision number.

B. Column 2: Substrate(s) and Surface Plane(s)

(a) *Chemical symbol or name.* Abbreviations: SAPPH—sapphire, alumina, or corundum; SPINEL—magnesium aluminate spinel.

(b) *Miller indices* (indicating type of surface plane): brackets have been omitted; also the distinction between the different faces in polar crystals has been ignored.

C. Column 3: Method(s) of Preparation

- HV evaporation in high vacuum (pressure up to 1×10^{-7} Torr during evaporation)
- UHV evaporation in ultrahigh vacuum (pressure below 1×10^{-7} Torr during evaporation)
- VG vapor growth, including reactive evaporation
- CVD chemical vapor deposition
- CG chemical growth
- SP sputtering
- ED electrodeposition (or electroplating)
- ELD electroless deposition or chemical displacement
- MG Melt growth
- LPE Liquid-phase epitaxy, molten-salt solution growth, or flux growth
- VLS vapor–liquid–solid growth or molten-drop growth
- SG solution growth (including solution growth of reacting chemicals)
- SPE solid-phase epitaxy

D. Column 4: Method(s) of Structural Observation

- OM optical microscopy
- RED reflection (high-energy) electron diffraction
- TED transmission electron diffraction
- TEM transmission electron diffraction and microscopy
- REM replica electron microscopy
- LED low-energy electron diffraction*
- AES Auger electron spectroscopy

* The abbreviation LED was used in the list, rather than the more common LEED, to conserve space.

ES	electron spectroscopy
EP	electron probe X-ray microanalysis
FEM	field electron microscopy or emission
FIM	field ion microscopy
XD	X-ray diffraction
XM	X-ray (diffraction) topography or microscopy
PCH	proton channeling, diffraction channeling of protons, or proton-scattering microscopy
ICH	ion channeling
IP	ion microprobe analysis

E. Column 5: References in Alphabetical Order

(a) The abbreviations for the names of journals are those used in "Chemical Abstracts," preceded by the year, and followed by the volume and page numbers.

(b) The abbreviations for the names of conference proceedings are those listed below; they are preceded by a star and the year of publication, and followed by the volume (if existing) and page numbers.

F. Abbreviations Used for References of Conference Proceedings

17ACMMM	*Proc. Ann. Conf. Magnetism and Magnetic Materials 17th, Chicago 1971* (C. D. Graham, Jr., and J. J. Rhyne, eds.). AIP, New York.
AIMEMAEM	*Proc. AIME Conf. Metallurgy of Advanced Electronic Materials, Philadelphia 1962.* Wiley (Interscience), New York.
AIMEMSM	*Proc. AIME Conf. Metallurgy of Semiconductor Materials, Los Angeles 1961* (J. B. Schroeder, ed.). Wiley (Interscience), New York.
CMPSS	"Molecular Processes on Solid Surfaces," *Proc. 3rd Battelle Institute Materials Science Coll., Kronberg, Germany 1968* (E. Drauglis, R. D. Gretz, and R. I. Jaffee, eds.). McGraw-Hill, New York.
2CTF	*Proc. 2nd Coll. Thin Films, Budapest 1967.* Hungarian Academy of Science, Budapest.
25EMSA	*Proc. Ann. Meeting Electron Microscopy Soc. Amer. 25th, Boston 1967* (C. J. Arceneaux, ed.). Claitor's, Baton Rouge, Louisiana.
27EMSA	*Proc. Ann. Meeting Electron Microscopy Soc. Amer. 27th, St. Paul, Minnesota 1969* (C. J. Arceneaux, ed.). Claitor's, Baton Rouge, Louisiana.
28EMSA	*Proc. Ann. Meeting Electron Microscopy Soc. Amer. 28th, Houston, Texas 1970* (C. J. Arceneaux, ed.). Claitor's, Baton Rouge, Louisiana.
29EMSA	*Proc. Ann. Meeting Electron Microscopy Soc. Amer. 29th, Boston 1971* (C. J. Arceneaux, ed.). Claitor's, Baton Rouge, Louisiana.
30EMSA	*Proc. Ann. Meeting Electron Microscopy Soc. Amer. 30th and 1st Pacific Regional Conf. Electron Microscopy, Los Angeles, California 1972* (C. J. Arceneaux, ed.). Claitor's, Baton Rouge, Louisiana.
31EMSA	*Proc. Ann. Meeting Electron Microscopy Soc. Amer. 31st, New Orleans, Louisiana 1973* (C. J. Arceneaux, ed.). Claitor's Baton Rouge, Louisiana.

2ERCEM — *Proc. 2nd European Regional Conf. Electron Microscopy, Delft 1960* (A. L. Houwink and B. J. Spit, eds.). De Nederlandse Verenigung voor Electronenmicroscopie, Delft.

3ERCEM — *Proc. 3rd European Regional Conf. Electron Microscopy, Prague 1964* (M. Titlbach, ed.). Czecheslovak Academy of Science, Prague.

4ERCEM — *Proc. 4th European Regional Conf. Electron Microscopy, Rome 1968* (D. S. Bocciarelli, ed.). Tipografia Poliglotta Vaticana, Rome.

1ICCG — "Crystal Growth," *Proc. Internat. Conf., Boston 1966* (H. S. Peiser, ed.). Pergamon Press, Oxford. (Also appeared as *J. Phys. Chem. Solids* **Suppl. 1.**)

5ICEM — *Proc. 5th Internat. Congr. Electron Microscopy, Philadelphia 1962* (S. S. Breese, Jr., ed.). Academic Press, New York.

6ICEM — *Proc. 6th Internat. Congr. Electron Microscopy, Kyoto 1966* (R. Uyeda, ed.). Maruzen, Tokyo.

7ICEM — *Proc. 7th Internat. Congr. Electron Microscopy, Grenoble 1970* (P. Favard, ed.). Socièté Francaise Microscopie Electronique, Paris.

ICPCSH — *Proc. Internat. Conf. Physics and Chemistry of Semiconductor Heterojunctions and Layer Structure, Budapest 1970* (G. Szigeti, ed.). Akad. Kiado, Budapest.

ICSPTF — "Structure and Properties of Thin Films," *Proc. Internat. Conf., Bolton Landing 1959* (C. A. Neugebauer, J. B. Newkirk, and D. A. Vermilyea, eds.). Wiley, New York.

ICTFBB — "Single Crystal Films," *Proc. Internat. Conf. Blue Bell, Penn. 1963* (M. H. Francombe and H. Sato, eds.). Pergamon Press, Oxford.

4IMATC — "The Structure and Chemistry of Solid Surfaces," *Proc. 4th Internat. Materials Symp., Berkeley 1968* (G. A. Somorjai, ed.). Wiley, New York.

ISCES — "Condensation and Evaporation of Solids," *Proc. Internat. Symp., Dayton 1962* (E. Rutner, P. Goldfinger, and J. P. Hirth, eds.). Gordon & Breach, New York.

1ISSI — "Semiconductor Silicon," *Proc. 1st Internat. Symp. Silicon Material Sci. and Technol., New York 1969* (R. R. Haberecht and E. L. Kern, eds.). Electrochem. Soc., New York.

ISTFCL — "Basic Problems in Thin Films Physics," *Proc. Internat. Symp. Clausthal-Göttingen 1965* (R. Niedermayer and H. Mayer, eds.). Vandenhoeck & Ruprecht, Göttingen.

2IVC — *Trans. 8th Nat. Vacuum Symp. and 2nd Internat. Congr. Vac. Sci. Technol., Washington 1961* (L. E. Preuss, ed.). Pergamon Press, Oxford.

3IVC — "Advances in Vacuum Science and Technology," *Proc. 3rd Internat. Congr. Vacuum Techniques, Stuttgart 1965* (H. Adams, ed.). Pergamon Press, Oxford.

4IVC — *Proc. 4th Internat. Vacuum Congr., Manchester 1968*. Institute of Physics Conf. Ser. 6, London.

NATOUTF — "The Use of Thin Films in Physical Investigations," NATO Advanced Study Inst., 1965 (J. C. Anderson, ed.). Academic Press, New York and London.

9NVSAVS — *Trans. 9th Nat. Vacuum Symp. Amer. Vac. Soc., Los Angeles 1962* (G. H. Bancroft, ed.). MacMillan, New York.

10NVSAVS — *Trans. 10th Nat. Vacuum Symp. Amer. Vac. Soc., Boston 1963* (G. H. Bancroft, ed.). MacMillan, New York.

SGSI — "Fundamentals of Gas–Surface Interactions," *Proc. Symp., San Diego, California 1966* (H. Saltsburg, J. N. Smith, Jr., and M. Rogers, eds.). Academic Press, New York.

III. Index of Overgrowth (Deposit or Film) Materials

The following index is ordered by increasing atomic number Z of the single elements or of the first element of alloys and compounds.

ELEMENTS

ALLOYS

Group V Compounds

Z		Page	Z		Page	Z		Page	Z		Page
5	BP	619	27	Co–P	636		GaP–Ge	651	49	InP	661
13	AlN	623	28	Ni–P	639		Ga–InP	648		InAs	661
	AlP	623	31	GaN	643		Ga–InAs	648		InSb	661
	AlAs	623		GaP	644		Ga–InSb	649		InP–As	662
	AlSb	623		GaAs	645		Ga–InP–As	649		InAs–Sb	662
	Al–GaAs	623		GaSb	647		GaSb–InAs	649		In–GaP	648
	Al–GaSb	623		GaP–As	648		GaP–ZnS	643		In–GaAs	648
	Al–GaP–As	623		GaP–GaSb	648		GaP–ZnSe	643		In–GaSb	649
	Al–GaAs–Sb	624		GaAs–Sb	648		GaAs–ZnSe	643		In–GaP–As	649
21	ScN	630		Ga–AlAs	623	32	Ge–GaP	651		InAs–GaSb	649
	ScP	631		Ga–AlSb	623	41	NbN	653		InP–CdS	661
	ScAs	631		Ga–AlP–As	623		Nb_4N_5	653	50	$SnCdP_2$	658
				Ga–AlAs–Sb	624	48	$CdSnP_2$	658	71	LuN	664

Chalcogenides

Z		Page	Z		Page	Z		Page	Z		Page
24	CrSe	631	47	Ag_2Se	658		Sn–PbSe	672	81	Tl_2Se	670
	CrTe	631		Ag_2Te	658		Sn–PbTe	672		Tl_2Te	670
	Cr_2CdS_4	661		$AgBiTe_2$	658	51	Sb_2S_3	663		$TlBiSe_2$	670
27	Co_9S_8	636	48	CdS	658		Sb_2Te_3	663	82	PbS	670
	CoSe	636		CdSe	660	62	SmSe	664		PbSe	671
29	CuSe	641		CdTe	660		SmTe	664		PbTe	671
30	ZnS	642		CdS–Se	661	63	EuSe	664		PbSe–Te	672
	ZnSe	642		$CdCr_2S_4$	661		EuTe	664		$PbSnS_2$	672
	ZnTe	643		CdS–InP	661		Eu–PbTe	664		Pb–SnSe	672
	ZnSe–Te	643		Cd–HgTe	661	70	YbSe	664		Pb–SnTe	672
	ZnS–GaP	643	49	In_2Te_3	662		YbTe	664		Pb–EuTe	664
	ZnSe–GaP	643	50	SnS	662	80	HgS	669	83	Bi_2S_3	673
	ZnSe–GaAs	643		SnSe	662		HgSe	669		Bi_2Se_3	673
32	GeSe	651		SnTe	662		HgTe	669		Bi_2Te_3	673
	GeTe	651		Sn–GeTe	652		Hg–CdTe	661		$BiAgTe_2$	658
	Ge–SnTe	652		$SnPbS_2$	672					$BiTlSe_2$	670

HALIDES

Z		Page	Z		Page	Z		Page	Z		Page
3	LiF	619		NaI	621	32	GeI_2	652	48	$CdCl_2$	661
	LiCl	619		Na_3AlF_6	621		GeI_4	652		CdI_2	661
	LiBr	619	12	MgF_2	622	37	RbF	652	55	CsF	663
	LiI	619	19	KF	629		RbCl	652		CsCl	663
7	NH_4Cl	620		KCl	629		RbBr	652		CsBr	664
	NH_4Br	620		KBr	629		RbI	653		CsI	664
	NH_4I	620		KI	630	47	AgCl	658	56	BaF_2	664
11	NaF	620	20	CaF_2	630		AgBr	658	81	TlCl	670
	NaCl	620	28	$NiBr_2$	639		AgI	658		TlBr	670
	NaBr	621	29	CuI	641					TlI	670

OTHER COMPOUNDS

Z		Page	Z		Page	Z		Page
1	Ice	619	25	Manganese oxides	631	50	Tin oxides	663
3	Lithium niobate	619	26	Iron oxides	632	56	Barium titanate	664
	Lithium lanthanide–molybdate	619		Orthoferrites	632		Barium stannate	664
				Spinel ferrites	633		Barium cerate	664
6	Carbon oxides	619		Garnets	633		Barium oxide	664
11	Sodium cyanide	621	27	Cobalt oxides	636	63	Europium oxides	664
	Sodium nitrate	621	28	Nickel oxides	639	73	Tantalum carbides	664
	Sodium niobate	621	30	Zinc oxides	643	74	Tungsten carbides	665
12	Magnesium oxides	622	32	Germanium oxides	652		Tungsten oxides	665
13	Alumina	624	33	Arsenic oxides	652	83	Bismuth silicate	673
14	Silicon carbides	628	37	Rubidium cyanide	653		Bismuth ferrite	673
19	Potassium cyanide	630	38	Strontium titanate	653		Bismuth zincate	673
	Potassium chlorate	630		Strontium stannate	653		Bismuth gallate	673
20	Calcium titanate	630	42	Molybdenum carbides	653		Bismuth titanates	673
22	Titanium oxides	631	47	Silver cyanide	658		Bismuth oxides	673
24	Chromium oxides	631	49	Indium oxides	662		Organic	673

IV. List of Epitaxial Systems

On pp. 619–673, 304 materials (elements, alloys, and compounds) grown as two-dimensionally oriented deposits or single-crystal films on the specified substrates have been listed, together with the methods of preparation and observation and the corresponding references (including those available as of July, 1974).

ACKNOWLEDGMENTS

Many thanks are due to Dr. L. May for advice in the organization of this work, to the staff of the Computer Center of Tel-Aviv University for their assistance in the computer program and printing of the list, and to my wife for encouragement and great assistance with the preparation and revision of the printed list.

Z= 1- 6	OVERGROWTH		SUBSTRATE	PREP	OBSERV	REFERENCE
1	ICE	4	AGI 111,00.1	VG	RED	LISGARTEN,ND 1958 PHIL.MAG.3,1306
3	LITHIUM	1	W 112	UHV	LED	MEDVEDEV,VK+NAUMOVETS,AG+SMEREKA,TP 1973 SURFACE SCI.34,368
	LITHIUM FLUORIDE	1	AG 111	HV	RED	SCHULZ,LG 1952(C) ACTA CRYST.5,266
		2	PBS 001,ZNS 110, MOS2 00.1	HV	RED	SHIGETA,J 1956 J.PHYS.SOC.JAPAN 11,206
		3	NACL 001	HV	TEM,REM	BANDYOPADHYAY,T+GRANZER,F+GUTLICH,KF 1972 J.CRYST.GROWTH 17,360
			LIF 001,KCL 001, NACL 001,110,111	HV	RED,TEM, REM	BAUER,E 1956ZKRISTALLOGR.107,72/1958IBID.110,395/ SEE ALSO BAUER,E+GREEN,AK+KUNZ,KM+POPPA,H 1966*ISTFCL, 135
			LIF 001,NACL 001, KCL 001,KBR 001, KI 001	HV	RED,REM	LUDEMANN,H 1954 Z.NATURFORSCH.9A,252
			NAF 001,NACL 001, KCL 001,KBR 001, KI 001	HV	RED	LUDEMANN,H 1956 Z.NATURFORSCH.11A,935
			LIF 001	HV	RED	LUDEMANN,H+RAETHER,H 1953 ACTA CRYST.6,873
			NACL 001,111	HV	TEM	MIDDLETON 1972.SEE CO 3
			LIF 001,110,111	HV	OM	NEUHAUS,A 1952 Z.ELEKTROCHEM.56,453
			NACL	HV	TEM	REICHELT,K 1973 J.CRYST.GROWTH 19,258
			LIF 001,NACL 001, KCL 001,KBR 001	HV	RED	SCHULZ,LG 1952(A) ACTA CRYST.5,130
			NACL 001	HV	RED	SHIGETA 1956.SEE 2
		4	MICA 00.1	HV	RED,TEM, REM	BAUER 1956.SEE 3
			MGO	UHV	TEM	REICHELT 1973.SEE 3
			MICA 00.1	HV, SG	RED	SCHULZ,LG 1951(B) ACTA CRYST.4,483
			NANO3 100 CACO3 100	HV, SG	RED	SCHULZ,LG 1952(B) ACTA CRYST.5,264
	LITHIUM CHLORIDE	2	MNS 001	SG	OM	ROYER,L 1928 BULL.SOC.FRANC.MINERAL.51,7
		3	NAF 001	HV	RED	LUDEMANN 1956.SEE LIF 3
			NACL 001	SG	OM	ROYER 1928.SEE 2
			LIF 001,NACL 001, KCL 001,KBR 001	HV	RED	SCHULZ 1952A.SEE LIF 3
			NACL 001	SG	OM	SLOAT,CA+MENZIES,AWC 1931 J.PHYS.CHEM.35,2005
		4	NANO3 100	SG	OM	HEINTZE,W 1937 Z.KRISTALLOGR.97,241
	LITHIUM BROMIDE	2	MNS 001	SG	OM	ROYER 1928.SEE LICL 2
		3	NACL 001,KCL 001	SG	OM	ROYER 1928.SEE LICL 2
			NACL 001,KCL 001	SG	OM	SLOAT+MENZIES 1931.SEE LICL 3
		4	NANO3 100	SG	OM	HEINTZE 1937.SEE LICL 4
			CAF2 111	SG	OM	KRASTANOW,L+STRANSKI,IN 1938 Z.KRISTALLOGR.99,444
	LITHIUM IODIDE	2	PBS 001	SG	OM	ROYER 1928.SEE LICL 2
		3	NABR 001	HV	RED,REM	LUDEMANN 1954.SEE LIF 3
			NACL 001,KCL 001, KBR 001	SG	OM	ROYER 1928.SEE LICL 2
	LITHIUM NIOBATE	4	LITAO3	MG	---	FUKUNISHI,S+UCHIDA,N+MIYAZAWA,S+NODA,J 1974 APPL.PHYS.LETT.24,424
			LITAO3	MG	OM,XD, EP	MIYAZAWA,S 1973 APPL.PHYS.LETT.23,198
	LITHIUM LANTHANIDE-MOLYBDATE	4	GELN2MO08	SG	OM,EP, XD	CLARK,GW+FINCH,CB+HARRIS,LA+YAKEL,HL 1972 J.CRYST.GROWTH 16,110
4	BERYLLIUM	3	NACL 001,KCL 001, KBR 001	HV	TED	CONJEAUD 1956.SEE TI 3
5	BORON	2	SI 111	CVD	OM,RED, XD	ARMINGTON,AF+POTTER,WD+TANNER,LE 1964 J.APPL.PHYS.35,730
			SI 111	CVD	OM,RED	PETERS,ED+POTTER,WD 1965 TRANS.AIME 233,473
	BORON PHOSPHIDE	2	SIC 00.1	CVD	OM,RED, XD	CHU,TL+JACKSON,JM+HYSLOP,AE+CHU,SC 1971 J.APPL.PHYS.42,420
6	CARBON OXIDES	1	CU 001	UHV	LED	CHESTERS,MA+PRITCHARD,J 1971 SURFACE SCI.28,460
			NI 110	UHV	LED,AES	TAYLOR,TN+ESTRUP,PJ 1973 J.VAC.SCI.TECHNOL.10,26

Z= 7-11 OVERGROWTH		SUBSTRATE	PREP	OBSERV	REFERENCE
7 AMMONIUM CHLORIDE	3	NACL 001+COATS	SG	OM	MOTOC,C+BADEA,M 1972 J.CRYST.GROWTH 17,337
		NACL 001,KCL 001	VG	OM	SLOAT+MENZIES 1931.SEE LICL 3
	4	MICA 00.1	SG	OM	LISGARTEN,ND 1954 TRANS.FARADAY.SOC.50,684
		MICA 00.1+COATS	SG	OM	MOTOC+BADEA 1972.SEE 3
		MICA 00.1	SG	OM	ROYER 1928.SEE LICL 2
AMMONIUM BROMIDE	1	AG 001	SG	OM	SLOAT+MENZIES 1931.SEE LICL 3
	3	NACL 001,KCL 001	VG	OM	SLOAT+MENZIES 1931.SEE LICL 3
	4	MICA 00.1	SG	OM	LISGARTEN 1954.SEE NH4CL 4
		MICA 00.1	SG	OM	ROYER 1928.SEE LICL 2
AMMONIUM IODIDE	3	KCN 001,KBR 001, KI 001,RBCN 001, RBBR 001,RBI 001	SG	OM	BARKER,TV 1906 J.CHEM.SOC.TRANS.89,1120/ 1907 MINERAL.MAG.14,235/ 1908 Z.KRISTALLOGR.45,1
		KBR 001,RBBR 001, RBI 001	SG	OM	ROYER 1928.SEE LICL 2
		NACL 001,KCL 001	VG, SG	OM	SLOAT+MENZIES 1931.SEE LICL 3
	4	MICA 00.1	SG	OM	DUNNING,WJ+FOX,PG+PARKER,DW 1967*1ICCG, 509
		MICA 00.1	SG	OM	DUNNING,WJ+SAVVA,M 1968 J.CRYST.GROWTH 3,4,350
		MICA 00.1	SG	OM	LISGARTEN 1954.SEE NH4CL 4
		MICA 00.1	SG	---	NEWKIRK,JB+TURNBULL,D 1955 J.APPL.PHYS.26,579
		MICA 00.1	SG	OM	ROYER 1928.SEE LICL 2
		MICA 00.1	HV, SG	RED	SCHULZ 1951(B).SEE LIF 4
		MICA 00.1	SG	OM	UPRETI,MC+WALTON,AG 1966 J.CHEM.PHYS.44,1936
11 SODIUM	1	W 110	UHV	---	CARROLL,CE+MAY,JE 1972 SURFACE SCI.29,60
		W 112 + O	UHV	LED	CHEN,JM 1970 J.APPL.PHYS.41,5008
		W 112,W 112 + O	UHV	LED	CHEN,JM+PAPAGEORGOPOULUS,CA 1970 SURFACE SCI.21,377/ 1971 IBID.26,499
		NI 001,110,111	UHV	LED	GERLACH,RL+RHODIN,TN 1968 SURFACE SCI.10,446/ 1969(A)*4IMATC, 55-1/ 1969(B) SURFACE SCI.17,32
		W 110	UHV	LED	MEDVEDEV,VK+NAUMOVETS,AG+FEDORUS,AG 1970 SOVIET PHYS-SOLID STATE 12,301
SODIUM FLUORIDE	2	PBS 001,ZNS 110, MOS2 00.1	HV	RED	SHIGETA 1956.SEE LIF 2
	3	LIF 001, NACL 001,110,111	HV	RED,TEM, REM	BAUER 1956.SEE LIF 3
		LIF 001,NACL 001, KI 001	HV	RED	LUDEMANN 1956.SEE LIF 3
		NACL 001	HV	RED	RAMOS,F+BRU,L 1963 ANAL.REAL SOC.ESP.FIS.QUIM.,SER.A 59,39
		LIF 001,NACL 001, KCL 001,KBR 001	HV	RED	SCHULZ 1952A.SEE LIF 3
		NACL 001	HV	RED	SHIGETA 1956.SEE LIF 2
	4	MICA 00.1	HV	RED,TEM, REM	BAUER 1956.SEE LIF 3
		NANO3 100, CACO3 100	HV, SG	RED	SCHULZ 1952(B).SEE LIF 4
SODIUM CHLORIDE	1	AG 001	HV, SG	TED	BRUCK 1936.SEE AL 3
		AG 001	SG	RED,TED	GOCHE+WILMAN 1939.SEE AG 3
		AG 001,111	SG	OM	JOHNSON,GW 1950(A,B) J.APPL.PHYS.21,449+1057/ 1950(C,D) J.CHEM.PHYS.18,154+560
		CU 001,111,ZN, AG 001,111,SB111, AU111,PB111,BI111	SG	OM	JOHNSON,GW 1950(E) NATURE 166,189/ 1951 J.APPL.PHYS.22,797
		AG 111	UHV	RED	RABBIT,LI+HAMPSHIRE,MJ+TOMLINSON,RD+CALDERWOOD,JH 1971CONDUCTION IN LOW-MOBILITY MATER.PROC.2.INT. CONF.EILAT,ISRAEL 1971(KLEIN,H+TANNHAUSER,DS+ POLLAK,M,EDS.),419.TAYLOR+FRANCIS,LONDON
		AG 001,111	HV, SG	RED	SCHULZ 1952(C).SEE LIF 1
	2	PBS 001	SG		BEUKERS,MCF 1939 REC.TRAV.CHIM.58,435
		PBTE 001	SG	OM	PALATNIK ET AL 1965.SEE PBTE 3
		MNS 001,PBS 001	SG	OM	ROYER 1928.SEE LICL 2
		PBS 001	SG	OM	SLOAT+MENZIES 1931.SEE LICL 3
	3	KCL 001,KBR 001, KI 001,RBCL 001	SG	OM	BARKER 1906,1907,1908.SEE NH4I 3
		NACL 001	HV	OM,RED	GREEN,AK+BAUER,E 1966 J.APPL.PHYS.37,917
		NACL 001	HV	OM,REM	KOSEVICH,VM+MOSKALEV,VM+PALATNIK,LS 1968 SOVIET.PHYS.-CRYST.12,581

Z=11 OVERGROWTH		SUBSTRATE	PREP	OBSERV	REFERENCE
SODIUM CHLORIDE(CONT.)	3	LIF 001,NACL 001, KCL 001,KBR 001, KI 001	HV	RED,REM	LUDEMANN 1954.SEE LIF 3
		LIF 001,NAF 001	HV	RED	LUDEMANN 1956.SEE LIF 3
		NACL 001	HV	RED	LUDEMANN+RAETHER 1953.SEE LIF 3
		C REPLICA KCL 001	HV	OM	MIETZ,I 1965 NATURWISSENSCHAFTEN 52,537
		NACL 001	HV	OM	NEUHAUS 1952.SEE LIF 3
		KCL 001	SG	OM	ROYER 1928.SEE LICL 2
		LIF 001,NACL 001, KCL 001,KBR 001	HV	RED	SCHULZ 1952(A).SEE LIF 3
		NACL 001,KCL 001	SG	OM	SLOAT+MENZIES 1931.SEE LICL 3
	4	MICA 00.1	SG	OM	DEICHA,G 1947(A,B) BULL.SOC.FRANC.MINERAL.70,177 +318/ 1947(C) LA NATURE 3143, 279
		NANO3 100	SG	OM	HEINTZE 1937.SEE LICL 4
		MICA 00.1	HV	RED,REM	KOCH+VOOK 1971.SEE AG 3
		MICA 00.1	UHV	RED,REM	KOCH,FA+VOOK,RW 1971*29EMSA,222
		MICA 00.1	HV, UHV	RED,REM	KOCH,FA+VOOK,RW 1972 THIN SOLID FILMS 14,231
		CAF2 111	SG	OM	KRASTANOV+STRANSKI 1968.SEE LIBR 4
		MICA 00.1	SG	OM	LISGARTEN 1954.SEE NH4CL 4
		MICA 00.1	HV	TEM	MISSIROLI,GF 1972 THIN SOLID FILMS 12,S35
		MICA 00.1	UHV	RED	RABBIT ET AL 1971.SEE 1
		MICA 00.1	HV, SG	RED	SCHULZ 1951(B).SEE LIF 4
		NANO3 100, CACO3 100	HV, SG	RED	SCHULZ 1952(B).SEE LIF 4
		MICA 00.1	HV	RED,TEM, REM	SHIMAOKA,G+HEROLD,JF 1967*25EMSA, 348
		TRIGLYCINE SULPH.	HV	TEM	VLASOV ET AL 1970.SEE AGCL 4
SODIUM BROMIDE	1	AG 001,111	HV, SG	RED	SCHULZ 1952(C).SEE LIF 1
	2	PBS 001	SG	OM	ROYER 1928.SEE LICL 2
		PBS 001	SG	OM	SLOAT+MENZIES 1931.SEE LICL 3
	3	NACL 001,KCL 001, KBR 001,KI 001, RBCL 001,RBBR 001, RBI 001	SG	OM	BARKER 1906,1907,1908.SEE NH4I 3
		LIF 001,NACL 001, KCL 001,KBR 001	HV	RED	SCHULZ 1952(A).SEE LIF 3
		NACL 001,KCL 001	SG	OM	SLOAT+MENZIES 1931.SEE LICL 3
	4	NANO3 100	SG	OM	HEINTZE 1937.SEE LICL 4
		CAF2 111	SG	OM	KRASTANOV+STRANSKI 1938.SEE LIBR 4
		MICA 00.1	SG	OM	LISGARTERN 1954.SEE NH4CL 4
		CACO3 100	SG	OM	ROYER, 1937 C.R.ACAD.SCI.205,1418
		MICA 00.1	HV, SG	RED	SCHULZ 1951(B).SEE LIF 2
SODIUM IODIDE	2	PBS 001	SG	OM	ROYER 1928.SEE LICL 2
		PBS 001	SG	OM	SLOAT+MENZIES 1931.SEE LICL 3
	3	NH4I 001, NACL 001,NABR 001, KCL 001,KCN 001, KBR 001,KI 001, RBCL 001,RBBR 001	SG	OM	BARKER 1906,1907,1908.SEE NH4I 3
		KBR 001	HV	RED,REM	LUDEMANN 1954.SEE LIF 3
		NACL 001	HV	TED	RAMOS,F+CASTRO,C+BRU,L 1965*3ERCEM A,403
		KI 001	SG	OM	ROYER 1928.SEE LICL 2
		NACL 001,KCL 001	SG	OM	SLOAT+MENZIES 1931.SEE LICL 3
	4	NANO3 100	SG	OM	HEINTZE 1937.SEE LICL 4
		MICA 00.1	SG	OM	LISGARTEN 1954.SEE NH4CL 4
		MICA 00.1	SG	OM	ROYER 1928.SEE LICL 2
		CACO3 100	SG	OM	ROYER 1937.SEE NABR 4
		MICA 00.1	HV, SG	RED	SCHULZ 1951(B).SEE LIF 4
		MICA 00.1	MG	OM	WEST,CD 1945 J.OPT.SOC.AMER.35,26
CRYOLITE	2-3	PBS 001,ZNS 110, MOS2 00.1,NACL001	HV	RED	SHIGETA 1956.SEE LIF 2
SODIUM CYANIDE	2-3	PBS 001,NACL 001, KCL 001	SG	OM	SLOAT+MENZIES 1931.SEE LICL 3
SODIUM NITRATE	4	CACO3	SG	OM,RED	FINCH,GI+WHITMORE,EJ 1938TRANS.FARADAY SOC.34,640
		MICA 00.1	MG	OM	WEST 1945.SEE NAI 4
SODIUM NIOBATE	3	LIF 001,NAF 001	HV	RED	MULLER ET AL 1963.SEE CATIO3 3

Z= 12-13 OVERGROWTH		SUBSTRATE	PREP	OBSERV	REFERENCE
12 MAGNESIUM	2	MOS2 00.1	UHV	TEM	HEYRAUD,JC+PERRIN,J THIN SOLID FILMS 12,531
		MOS2 00.1	UHV	TEM,SEM	HEYRAUD,JC+MAURISSEN-RUGLIONI,Y+PERRIN,J+ CAPELLA,J 1973 THIN SOLID FILMS 18,213
MAGNESIUM-GOLD ALLOYS	4	MICA 00.1	HV	TED	VERDERBER,RR 1963 J.APPL.PHYS.34,2103
MAGNESIUM FLUORIDE	3-4	LIF 001,NACL 001, 110,111,MICA 00.1	HV	RED,TEM, REM	BAUER 1956.SEE LIF 3
MAGNESIUM OXIDES	3	LIF	HV	TEM	REICHELT 1973.SEE LIF 3
	4	MGO 001	VG	OM,RED, SEM, XD,EP	COCKAYNE,B+FILBY,JD+GASSON,DB 1971 J.CRYST.GROWTH 9,340
		MNFE2O4 001, MN-ZNFE2O4	CG	XD	HANAK,JJ+JOHNSON,D 1968 J.APPL.PHYS.39,1161
		MGO	CVD	OM	MEE,JB+PULLIAM,GR 1967*1ICCG,333
		MGO 001	CVD	OM,XD, XM	SPOONER,FJ+VERNON,MW 1970 J.MATER.SCI.5,731
13 ALUMINUM	1	MO 110	UHV	LED,AES	JACKSON,AG+HOOKER,MP 1971 SURFACE SCI.28,373
		NB 110	UHV	LED,AES	JACKSON,AG+HOOKER,MP 1972 J.VAC.SCI.TECHN.9,784
		TA 110	UHV	LED	JACKSON,AG+HOOKER,MP+HAAS,TW 1967 J.APPL.PHYS.38,4998
		AL 001,110,111	UHV	LED	JONA,F 1967 J.PHYS.CHEM.SOLIDS 28,2155
	2	SI 111	UHV	LED	LANDER,JJ+MORRISON,J 1964 SURFACE SCI.2,553/ 1965 J.APPL.PHYS.36,1706
		GAAS 001,110,111	UHV	RED	LUDEKE,R+CHANG,LL+ESAKI,L 1973APPLPHYSLETT.23,201
		SI 111	UHV	LED	STRONGIN,M+KAMMERER,OF+FARRELL,HH+MILLER,DL. 1973 PHYS.REV.LETT.30,129
	3	NACL 001	HV	TED	BRUCK,L 1936 ANN.PHYSIK (LEIPZIG) 26,233
		NACL 001,KBR 001	UHV	TEM	DUMPICH,G+WASSERMANN,EF 1972 SURFACE SCI.33,203
		LIF 001,NACL 001, KCL 001	HV	RED	GOTTSCHE,H 1956 Z.NATURFORSCH.11A,55
		LIF 001	UHV	RED	GREEN,AK+DANCY,J+BAUER,E 1970 J.VAC.SCI.TECHN.7,159
		NACL 001,KCL 001	UHV	TEM	INO,S+OGAWA,S 1966*6ICEM 1,521
		NACL 001	UHV	TEM	INO,S+WATANABE,D+OGAWA,S 1964 J.PHYS.SOC.JAPAN 19,881
		NACL 001	HV	TED	JAHRREISS,H+ISKEN,HJ 1966 PHYS.STAT.SOLIDI 17,619
		NACL 001	HV	TEM	JAUNET,J+SELLA,C 1965*3ERCEM A,381
		NACL 001	HV	TEM	JAYADEVAIAH,TS+KIRBY,RE 1969APPL.PHYS.LETT.15,150
		KCL 001,KBR 001, KI 001	HV	TED	KATO,T 1968 JAPAN.J.APPL.PHYS.7,1162
		NACL 001	HV	TEM	KOMODA,T 1966 JAP.J.APPL.PHYS.5,419
		NACL 001	UHV	RED,TEM	KUNZ,KM+GREEN,AK+BAUER,E 1966 PHYS.STATUS.SOLIDI 18,441
		KBR 001,110,111, 210,211	HV	TEM	LANDRY,JD+MITCHELL,EN 1969 J.APPL.PHYS.40,3400
		KCL 123	HV	TEM	MADER,S+CHAUDHARI,P 1969 J.VAC.SCI.TECHN.6,615
		KCL 001	UHV	TEM	METOIS,JJ+GAUCH,M+MASSON,A+KERN,R 1972(A) THIN SOLID FILMS 11,205/ 1972(B) SURFACE SCI.30,43
		NACL 001	HV	XD	MIWA,M+ANNAKA,S 1954 J.PHYS.SOC.JAPAN 9,302
		NACL 001	HV, UHV	TEM	MURR,LE+INMAN,MC 1966 PHIL.MAG.14,135
		NACL 001,KCL 001	UHV	TEM	OGAWA+INO 1969.SEE AU 3
		NACL 001,KCL 001	UHV	TED	OGAWA,S+INO,S+KATO,T+OTA,H 1966 J.PHYS.SOC.JAPAN 21,1963
		NACL	HV	TEM	POSTNIKOV,VS+IEVLEV,VM+BELONOGOV,VK+ZOLATUKHIN,IV 1968 SOVIET PHYS.SOLID STATE 10,722
		LIF 001,111, NACL 001,110	HV	XD	RHODIN,JR.TN 1949 DISC.FARADAY SOC.5,215
		NACL 001	HV	TED	SELLA,C+TRILLAT,J 1964*ICTFBB 201
		NACL 001,KCL 001	HV	TED	SHARMA,SK 1963 CURRENT SCI.(INDIA) 32,107
	4	MGO 001	UHV	RED	GREEN ET AL 1970.SEE 3
		DIAMOND 111	UHV	LED	LANDER,JJ+MORRISON,J 1966 SURFACE SCI.4,241
		MICA 00.1	HV	TEM	POSTNIKOV ET AL 1968.SEE 3
		MICA 00.1	HV	XD	RHODIN 1949.SEE 3
ALUMINUM-NICKEL ALLOYS	3	NACL 001	HV	TED	SHIRAI ET AL 1961.SEE NI 1
ALUMINUM-COPPER ALLOYS	3	KCL 001	HV, SP	TEM	MADER,S+HERD,S 1972 THIN SOLID FILMS 10,377
		NACL 001	HV	TED	RAETHER,H 1950 C.R.ACAD.SCI.231,653
ALUMINUM-PALLADIUM ALL.	3	NACL 001	HV	TED	SHIRAI ET AL 1961.SEE NI 1
ALUMINUM-SILVER ALLOYS	3	NACL 001	HV	TEM	FUKANO,Y+OGAWA,S 1956 ACTA CRYST.9,971/ 1959 J.PHYS.SOC.JAPAN 14,1671
		NACL 001	HV	RED,TED	GOTTSCHE,H 1953 Z.PHYSIK 134,504

Z= 13 OVERGROWTH		SUBSTRATE	PREP	OBSERV	REFERENCE
		NACL 001	HV	TED	SHIRAI ET AL 1961.SEE NI 1
		NACL 001	HV	TED	WATANABE,Y 1956 J.PHYS.SOC.JAPAN 11,1072
	4	MICA 00.1	HV	TEM	POSTNIKOV ET AL 1968.SEE AL 3
ALUMINUM-GOLD ALLOYS	1-3	AU 001, NACL 001,KCL 001	HV	TED,XD	FRANCOMBE,MH+NOREIKA,AJ+TAKEI,WJ 1967/1968 THIN SOLID FILMS 1,353
ALUMINUM NITRIDE	2	SIC 00.1,ALN	CVD	OM,RED	CHU,TL+ING,DW+NOREIKA,AJ 1967 SOLID-ST.ELECTRON.10,1023
		SI 111,SIC	CVD	OM,RED	MANASEVIT,HM+ERDMANN,FM+SIMPSON,WI 1971 J.ELECTROCHEM.SOC.118,1864
		SI 111	CVD	TEM	NOREIKA,AJ+ING,DW 1968 J.APPL.PHYS.39,5578
		SI 111,SIC 00.1	CVD	OM	PASTERNAK,J+ROSKOVCOVA,L 1965PHYSSTATSOLIDI 9,K73
	4	SAPPHIRE	CVD	OM,RED	COX,GA+CUMMINS,DO+KAWABE,K+TREDGOLD,RH 1967 J.PHYS.CHEM.SOLIDS 28,543
		SAPPH. 10.2,11.0, 00.1,SPINEL 110	CVD	RED	MANASEVIT ET AL 1971.SEE 2
		SAPPH. 10.2,00.1	SP	ED,XD	SHUSKUS,AJ+REEDER,TM+PARADIS,EL 1974 APPL.PHYS.LETT.24,155
		SAPPH. 10.2,00.1	CVD	OM,RED	YIM,WM+STOFKO,EJ+ZANZUCCHI,PJ+PANCOVE,JI+ ETTENBERG,M+GILBERT,SL 1973 J.APPL.PHYS.44,292
ALUMINUM PHOSPHIDE	2	SI 111,GAAS	CVD	---	REID,FJ+MILLER,SE+GOERING,HL 1966 J.ELECTROCHEM.SOC.113,467
		SI	CVD	XD	RICHMAN,D 1968 J.ELECTROCHEM.SOC.115,945
ALUMINUM ARSENIDE	2	GAAS	CVD	---	BOLGER,DE+BARRY.BE 1963 NATURE(LONDON)199,1287
		GAAS 001	CVD	XD	ETTENBERG,M+SIGAI,AG+DREEBEN,A+GILBERT,S 1971 J.ELECTROCHEM.SOC.118,1355
		GAAS 001,111	CVD	OM	MINDEN,HT 1970 APPL.PHYS.LETT.17,358
		GAAS 001	CVD	TEM,EP, XD	SIGAI,AG+ABRAHAMS,MS+BLANC,J 1972 J.ELECTROCHEM.SOC.119,952
	4	SAPPH. 00.1	CVD	OM,RED, XD	MANASEVIT,HM 1971 J.ELECTROCHEM.SOC.118,647
ALUMINUM ANTIMONIDE	2	GE 001,110,111	HV	OM,RED, XD	RICHARDS,JL+HART,PB+GALLONE,LM 1963 J.APPL.PHYS.34,3418
		GE 001,110,111	HV	OM,RED, XD	RICHARDS,JL+HART,PB+MULLER,EK 1964*ICTFBB,241
ALUMINUM-GALLIUM ARSENIDE	2	GAAS 110	CVD	OM,EP, XD	BLACK,JK+KU,SM 1966 J.ELECTROCHEM.SOC.113,249
		GAAS	LPE	OM,EB	BLUM,JM+SHIH,KK 1972 J.APPL.PHYS.43,1394
		GAAS 001	UHV	RED,TEM, EP,ICH	CHANG,LL+ESAKI,L+HOWARD,WE+LUDEKE,R 1973 J.VAC.SCI.TECHN.10,11
		GAAS 001	UHV	OM	CHO,AY+CASEY,JR.HC 1974 J.APPL.PHYS.45,1258
		GAAS 001	UHV	RED,SEM	CHO,AY+STOKOWSK,SE 1971 SOLID-ST.COMMUN.9,565
		GAAS 001,111	LPE	---	HAYASHI,I+PANISH,MB 1970 J.APPL.PHYS.41,150
		GAAS 001,111	LPE	---	HAYASHI,I+PANISH,MB+FOY,PW+SUMSKI,S 1970 APPL.PHYS.LETT.17,109
		GAAS 001	LPE	OM	HOVEL,HJ+WOODALL,JM 1973 JELECTROCHEMSOC.120,1246
		GAAS 001	UHV	AES	LUDEKE,R+ESAKI,L+CHANG,LL 1974APPLPHYSLETT.24,417
		GAAS	CVD	EP,XD	MANASEVIT 1971.SEE ALAS 4
		GAP 111	LPE	OM	MOON,RL+ANTYPAS,GA 1973 J.CRYST.GROWTH 19,109
		GAAS 001	LPE	SEM	NAKAMURA,M+AIKI,K+UMEDA,J+YARIV,A+YEN,HW+ MORIKAWA,T 1974 APPL.PHYS.LETT.24,466
		GAAS 001	LPE	OM,EP	POTEMSKI,RM+WOODALL,JM 1972 J.ELECTROCHEM.SOC.119,277
		GAAS 111	LPE	OM,ER	RADO,WG+JOHNSON,WJ+CRAWLEY,RL 1972 J.ELECTROCHEM. SOC.119,652/ 1972 J.APPL.PHYS.43,2763
		GAP 001,111	LPE	OM,XM	VAN OIRSCHOT,TGJ 1974 APPL.PHYS.LETT.24,211+577
		GAAS 001	LPE	OM,REM, SEM	WOODALL,JM 1972 J.CRYST.GROWTH 12,32
		GAAS 001,111	LPE	OM,EP	WOODALL,JM+RUPPRECHT,H+REUTER,W 1969 J.ELECTROCHEM.SOC.116,899
	4	SAPPH. 00.1	CVD	EP,XD	MANASEVIT 1971.SEE ALAS 4
ALUMINUM-GALLIUM ANTIMONIDE	2	GASB 111	LPE	EP	GOLUBEV,LV+KHACHATURYAN,OA+SHIK,A.YA+ SHMARTSEV,YU.V 1974 PHYS.STAT.SOL.A22, K203
ALUMINUM-GALLIUM PHOSPHIDE-ARSENIDE	2	GAP-AS	LPE	OM	BURNHAM,RD+HOLONYAK,JR.N+SCIFRES,DR 1970 APPL.PHYS.LETT.17,455
		GAP-AS,AL-GAP-AS	LPE	OM,EP	BURNHAM,RD+HOLONYAK,JR.N+KORB,HW+MACKSEY,HM+ SCIFRES,DR+WOODHOUSE,JB+ALFEROV,ZH.I 1972 SOVIET PHYS.-SEMICOND.6, 77
		GAAS 001	LPE	XD,XM	ROZGONYI,GA+PANISH,MB 1973 APPL.PHYS.LETT.23,533
		GAAS	LPE	TEM,XM	ROZGONYI,GA+PETROFF,PM+PANISH,MB 1974 APPL.PHYS.LETT.24,251

Z= 13-14 OVERGROWTH		SUBSTRATE	PREP	OBSERV	REFERENCE
ALUMINUM-GALLIUM ARSENIDE-ANTIMONIDE	2	GAAS-SB 001	LPE	EP	SUGIYAMA,K+SAITO,H 1972 JAPAN.J.APPL.PHYS.11,1057
ALUMINA	4	SAPPH. 00.1,11.0, 10.2,21.3	CVD	OM,XD	MESSIER,DR+WONG,P 1971 J.ELECTROCHEM.SOC.118,772
		SAPPH. 00.1	CVD	OM	SCHAFFER,PS 1965 J.AMER.CERAMIC SOC.48,508
14 SILICON	1	W 001	UHV	FEM	COLLINS,RA 1971 SURFACE SCI.26,624
	2A	SI 111	UHV	REM	ABBINK,HC+BROUDY,RM+MCCARTY,GP 1968 J.APPL.PHYS.39,4673
		SI 111	CVD	OM	AVIGAL,Y+SCHIEBER,M 1970 JELECTROCHEMSOC.117,1585
		SI 111	CVD	OM,TEM, SEM	AVIGAL,Y+SCHIEBER,M 1971 J.CRYST.GROWTH 9,127
		SI 111	CVD	OM	BATSFORD,KD+THOMAS,DJD 1962SOLID-ST.ELECTRON5,353
		SI 111	UHV	OM,SEM	BENNETT,RJ+GALE,RW 1970 PHIL.MAG.22,135
		SI	UHV	---	BENNETT,RJ+PARISH,C 1973 SOLID-ST.ELECTRON.16,497
		SI 111	CVD	OM,XD	BHOLA,SR+MAYER,A 1963 RCA REV.24,511
		SI 111	CVD	OM	BLOEM,J 1971 J.ELECTROCHEM.SOC.118,1837
		SI 111	CVD	---	BLOEM,J 1973 J.CRYST.GROWTH 18,70
		SI 111	CVD	OM	BLOEM,J+GOEMANS,AH 1972 J.APPL.PHYS.43,1281
		SI 001,111	HV, CVD	TEM	BOOKER,GR 1964 DISCUSS.FARADAY SOC.38,298
		SI	CVD	TEM	BOOKER,GR 1965*3ERCEM A,383
		SI 111	HV, CVD	TEM	BOOKER,GR 1965 PHIL.MAG.11,1007
		SI 001,111	CVD	TEM	BOOKER,GR 1966 J.APPL.PHYS.37,441
		SI	HV, CVD	TEM	BOOKER,GR+HOWIE,A 1963 APPL.PHYS.LETT.3,156
		SI 111	CVD+ UHV	OM,RED, TEM,REM	BOOKER,GR+JOYCE,BA 1966 PHIL.MAG. 14,301
		SI 110	CVD	OM,RED, TEM	BOOKER,GR+JOYCE,BA+BRADLEY,RR 1964PHILMAG.10,1087
		SI 001,111,311	CVD+ UHV	OM,TEM, REM,SEM	BOOKER,GR+JOYCE,BA+BRADLEY,RR+WATTS,BE 1968*4ERCEM 1,537
		SI 111	CVD	OM,TEM	BOOKER,GR+STICKLER,R 1962 J.APPL.PHYS.33,3281
		SI	HV, CVD	TEM	BOOKER,GR+STICKLER,R 1963 APPL.PHYS.LETT.3,158
		SI 111	HV	OM,RED	BOOKER,GR+UNVALA,BA 1963 PHIL.MAG.8,1597
		SI 111	HV	TEM	BOOKER,GR+UNVALA,BA 1965 PHIL.MAG.11,11
		SI001,110,111,211		TEM	BOOKER,GR+VALDRE,U 1966 PHIL.MAG.13,421
		SI 111	UHV	LED	BROUDY,RM+ABBINK,HC 1968 APPL.PHYS.LETT.13,212
		SI 001,111	CVD	OM	BURMEISTER,J 1971 J.CRYST.GROWTH 11,131
		SI 111,110	CVD	OM	BYLANDER,EG 1962 J.ELECTROCHEM.SOC.109,1171
		SI 001 + PD	SPE	SEM,ICH	CANALI,C+MAYER,JW+OTTAVIANI,G+SIGURD,D+ VAN DER WEG,W 1974 APPL.PHYS.LETT.25, 3
		SI 111	CVD	OM,XD	CAVE,EF+CZORNY,BR 1963 RCA.REV.24,523
		SI 001,110,111	CVD	RED,REM	CHARIG,JM+JOYCE,BA 1962 J.ELECTROCHEM.SOC.109,957
		SI 001,110,111	CVD	OM,RED, TEM,REM	CHARIG,JM+JOYCE,BA+STIRLAND,DJ+BICKNELL,RW 1962 PHIL.MAG.7,1847
		SI 111	CVD, UHV	LED,AES	CHARIG,JM+SKINNER,DK 1969 SURFACE SCI.15,277
		SI001,110,111,211	CVD	OM	CHU,TL 1966 J.ELECTROCHEM.SOC.113,717
		SI001,110,111,211	CVD	OM	CHU,TL+GAVALER,JR 1963 J.ELECTROCHEM.SOC.110,388
		SI	CVD	OM,TEM	CHU,TL+GAVALER,JR 1963*AIMEMAEM 19,209
		SI 111	CVD	OM	CHU,TL+GAVALER,JR 1964 PHIL.MAG.9,993
		SI 111	SP	TED	CLARK,AH+ALIBOCEK,RG 1968 J.APPL.PHYS.39,2156
		SI 111	UHV	OM,TEM, REM,SEM	CULLIS,AG+BOOKER,GR 1971 J.CRYST.GROWTH 9,132
		SI	LPE	OM,XD	D≠ASARO,LA+LANDORF,RW+FURNANAGE,RA 1969*ISSI,233
		SI 001,111	CVD	OM,PCH	DEXTER,RJ+WATELSKI,SB+PICRAUX,ST 1973 APPL.PHYS.LETT.23,455
		SI 111	UHV	TEM,EP	DITCHFIELD,RW+CULLIS,AG 1970*7ICEM 2,125
		SI 001,111	CVD	OM,SEM	DUMIN,DJ 1971 J.CRYST.GROWTH 8,33
		SI 110,112	CVD	OM	DYER,LD 1971 J.ELECTROCHEM.SOC.118,957
		SI OFF 001,OFF111	CVD	---	EVERSTEYN,FC+VAN DEN HEUVEL,GJPM 1973 J.ELECTROCHEM.SOC.120, 699
		SI 111	CVD+ UHV	---	FARROW,RFC+FILBY,JD 1971J.ELECTROCHEM.SOC.118,149
		SI 111 + AU,CU, IN,SN	UHV+ VLS	OM,RED	FILBY,JD+NIELSEN,S 1965J.ELECTROCHEM.SOC.112,535/ 1966 BRIT.J.APPL.PHYS.17,81
		SI			FILBY,JD+NIELSEN,S+RICH,GJ 1966*NATOUTF,233
		SI 111+AU	HV+ VLS	OM,RED	FILBY,JD+NIELSEN,S 1967*3IVC 2,295
		SI	HV, CVD	OM,TEM, SEM	FILBY,JD+NIELSEN,S+RICH,GJ+BOOKER,GR+LARCHER,JM 1967 PHIL.MAG.16,565
		SI 001,110,111	CVD	OM,TEM	FINCH,RH+QUEISSER,HJ+THOMAS,G+WASHBURN,J 1963 J.APPL.PHYS.34,406
		SI 111	CVD	OM,RED	FRIESER,RG 1968 J.ELECTROCHEM.SOC.115,401
		SI	CVD	OM,RED	GITTLER,FL 1972 J.CRYST.GROWTH 17,271
		SI 001,110,111	CVD	OM	GIVARGIZOV,EI 1964 SOVIET PHYS-SOLID STATE 6,1415
		SI	CVD	---	GLANG,R+KIPPENHAN,BW 1960 IBM.J.RES.DEVELOP.4,299
		SI 111	CVD		GLANG,R+HAJDA,ES 1962*AIMEMSM 15,27
		SI 111	CVD	OM	GUPTA,DC+YEE,R 1969 J.ELECTROCHEM.SOC.116,1561

Z= 14 OVERGROWTH		SUBSTRATE	PREP	OBSERV	REFERENCE
SILICON (CONT.)	2A	SI 111,115	HV	OM,XD	HALE,AP 1963 VACUUM 13,93
		SI	UHV	OM,RED	HANDELMAN,E+POVILONIS,EI 1964 J.ELECTROCHEM.SOC.111,201
		SI 111	CVD	OM,RED, SEM,AES	HENDERSON,RC+HELM,RF 1972 SURFACE SCI.30,310
		SI	CVD		HOLONYAK,JR.N+JILLSON,DC+BEVACQUA,SF 1961 AIME CONF.METALL.ELEM+COMP.SEMICOND.12,81,INTERSC.PUB
		SI 111	CVD	OM,XD	INOUE,M 1965 J.ELECTROCHEM.SOC.112,189
		SI	CVD	---	JACCODINE,RJ 1963 APPL.PHYS.LETT.2,201
		SI 111	CVD	TEM	JACCODINE,RJ 1965 J.APPL.PHYS.36,2811
		SI 111	CVD+ VLS	OM	JAMES,DWF+LEWIS,C 1965 BRIT.J.APPL.PHYS.16,1089
		SI 111,001	UHV	LED	JONA,F 1966 APPL.PHYS.LETT.9,235
		SI 110	CVD	OM,RED, TEM	JOYCE,BA+BRADLEY,RR 1963 J.ELECTROCHEM.SOC.110,1235
		SI 111	CVD+ UHV	OM,RED, REM	JOYCE,BA+BRADLEY,RR 1966 PHIL.MAG.14,289
		SI 111	CVD+ UHV	OM,RED, TEM,REM, SEM	JOYCE,BA+BRADLEY,RR+BOOKER,GR 1967 PHIL.MAG.15,1167
		SI 001	CVD+ UHV	OM,RED, TEM, REM,SEM	JOYCE,BA+BRADLEY,RR+WATTS,BE+BOOKER,GR 1969 PHIL.MAG.19,403
		SI 111	CVD+ UHV	OM,RED, TEM, REM,LED	JOYCE,BA+NEAVE,JH+WATTS,BE 1969 SURFACE SCI.15,1
		SI 001,110,111	LPE	OM	KIM,HJ 1972 J.ELECTROCHEM.SOC.119,1394
		SI	CVD	OM,EP	KORMANY,T+KOSZA,G 1972 J.MATER.SCI.7,1080
		SI 111	CVD	OM,XD	LAWRENCE,JE+TUCKER,RN 1965 J.APPL.PHYS.36,3095
		SI	CVD	OM	LEKHOLM,A 1972 J.ELECTROCHEM.SOC.119,1122
		SI	CVD		LI,CH 1966 PHYS.STAT.SOLIDI 15,419
		SI 111	CVD	OM	LIETH,RMA+EGGELS,AGM 1964 J.APPL.PHYS.35,3015
		SI	CVD	OM,TEM	LIGHT,TB 1962*AIMEMSM 15,137
		SI	CVD	OM,RED	MARK,A 1961 J.ELECTROCHEM.SOC.108,880
		SI	CVD	---	MATTHEWS,JW+MADER,S+LIGHT,TB 1970 J.APPL.PHYS.41,3800
		SI 001,110,111, 112,114,221,334	CVD	OM	MENDELSON,S 1964*ICTFBB,251/ 1964 J.APPL.PHYS.35,1570/ 1964 J.METALS 16,108
		SI 111	CVD	OM	MENDELSON,S 1965 J.APPL.PHYS.36,2525
		SI	CVD	OM	MENDELSON,S 1965 ACTA MET.13,555
		SI 111	CVD	OM	MENDELSON,S 1967 J.APPL.PHYS.38,1573
		SI		OM	MENDELSON,S 1968 J.APPL.PHYS.39,2477
		SI 111	CVD	OM,RED, REM	MILLER,DP+WATELSKI,SB+MOORE,CR 1963 J.APPL.PHYS.34,2813
		SI	CVD	---	NAKANUMA,S 1966IEEETRANSELECTRONDEVICES ED-13,578
		SI 111	UHV	RED	NANNICHI,Y 1963 NATURE(LONDON) 200,1087
		SI 001,111	CVD	OM,RED	NEWMAN,RC+WAKEFIELD,J 1963 J.ELECTROCHEM.SOC.110,1068
		SI	HV, UHV	RED,REM	NIELSEN,S+COATES,DG+MAINES,JE 1964*ISCES,685
		SI 111	HV, CVD	OM	NIELSEN,S+RICH,GJ 1964 MICROELECTRON.RELIABILITY 3,165+171
		SI 111	CVD	OM	NIELSEN,S+RICH,GJ 1965 J.APPL.PHYS.36,3360
		SI	CVD	OM	NIELSEN,S+RICH,GJ+FAIRHURST,K 1964 MICROELECTRON.RELIABILITY 3,233
		SI 001,110,111, 112,113	CVD	OM	NISHIZAWA,J+TERASAKI,T+SHIMBO,M 1972 J.CRYST.GROWTH 13/14,297/ 1972 IBID.17,241
		SI 111	CVD	OM	NOTIS,MR+CONARD,GP 1964 J.APPL.PHYS.35,695
		SI 111	SP	RED, TEM, REM	PCHELYAKOV,OP+LOVYAGIN,RN+KRIVOROTOV,EA+ TOROPOV,AI+ALEKSANDROV,LN+STENIN,SI 1973 PHYS.STATUS SOLIDI A17,339
		SI 111	SP	TEM,REM	PCHELYAKOV,OP+LOVYAGIN,RN+TOROPOV,AI+STENIN,SI 1973 PHYS.STAT.SOLIDI A17,547
		SI 111	HV	OM,RED	PETRIN,AI+KUROV,SA 1966 SOVIET PHYS.CRYST.10,634
		SI 110	CVD	TEM	PHILLIPS,VA 1970*28EMSA,508/ 1972ACTA MET.20,1143
		SI 111	CVD	OM	POMERANTZ,D 1967 J.APPL.PHYS.38,5020
		SI 111	CVD	OM	POMERANTZ,DI 1972 J.ELECTROCHEM.SOC.119,255
		SI 111		OM,TEM	QUEISSER,HJ+FINCH,RH+WASHBURN J 1962 J.APPL.PHYS.33,1536
		SI 111	CVD	OM,RED, TEM	RAI-CHOUDHURY,P 1971 J.CRYST.GROWTH 8,165
		SI 001,111	CVD	OM,SEM	RAI-CHOUDHURY,P 1971 J.ELECTROCHEM.SOC.118,1183
		SI 001,111	CVD	OM,SEM	RAI-CHOUDHURY,P+SCHRODER,DK 1971 J.ELECTROCHEM.SOC.118,107
		SI OFF 111	CVD	---	RAI-CHOUDHURY,P+SCHRODER,DK 1972 J.ELECTROCHEM.SOC.119,1580
		SI001,110,111,115	CVD	OM	RAI-CHOUDHURY,P+SCHRODER,DK 1973 J.ELECTROCHEM.SOC.120, 664
		SI 111	CVD	XM	RAI-CHOUDHURY,P+TAKEI,WJ 1969 JAPPL.PHYS.40,4980
		SI 111	CVD	OM,TEM, XM	RAI-CHOUDHURY,P+NOREIKA,AJ+THEODORE,ML 1969 J.ELECTROCHEM.SOC.116,97
		SI 111	CVD	OM,RED, REM,XD	REVESZ,AG+EVANS,RJ 1964 TRANS.AIME 230,581
		SI 111	CVD	XD	RICHMAN,D+ARLETT,RH 1969J.ELECTROCHEM.SOC.116,872

Z= 14 OVERGROWTH		SUBSTRATE	PREP OBSERV	REFERENCE
SILICON (CONT.)	2A	SI 111	CVD OM,SEM	RICHMAN,D+ARLETT,RH 1969*1ISSI,200
		SI	CVD OM	ROY,K 1971 J.CRYST.GROWTH 9,139
		SI		RUNYAN,WR 1969*1ISSI,169
		SI	CVD OM,XD	SANGSTER,RC+MAVERICK,EF+CROUTCH,ML 1957 J.ELECTROCHEM.SOC.104,317
		SI 111+AL	SPE SEM	SANKUR,H+MCCALDIN,JO+DEVANEY,J 1973 APPL.PHYS.LETT.22,64
		SI 111	CVD OM,XM	SCHWUTTKE,GH 1962 J.APPL.PHYS.33,1538
		SI	XM	SCHWUTTKE,GH 1967*3IVC 2,301
		SI 111	CVD OM,XM	SCHWUTTKE,GH+SILS,V 1963 J.APPL.PHYS.34,3127
		SI 111	CVD OM,XM	SEKI,Y+TANNO,K+MATSUI,J+KAWAMURA,T 1969*1ISSI,653
		SI 111,113	CVD OM	SHIMBO,M+TERASAKI,T+NISHIZAWA,J 1971 J.APPL.PHYS.42,487
		SI 111	CVD OM	SIRTL,E 1963 J.PHYS.CHEM.SOLIDS 24,1285
		SI 001,111	CVD ---	SKELLY,G+ADAMS,AC 1973 J.ELECTROCHEM.SOC.120,116
		SI	CVD TEM	SOROKIN,LM+PROCHOROV,VI 1970*7ICEM 2,401
		SI 001,111	CVD XM	SUGITA,Y+TAMURA,M+SUGAWARA,K 1969 J.APPL.PHYS.40,3089/ 1969JVACSCITECHN.6,585/ 1970 J.APPL.PHYS.41,1877
		SI 111	CVD OM	SUNAMI,H+TERASAKI,T+MIYAMOTO,N+NISHIZAWA,J 1969 J.APPL.PHYS.40,4670
		SI 111	CVD OM	SUZUKI,T+URA,M+OGAWA,T 1972 JAPANJAPPLPHYS.11,667
		SI 001,110,111	CVD OM	TAFT,EA 1971 J.ELECTROCHEM.SOC.118,1535
		SI 110,112,113	CVD XM	TAMURA,M+SUGITA,Y 1973 J.APPL.PHYS.44,3442
		SI 111	CVD OM,RED	THEURER,HC 1961 J.ELECTROCHEM.SOC.108,649
		SI 111	HV RED,TEM	THOMAS,DJD 1966 PHYS.STATUS SOLIDI 13,359
		SI 001,111	UHV LED	THOMAS,RN+FRANCOMBE,MH 1967 APPL.PHYS.LETT.11,108
		SI 111	UHV TEM,LED	THOMAS,RN+FRANCOMBE,MH 1967 APPL.PHYS.LETT.11,134
		SI 001,111	UHV RED	THOMAS,RN+FRANCOMBE,MH 1968 APPL.PHYS.LETT.13,270
		SI 001,111	UHV RED,TEM	THOMAS,RN+FRANCOMBE,MH 1969SOLIDSTELECTRON.12,799
		SI 111	UHV TEM,LED	THOMAS,RN+FRANCOMBE,MH 1971 SURFACE SCI.25,357
		SI 111	CVD+SEM VLS	THORNTON,PR+JAMES,DWF+LEWIS,C+BRADFORD,A 1966 PHIL.MAG.14,165
		SI	CVD OM,XD	TOWNSEND,WG+UDDIN,ME 1973 SOLID-ST.ELECTRON.16,39
		SI 111	CVD	TUNG,SK 1962*AIMEMSM 15,87
		SI	CVD OM,XD	TUNG,SK 1965 J.ELECTROCHEM.SOC.112,436
		SI	HV RED	UNVALA,BA 1962 NATURE (LONDON) 194,966
		SI 111	HV OM,RED	UNVALA,BA 1963 LE VIDE NO.104,109
		SI 111	HV OM,RED	UNVALA,BA+BOOKER,GR 1964 PHIL.MAG.9,691
		SI 111	SP TEM	UNVALA,BA+PEARMAIN,K 1970 J.MATER.SCI.5,1016
		SI	VLS ---	WAGNER,RS+ELLIS,WC 1964 APPL.PHYS.LETT.4,89/ 1965 TRANS.AIME 233,1053
		SI 001,111	CVD OM,RED, XD	WAJDA,ES+KIPPENHAN,BW+WHITE,WH 1960 IBM J.RES.DEVELOP.4,288
		SI 110	CVD TEM	WASHBURN,J+THOMAS,G+QUEISSER,HJ 1963 APPL.PHYS.LETT.3,44
		SI 001	CVD+OM,RED, UHV TEM,SEM	WATTS,BE+BRADLEY,RR+JOYCE,BA+BOOKER,GR 1968 PHIL.MAG.17,1163
		SI 001,111	UHV RED,XD	WEISBERG,LR 1967 J.APPL.PHYS.38,4537
		SI 111	UHV RED,TEM	WIDMER,H 1964APPL.PHYS.LETT.5,108/1967*3IVC 2,309
		SI 111	CVD TEM	WILHELM,FJ,JOSHI,ML 1966 J.APPL.PHYS.37,1933
		SI 111	CVD OM	WILLIAMS,JH+JACKSON,JR.DM 1969*1ISSI, 206
		SI	CVD OM,TEM, REM	WOLLEY,ED+STICKLER,R+CHU,TL 1968 J.ELECTROCHEM.SOC.115,409
		SI 111	CVD OM,RED, XD	ZEVEKE,TA+KORNEV,LN+TOLOMASOV,VA 1968 SOVIET PHYS.CRYST.12,919
	2B	ZNSIP2	CVD OM,RED, XD	BERTOTI,I 1970 J.MATER.SCI.5,1073
		ZNS 110	HV RED	SEGMULLER,A 1956 Z.KRIST.107,18
		GE 111	UHV LED	TAKEISHI,Y+SASAKI,I+HIRABAYASHI,K 1967 APPL.PHYS.LETT.11,330
		SIC 00.1	CVD OM,RED	TALLMAN,RL+CHU,TL+GRUBER,GA+OBERLEY,JJ+WOLLEY,ED 1966 J.APPL.PHYS.37,1588
		SIC 00.1	CVD OM,RED	TALLMAN,RL+CHU,TL+OBERLEY,JJ 1966 SOLID-ST.ELECTRON.9,327
	3	NACL 001	UHV TEM	SHIMAOKA,G+CHANG,SC 1971*29EMSA, 224
		NACL 001	HV, RED,TEM UHV	SHIMAOKA,G+CHANG,SC 1972 J.VAC.SCI.TECHN.9,235
	4	SAPPHIRE 00.1	CVD OM	ANG,CY+MANASEVIT,HM 1965 SOLID-ST.ELECTRON.8,994
		QUARTZ 10.0,11.0, 00.1	CVD RED,REM, TEM	BICKNELL,RW+CHARIG,JM+JOYCE,BA-+STIRLAND,DJ 1964 PHIL.MAG.9,965
		SAPPH. 00.1,11.2	CVD RED,TEM, REM,XD	BICKNELL,RW+JOYCE,BA+NEAVE,JH+SMITH,GV 1966 PHIL.MAG.14,31
		SAPPHIRE 22.3	CVD OM,XD	BLANK,JM+RUSSEL,VA 1966 TRANS.AIME 236,291
		SAPPH. 10.2,11.3, 00.1	UHV LED	CHANG,CC 1969*4IMATC, 77-1
		SAPPH. 10.2,11.3, 00.1,11.0	UHV LED,AES	CHANG,CC 1971 J.VAC.SCI.TECHN.8,500
		SPINEL 001,111, SAPPH. 10.2,00.1, 11.3	CVD ---	CHIANG,YS+LOONEY,GW 1973 JELECTROCHEM.SOC.120,550

Z= 14 OVERGROWTH		SUBSTRATE	PREP	OBSERV	REFERENCE
SILICON (CONT.)	4	SAPPHIRE,SPINEL	CVD	OM,SEM	CULLEN,GW 1971 J.CRYST.GROWTH 9,107
		SPINEL 111	CVD	SEM	CULLEN,GW+DOUGHERTY,FC 1972 J.CRYST.GROWTH 17,230
		SPINEL 111, SAPPHIRE 10.2	CVD	SEM	CULLEN,GW+WANG,CC 1971 J.ELECTROCHEM.SOC.118,640
		SPINEL 111	CVD	---	CULLEN,GW+GOTTLIEB,GE+WANG,CC+ZAININGER,KH 1969*1ISSI,291/ 1969 J.ELECTROCHEM.SOC.116,1444
		SAPPH.,SPINEL 111	CVD	---	CULLEN,GW+GOTTLIEB,GE+WANG,CC 1970 RCA REV.31,355
		SAPPHIRE 10.2, SPINEL 001	CVD	OM	DRUMINSKI,M+SCHLOTTERER,H 1972 J.CRYST.GROWTH 17,249
		SAPPHIRE	CVD	OM,RED, XD	DUMIN,DJ 1965 J.APPL.PHYS.36,2700
		SAPPH. 00.1,11.2	CVD	OM	DUMIN,DJ 1967 J.APPL.PHYS.38,1909
		SAPPH. 00.1,10.2	CVD	OM	DUMIN,DJ+ROBINSON,PH 1966 JELECTROCHEMSOC.113,469
		SAPPH. 10.2 + AU	HV+ VLS	OM,RED	FILBY+NIELSEN 1967.SEE 2A
		SAPPH. 00.1,11.3	CVD, UHV	OM,RED	FRAIMBAULT,JL+GYOMLAI,I+MONTMORY,R+VUILLOD,J 1966*ISTFCL,638
		SPINEL 111	UHV	XD	GASSMANN,F+DELLA CASA,A+AESCHLIMANN,R 1971 MATER.RES.BULL.6,817
		SPINEL 111	UHV	OM,XD	GASSMANN,F+AESCHLIMANN,R+BANZIGER,U 1972 MATER.RES.BULL.7,1493
		SPINEL 111	CVD	---	GOTTLIEB,GE 1972 J.CRYST.GROWTH 12,327
		SAPPHIRE 10.2	CVD	RED	GOTTLIEB,GE+CORBOY,JR.JF 1972 JCRYSTGROWTH 17,261
		SAPPHIRE 10.2	CVD	OM,XD	HART,PB+ETTER.PJ+JERVIS,BW+FLANDERS,JM 1967 BRIT.J.APPL.PHYS.18,1389
		SPINEL 001,111	CVD	OM,TEM, REM,XD	HEYWANG,W 1968 MATER.RES.BULL.3,315
		SAPPH. 00.1,10.2	HV, UHV	RED	ITOH,T+HASEGAWA,S+KAMINAKA,N 1968 J.APPL.PHYS.39,5310
		SAPPH. 00.1,10.2	HV, UHV	RED,REM	ITOH,T+HASEGAWA,S+WATANABE,H 1968 J.APPL.PHYS.39,2969
		SPINEL 111	UHV	RED	ITOH,T+HASEGAWA,S+KAMINAKA,N 1969 J.APPL.PHYS.40,2597
		SAPPH. 10.0,11.0, 00.1-QUARTZ10.0, 11.0,00.1	CVD	RED,XD	JOYCE,BA+BENNET,RJ+BICKNELL,RW+ETTER,PJ 1965 TRANS.AIME 233,556
		QUARTZ 11.0,00.1	CVD	RED,TEM	JOYCE,BA+BICKNELL,RW+CHANG,JM+STIRLAND,DJ 1963 SOLID-ST.COMMUN.1,107
		SAPPHIRE 10.2	CVD	RED	KROKO,LJ+SHAW,GL 1969*1ISSI,316
		SAPPH. 11.0,10.2, 00.1	CVD	XD	LARSSEN,PA 1966 ACTA CRYST.20,599
		SAPPHIRE	UHV	TEM	LININGTON,PF 1970*7ICEM 2,447
		SPINEL001,110,111	CVD	OM,XD	MANASEVIT,HM+FORBES,DH 1966 J.APPL.PHYS.37,734
		SAPPHIRE	CVD	OM,XD	MANASEVIT,HM+SIMPSON,WI 1964 J.APPL.PHYS.35,1349
		SAPPH. 10.2,11.0, 11.3,11.4,00.1	CVD	OM,REM, XD	MANASEVIT,HM+MILLER,A+MORRITZ,FL+NOLDER,RL 1965 TRANS.AIME 233,540
		BEO 10.0,10.1, 20.1,00.1	CVD	OM,XD	MANASEVIT,HM+FORBES,DH+CADOFF,IB 1966 TRANS.AIME 236,275
		SAPPHIRE	CVD	OM,RED, XD	MANASEVIT,HM+NOLDER,RL+MOUDY,LA 1968 TRANS.AIME 242,465
					MATARE,HF 1969*1ISSI,249
		SAPPHIRE 10.2	CVD	XD,XM	MERCIER,J 1970 J.ELECTROCHEM.SOC.117,666
		SAPPHIRE	CVD	XM	MERCIER,J 1970 J.ELECTROCHEM.SOC.117,812
		SAPPHIRE 10.2	CVD	RED,REM, SEM,XD, IP	MERCIER,J 1971 J.ELECTROCHEM.SOC.118,962
		SPINEL001,110,111, SAPPH. 10.2,11.0, 11.3,11.4,00.1, BEO 10.0,10.1, 20.1,00.1	CVD	OM,REM, XD	MILLER,A+MANASEVIT,HM 1966 J.VAC.SCI.TECHN.3,68
		SAPPHIRE	CVD	OM,RED XD	MUELLER,CW+ROBINSON,PH 1964 PROC.IEEE 52,1487
		SAPPH. 10.2,00.1	UHV	OM,RED	NABER,CT+O≠NEAL,JE 1968 TRANS.AIME 242,470
		SAPPH. 10.2,11.0, 11.4,00.1	CVD	XD	NOLDER,RL+CADOFF,I 1965 TRANS.AIME 233,549
		SAPPH. 10.2,11.0, 11.3,11.4,00.1	CVD	REM,XD	NOLDER,RL+KLEIN,DI+FORBES,DH 1965 J.APPL.PHYS.36,3444
		SPINEL 001,111	CVD	PCH	PICRAUX,ST 1972 APPL.PHYS.LETT.20,91/ 1973 J.APPL.PHYS.44,587
		SPINEL 001,111, SAPPHIRE 10.2	CVD	OM,TEM, SEM,PCH	PICRAUX,ST+THOMAS,GJ 1973 J.APPL.PHYS.44,594
		SAPPH. 00.1,26.9	CVD	OM,XD	PORTER,JL+WOLFSON,RG 1965 J.APPL.PHYS.36,2746
		SAPPH. 00.1	HV	OM,RED	REYNOLDS,FH+ELLIOT,AB 1967SOLIDSTELECTRON.10,1093
		SPINEL001,110,111	CVD	OM,RED	ROBINSON,PH+DUMIN,DJ 1968J.ELECTROCHEM.SOC.115,75
		SAPPH. 10.2,00.1	CVD	OM,RED, REM,XD	ROBINSON,PH+MUELLER,CW 1966 TRANS.AIME 236, 268
		SAPPH. 00.1	HV	OM,RED	SALAMA,CAT+TUCKER,TW+YOUNG,L 1967 SOLID-ST.ELECTRON.10,339
		SAPPH. 00.1,10.2, SPINEL 001,111	CVD	---	SCHLOTTERER,H 1968 SOLID-ST.ELECTRON.11,947
		SPINEL 001,111	CVD	TEM,REM	SCHLOTTERER,H+ZAMINER,CH 1966 PHYSSTATSOL.15,399
		SAPPHIRE 11.2	CVD	OM,RED, SEM	SCHRODER,DK+RAI-CHOUDHURY,P 1973 APPL.PHYS.LETT.22,455

Z= 14 OVERGROWTH		SUBSTRATE	PREP	OBSERV	REFERENCE
SILICON (CONT.)	4	SPINEL 001,111	CVD	OM,XD	SEITER,H+ZAMINER,CH 1965 Z.ANGEW.PHYS.20,158
		CAF2	HV	OM,RED	UNVALA 1963.SEE 2A
		SPINEL 001,110,111	CVD	OM,RED, REM,SEM, XD	WANG,CC+GOTTLIEB,GE+CULLEN,GW+MCFARLANE III,SH+ ZAININGER,KH 1969 TRANS.AIME 245,441
		SAPPH. 00.1,10.2	UHV	XD	WEISBERG,LR+MILLER,EA 1968 TRANS.AIME 242,479
		SPINEL 001	SP	RED,SEM	WEISSMANTEL,CHR+FIEDLER,O+HECHT,G+REISSE,G 1972 THIN SOLID FILMS 13,359
		SAPPH. 10.2,11.0, SPINEL 001,111,113	HV, UHV	RED,REM	YASUDA,Y 1971 JAP.J.APPL.PHYS.10,45
		SAPPH. 00.1	HV	RED,REM, XD	YASUDA,Y+OHMURA,Y 1969 JAP.J.APPL.PHYS.8,1098
		SPINEL 001	CVD	TEM,REM	ZAMINER,C 1968 Z.ANGEW.PHYS.24,223
		SAPPH. 00.1,11.0	CVD	OM,RED	ZEVEKE,TA+KORNEV,LN+TOLOMASOV,VA 1968 SOVIET PHYS.CRYST.13,493
		SAPPHIRE 00.1	CVD	XD,XM	ZEYFANG,R 1970 THIN SOLID FILMS 6,321
SILICON-GERMANIUM ALLOYS	2	SI	CVD	OM,SEM, EP,XD	AHARONI,H+BAR-LEV,A+MARGALIT,S 1972 J.CRYST.GROWTH 17,254
		SI 111,ETC.	CVD	OM,ER, XD	AHARONI,H+BAR-LEV,A+BLECH,IA+MARGALIT,S 1972 THIN SOLID FILMS 11,313
		SI 001,110,111	CVD	OM,RED, XD	MILLER,KJ+GRIECO,MJ 1962 J.ELECTROCHEMSOC.109,70
		GE 001,111	CVD	OM,RED, EP,XD	NEWMAN+WAKEFIELD 1963.SEE SI 2A
		GE 001,110	CVD	OM,XD	OLESZCH,GM+ANDERSON,RL 1973 J.ELECTROCHEM.SOC.120,554
SILICON-PALLADIUM ALL.	2	SI 111 + PD	SPE	RED,TED, SEM	BUCKLEY,WD+MOSS,SC 1972 SOLID-ST.ELECTRON.15,1331
		SI 111	CG	OM,RED, EP	HUTCHINS,GA+SHEPELA,A 1973 THINSOLIDFILMS 18,343
SILICON-GOLD ALLOYS	2	SI 111	UHV+ CG	TED,AES	NARUSAWA,T+SAKURAI,T+HAYAKAWA,K 1971 JAPAN.J.APPL.PHYS.10,280
		SI 001,111	HV+ CG	OM,SEM, XD	PHILOFSKY,E+RAVI,KV+BROOKS,J+HALL,E 1972 J.ELECTROCHEM.SOC.119,527
SILICON CARBIDES	2A	SIC 00.1	SG	OM,XD	BRANDER,RW+SUTTON,RP 1969J.PHYS.D=APPL.PHYS.2,309
		SIC 00.1	CVD	OM,XD, XM	CALLAGHAN,MP+BRANDER,RW 1972 J.CRYST.GROWTH 13/14,397
		SIC 00.1	CVD	OM	CAMPBELL,RB+CHU,TL 1966 J.ELECTROCHEMSOC.113,825
		SIC	CVD	OM,XD	GRAMBERG,G+KONIGER,M 1972SOLID-ST.ELECTRON.15,285
		SIC 00.1	CVD	OM	HARRIS,JM+GATOS,HC+WITT,AF 1971 J.ELECTROCHEM.SOC.118,335
		SIC 00.1	CVD	OM,XD	JENNINGS,VJ+SOMMER,A+CHANG,HC 1966 J.ELECTROCHEM.SOC.113,728
		SIC	VG	OM,XD	KNIPPENBERG,WF 1963 PHILIPS RES.REPT.18,161
		SIC 111	CVD	OM,RED	KUROIWA,K+SUGANO,T 1973 J.ELECTROCHEM.SOC.120,138
		SIC	VG	OM	LIEBMAN,WK 1964 J.ELECTROCHEM.SOC.111,885
		SIC 00.1	CVD	OM,SEM, XD	MINAGWA,S+GATOS,HC 1971 JAP.J.APPL.PHYS.10,1680
		SIC	CVD	OM,XD	POWELL,JA+WILL,HA 1973 J.APPL.PHYS.44,5177
		SIC	CVD	OM,RED, TEM,XM	RAI-CHOUDHURY,P+FORMIGONI,NP 1969 J.ELECTROCHEM.SOC.116,1440
		SIC 00.1	CVD	OM	SIRTL 1963.SEE SI 2A
		SIC 00.1	CVD	XD	SPIELMAN,W 1965 Z.ANGEW.PHYS.19,93
		SIC	VG	OM	SWIDERSKI,I+SZCZUTOWSKI,W+NIEMYSKI,T 1974 J.CRYST.GROWTH 21,125
		SIC 00.1	VG	OM,EP, XD	YAMADA,S+KUMAGAWA,M 1971 J.CRYST.GROWTH 9,309
	2B	SI001,110,111,311	CG	RED,TEM, REM	BROWN,AS+WATTS,BE 1970 J.APPL.CRYST.3, 172
		SI 111	UHV+ CG	OM,RED	HAQ,KE+KHAN,IH 1970 J.VAC.SCI.TECH.7,490
		SI 111	UHV+ CG	RED,LED	HENDERSON,RG+POLITO,WJ+SIMPSON,J 1970 APPL.PHYS.LETT.16,15
		SI	UHV+ CG	RED,TEM, REM	HENDERSON,RG+MARCUS,RB+POLITO,WJ 1971 J.APPL.PHYS.42,1208
		SI 001,111	CVD	OM,TED, XD	JACKSON,DM+HOWARD,RW 1965 TRANS.AIME 233,468
		SI001,110,111,211	UHV+ CG	RED,REM	KHAN,IH+SUMMERGRAD,RN 1967 APPL.PHYS.LETT.11,12
		SI 001,111	UHV+ CG	RED,LED, AES	KRAUSE,GO 1970 PHYS.STATUS SOLIDI 3,899
		SI	CG	SEM	LEAMY,HJ+MOGAB,CJ 1973*31EMSA, 36
		SI 111	CG	OM,RED, SEM	LEARN,AJ+KHAN,IH 1970 THIN SOLID FILMS 5,145
		SI 111	SP	RED	MATSUMOTO,S+SUZUKI,H+UEDA,R 1972 JAPAN.J.APPL.PHYS.11,607
		SI 111	CVD	OM,RED, TEM	MILLER ET AL 1963.SEE SI 2A

Z= 14-19 OVERGROWTH		SUBSTRATE	PREP	OBSERV	REFERENCE
SILICON CARBIDES(CONT)	2B	SI 111	CG	OM,TEM, SEM	MOGAB,CJ+LEAMY,HJ 1974 J.APPL.PHYS.45,1075
		SI	CG	RED	NAKASHIMA,H+SUGANO,T+YANAI,H 1966 JAP.J.APPL.PHYS.5,874
		SI 111	CG	---	NEWMAN,RC+WAKEFIELD,J 1960 SOLID STATE PHYSICS IN ELECTRONICS TELECOM. PROC.INT.CONF.BRUSSELS,1958 2,318.ACADEMIC PRESS
		SI 001,110,111	CVD	OM,RED, TEM,XM	RAI-CHOUDHURY+FORMIGONI 1969.SEE 2A
		SI 111	CG	OM,RED	SATO,K 1964 SOLID-ST. ELECTRON.7,743
		SI 001,111	CG	RED	TOMBS,NC+COMER,JJ+FITZGERALD,JF 1965 SOLID-ST.ELECTRON.8,839
18 ARGON	1	NB 001	VG	LED	DICKEY,JM+FARRELL,HH+STRONGIN,M 1970 SURFACE SCI.23,448
	2	GRAPHITE 00.1	VG	TEM	BALL,DJ+VENABLES,JA 1970*7ICEM 2,459
19 POTASSIUM	1	NI 001	UHV	LED	ANDERSSON,S+JOSTELL,U 1973 SOLID-ST.COMMUN.13,829
		NI 110	UHV	LED	GERLACH+RHODIN 1969(A,B).SEE NA 1
		W 001	UHV	LED,AES	THOMAS,S+HAAS,TW 1973 J.VAC.SCI.TECHN.10,218
	3	NACL 001,110,111, CSCL 001,110,111	HV	---	REALE,C 1972 PHYS.LETT.40A,71
POTASSIUM FLUORIDE	2	MNS 001	SG	OM	ROYER 1928.SEE LICL 2
	3	NACL 001	HV	TED	RAMOS ET AL 1964.SEE NAI 3
		NACL 001	SG	OM	ROYER 1928.SEE LICL 2
		LIF 001,NACL 001, KCL 001,KBR 001	HV	RED	SCHULZ 1952(A).SEE LIF 3
		NACL 001	SG	OM	SLOAT+MENZIES 1931.SEE LICL 3
	4	MICA 00.1	HV, SG	RED	SCHULZ 1951(B).SEE LIF 4
		CACO3 100	HV, SG	RED	SCHULZ 1952(B).SEE LIF 4
POTASSIUM CHLORIDE	1	AG 001,111,BI 111	SG	OM	JOHNSON 1950(A),1951.SEE NACL 1
		AG 001,111	HV, SG	RED	SCHULZ 1952(C).SEE LIF 1
	2	PBS 001	SG	OM	ROYER 1928.SEE LICL 2
		PBS 001	SG	OM	SLOAT+MENZIES 1931.SEE LICL 3
	3	NACL 001,KCN 001, KBR 001,RBCN 001, RBBR 001	SG	OM	BARKER 1906,1907,1908.SEE NH4I 3
		LIF 001,NACL 001, KCL 001,KBR 001, KI 001	HV	RED,REM	LUDEMANN 1954.SEE LIF 3
		LIF 001	HV	RED	LUDEMANN 1956.SEE LIF 3
		KCL 001	HV	RED	LUDEMANN+RAETHER 1953.SEE LIF 3
		NACL 001,RBCL 001	SG	OM	ROYER 1928.SEE LICL 2
		LIF 001,NACL 001, KCL 001,KBR 001	HV	RED	SCHULZ 1952(A).SEE LIF 3
		NACL 001,KCL 001	SG	OM	SLOAT+MENZIES 1931.SEE LICL 3
		KCL 001	HV, SG	RED,REM	SONKSEN,D 1956 Z.NATURFORSCH.11A,646
	4	MICA 00.1	SG	OM	DEICHA,G 1946 C.R.ACAD.SCI.223,1155/ 1947(A,B,C).SEE NACL 4
		MICA 00.1	HV	REM	EGERTON,RF 1970 THIN SOLID FILMS 5.R41
		MICA 00.1	HV	---	GREEN,M 1971 SURFACE SCI.26,549
		NANO3 100	SG	OM	HEINTZE 1937.SEE LICL 4
		CAF2 111	SG	OM	KRASTANOV+STRANSKI 1938.SEE LIBR 4
		MICA 00.1	SG	OM	LISGARTEN 1954.SEE NH4CL 4
		MICA 00.1	SG	OM	ROYER 1928.SEE LICL 2
		CACO3 100	SG	OM	ROYER 1937.SEE NABR 4
		MICA 00.1	HV, SG	RED	SCHULZ 1951(B).SEE LIF 4
		NANO3 100, CACO3 100	HV, SG	RED	SCHULZ 1952(B).SEE LIF 4
		MICA 00.1	SG	OM,REM	UPRETI+WALTON 1966.SEE NH4I 4
POTASSIUM BROMIDE	1	W 110	UHV	LED,AES	MORRISON,J+LANDER,JJ 1969 SURFACE SCI.18,428
	2	PBS 001	SG	OM	SLOAT+MENZIES 1931.SEE LICL 3
	3	NACL 001,KCL 001, KCN 001,KI 001, RBCL 001,RBCN 001, RBBR 001,RBI 001	SG	OM	BARKER 1906,1907,1908.SEE NH4I 3

Z= 19-21	OVERGROWTH		SUBSTRATE	PREP	OBSERV	REFERENCE
	POTASSIUM BROMIDE(CONT.)	3	LIF 001,NACL 001, KCL 001,KBR 001, KI 001	HV	RED,REM	LUDEMANN 1954.SEE LIF 3
			LIF 001	HV	RED	LUDEMANN 1956.SEE LIF 3
			KBR 001	HV	RED	LUDEMANN+RAETHER 1953.SEE LIF 3
			NH4I 001,KI 001, RBBR 001,RBI 001	SG	OM	ROYER 1928.SEE LICL 2
			LIF 001,NACL 001, KCL 001,KBR 001	HV	RED	SCHULZ 1952(A).SEE LIF 3
			NACL 001,KCL 001	SG	OM	SLOAT+MENZIES 1931.SEE LICL 3
		4	MICA 00.1	SG	OM	DEICHA 1947(A,B,C).SEE NACL 3
			NANO3 100	SG	OM	HEINTZE 1937.SEE LICL 4
			MICA 00.1	SG	OM	LISGARTEN 1954.SEE NH4CL 4
			MICA 00.1	SG	OM	ROYER 1928.SEE LICL 2
			CACO3 100	SG	OM	ROYER 1937.SEE NABR 4
			MICA 00.1	HV, SG	RED	SCHULZ 1951(B).SEE LIF 4
			NANO3 100, CACO3 100	HV, SG	RED	SCHULZ 1952(B).SEE LIF 4
			MICA 00.1	SG	OM,REM	UPRETI+WALTON 1966.SEE NH4I 4
			MICA 00.1	MG	OM	WEST 1945.SEE NAI 4
	POTASSIUM IODIDE	1	AG 111	HV, SG	RED	SCHULZ 1952(C).SEE LIF 1
		2	PBS 001	SG	OM	SLOAT+MENZIES 1931.SEE LICL 3
		3	NACL 001,NABR 001, KCN 001,KBR 001, RBCN 001,RBBR 001, RBI 001	SG	OM	BARKER 1906,1907,1908.SEE NH4I 3
			LIF 001,NACL 001, KCL 001,KBR 001, KI 001	HV	RED,REM	LUDEMANN 1954.SEE LIF 3
			LIF 001	HV	RED	LUDEMANN 1956.SEE LIF 3
			KI 001	HV	RED	LUDEMANN+RAETHER 1953.SEE LIF 3
			KBR 001,RBCL 001, RBBR 001,RBI 001	SG	OM	ROYER 1928.SEE LICL 2
			LIF 001,NACL 001, KCL 001,KBR 001	HV	RED	SCHULZ 1952(A).SEE LIF 3
			NACL 001,KCL 001	SG	OM	SLOAT+MENZIES 1931.SEE LICL 3
		4	MICA 00.1	SG	OM	DEICHA 1947(A,B,C).SEE NACL 3
			NANO3 100	SG	OM	HEINTZE 1937.SEE LICL 4
			MICA 00.1	SG	OM	LISGARTEN 1954.SEE NH4CL 4
			MICA 00.1	SG	OM	ROYER 1928.SEE LICL 2
			CACO3 100	SG	OM	ROYER 1937.SEE NABR 4
			MICA 00.1	HV, SG	RED	SCHULZ 1951(B).SEE LIF 4
			NANO3 100, CACO3 100	HV, SG	RED	SCHULZ 1952(B).SEE LIF 4
			MICA 00.1	SG	OM,REM	UPRETI+WALTON 1966.SEE NH4I 4
			MICA 00.1	MG	OM	WEST 1945.SEE NAI 4
			MICA 00.1	HV	TEM	ZOUCKERMANN,R 1961*2ERCEM 1,316
	POTASSIUM CYANIDE	2-3	PBS 001, NACL 001,KCL 001	SG	OM	SLOAT+MENZIES 1931.SEE LICL 3
		4	NH4I 001	SG	OM	BARKER 1906,1907,1908.SEE NH4I 3
	POTASSIUM CHLORATE	2	PBS 001	SG	OM	SLOAT+MENZIES 1931.SEE LICL 3
20	CALCIUM FLUORIDE	2	MOS2 00.1,ZN 110	HV	RED	SHIGETA 1956.SEE LIF 2
		3	LIF 001, NACL 001,110,111	HV	RED,TEM, REM	BAUER 1956.SEE LIF 3
			NACL 001	HV	TEM	BUJOR,M+VOOK,RW 1969 J.APPL.PHYS.40,5373
			NACL 001	HV, UHV	RED,TEM, REM	KOCH,FA+VOOK,RW 1973 J.APPL.PHYS.44,2475
			NACL 001	UHV	RED,TEM	KOCH,FA+VOOK,RW 1973 J.VAC.SCI.TECHN.10,313
			NACL	HV	TEM	REICHELT,K+REY,M 1973 J.VAC.SCI.TECHN.10,1153
			NACL 001	HV	TEM	VOOK,RW+BUJOR,M 1970 J.VAC.SCI.TECHN.7,115
		4	MICA 00.1	HV	RED,TEM, REM	BAUER 1956.SEE LIF 3
	CALCIUM TITANATE	1-3	AU 001,LIF 001, NAF 001	HV	RED	MULLER,EK+NICHOLSON,BJ+TURNER,G.L≠E 1963 J.ELECTROCHEM.SOC.110,969
21	SCANDIUM NITRIDE	4	SAPPHIRE 10.2	CVD	XD	DISMUKES,JP+YIM,WM+TIETJEN,JJ+NOVAK,RE 1970 RCA REV.31,680
			SAPPHIRE	CVD	RED,SEM, XD	DISMUKES,JP+YIM,WM+BAN,VS 1972 J.CRYST.GROWTH 13/14,365

Z= 21-26 OVERGROWTH		SUBSTRATE	PREP	OBSERV	REFERENCE
SCANDIUM PHOSPHIDE	2	SI 001,111	CVD	OM,XD	YIM,WM+STOFKO,EJ+SMITH,RT 1972J.APPL.PHYS.43,254
SCANDIUM ARSENIDE	2	SI 001,111	CVD	OM,XD	YIM ET AL 1972.SEE SCP 2
22 TITANIUM	1	W,RH	UHV	FEM	ANDERSON,JR+THOMPSON,N 1971 SURFACE SCI.26,397
	3	NAF 001,KCL 001	UHV	RED	ARNTZ,F+CHERNOW,F 1965 J.VAC.SCI.TECHN.2,20
		NACL 001,KCL 001, KBR 001	HV	TED	CONJEAUD,P 1955 J.RECHERCHES CNRS.32,273/ 1956 J.CHIM.PHYS.53,620
		NACL 001,110,111	UHV	TEM	LAWLESS,KR+WAWNER,FE 1967*25EMSA, 296
		NACL 001,110,111	UHV	TEM	WAWNER,FE+LAWLESS,KR 1969 J.VAC.SCI.TECHN.6,588
	4	CAF2 111	UHV	RED	ARNTZ+CHERNOW 1965.SEE 3
		MICA 00.1	UHV	TEM	EADES,J 1970*7ICEM 2,421
		MICA 00.1	UHV	RED,TEM	GRUNBAUM,E+SCHWARZ,R 1969 J.APPL.PHYS.40,3364
		SAPPHIRE 00.1	UHV	OM,RED, REM	O≠NEAL,JE+HYATT,RL+LEONHARD,FW 1970 J.CRYST.GROWTH 7,177
		MICA 00.1	UHV	TEM	STEEDS,JW+EADES,JA 1973 SURFACE SCI.38,187
TITANIUM OXIDES	4	RUTILE001,110,111, MGO001,SAPPH.00.1	CVD	RED	GHOSHTAGORE,RN+NOREIKA,AJ 1970 J.ELECTROCHEM.SOC.117,1310
23 VANADIUM	4	SAPPH. 10.2,00.1	UHV	OM,RED, REM	O≠NEAL,JE+HYATT,RL 1971 THIN FILMS 2,1
		SAPPH. 10.2,00.1	UHV	RED,REM	O≠NEAL,JE+BELLINA,JR.JJ+RATH,BB 1972*30EMSA, 492
		TIO2 100,110,001	SP	RED,REM	ROZGONYI,GA+HENSLER,DH 1968 JVACSCITECHN.5,194
24 CHROMIUM	1	W 001	UHV	LED	BONDARENKO,BV+MAKHOV,VI+KOZLOV,AM 1970 SOVIET PHYS.SOLID STATE 11,2991
		AG 111	HV	TEM	CINTI,R+DEVENYI,J+ESCUDIER,P+MONTMORY,R+YELON,A 1965 C.R.ACAD.SCI.,260,6849
		NI,CU	ED	TEM	CLEGHORN,WH+WARRINGTON,DH+WEST,JM 1968 ELECTROCHIM.ACTA 13,331
		CU 111	ED	RED	COCHRANE,W 1936 PROC.PHYS.SOC,LONDON 48,723
		CU 001,110,111	ED	RED	GOSWAMI,A 1958 TRANS.FARADAY SOC.54,821
		NI 001	UHV	TEM	JESSER,WA+MATTHEWS,JW 1968 PHIL.MAG.17,475
	3	NACL 001	HV	TED	BRUCK 1936.SEE AL 3
		NACL 001	UHV	TEM	FUKANO,Y+WAYMAN,CM 1972 J.CRYST.GROWTH 15,32
		NACL 001	HV	TED	SELLA,TRILLAT 1964.SEE AL 3
		NACL 001	HV	TED	SHIRAI,S 1939 PROC.PHYS.-MATH.SOC.JAPAN 21,800
		NACL 001	HV	TEM	TAKAHASHI,N+MARTINA,H 1966 C.R.ACAD.SCI.263B,203
CHROMIUM SELENIDE	3-4	NACL 001,110,111, MICA 00.1	HV	RED,TED	GOSWAMI,A+NIKAM,PS 1972 THIN SOLID FILMS 11,353
CHROMIUM TELLURIDE	3-4	NACL 001,110,111, MICA 00.1	HV	RED,TED	GOSWAMI+NIKAM 1972.SEE CRSE 3-4
CHROMIUM OXIDES	4	RUTILE001,110,210, AL2O3 00.1, FE2O3 00.1	CVD	OM,XD	DE VRIES,RC 1966 MATER.RES.BULL.1,83
25 MANGANESE	1	AL 111	UHV	LED	EDWARDS,IAS+THIRSK,HR 1973 SURFACE SCI.39,245
MANGANESE-BISMUTH ALLOY	4	MICA 00.1	HV	TEM,XD	CHEN,D 1966 J.APPL.PHYS.37,1486
		MICA 00.1	HV	TEM	UNGER,WK+STOLZ,M 1971 J.APPL.PHYS.42,1085
MANGANESE OXIDES	4	MGO 001	CVD	REM,EP, XD	CASLAVSKA,V+ROY,R 1970 J.APPL.PHYS.41,825
26 IRON	1	AG 111	HV	TEM	BIRAGNET,F+DEVENYI,J+ESCUDIER,P+MONTMERY,R+ PACCARD,D+YELON,A 1966*ISTFCL,447
		AU 001,111	HV	TED	CAHOREAU,M+GILLET,M 1965 J.MICROSCOPE 4,207
		AU 111	HV	TEM	CAHOREAU,M+GILLET,M 1969 C.R.ACAD.SCI.B268,1024/ 1970 IBID.B271,425+445/ 1970*7ICEM 2,453/ 1971 SURFACE SCI.26,415
		AG 111	HV	TEM	CINTI ET AL 1965.SEE CR 1
		PD 001,AU 001	ED	TED	FINCH,GI+SUN,CH 1936 TRANS.FARADAY SOC.32,852
		AU 111	CVD	TEM	GABOR,T+BLOCHER,JR.JM 1969 J.VAC.SCI.TECHN.6,815
		AU 001	HV	TED	GILLET,M+CAHOREAU,M+GILLET,E 1965*3ERCEM A,379
		CU 001,110,111	ED	RED	GOSWAMI 1958.SEE CR 1
		CU 111	HV, UHV	RED,TEM	GRADMANN,U 1969 J.APPL.PHYS.40,1182
		AU 111	UHV	TEM	GUEGUEN,P+CAHOREAU,M+GILLET,M 1973 THIN SOLID FILMS 16,27
		CU 111	HV	RED	HAASE,O 1956 Z.NATURFORSCH.11A,862
		CU 001,110,111	HV	RED	HAASE,O 1959 Z.NATURFORSCH.14A,920
		CU 001,110, PD 001,AU 001	HV	RED	HAASE,O 1961(A) Z.NATURFORSCH.16A,202
		CU 001	HV	RED	HAASE.O 1961(B) Z.NATURFORSCH.16A,206
		AL 001,NI 001, CU 001,RH 001, PD 001,AU 001	HV	TEM	HOTHERSALL,DC 1967 PHIL.MAG.15,1023

Z= 26 OVERGROWTH		SUBSTRATE	PREP	OBSERV	REFERENCE
IRON (CONT.)	1	CU 001	UHV	TEM	JESSER,WA+MATTHEWS,JW 1967 PHIL.MAG.15,1097
		CU 001	UHV	TEM	JESSER,WA+MATTHEWS,JW 1968 PHIL.MAG.17,595
		CU 001	HV, ED	TEM	JONES,GA 1967 PHYS.STATUS SOLIDI 19,811
		CU 001	ED	TEM	JONES,GA+OXLEY,DP+TEBBLE,RS 1965 PHIL.MAG.11,993
		CU 001,110,111	HV	RED	KORITKE,H 1961 Z.NATURFORSCH.16A,531
		NI 001	UHV	TEM,REM, SEM	MATTHEWS,JW+JESSER,WA 1969 PHIL.MAG.20,999
		W	UHV	FEM	MELMED,AJ 1967 SURFACE SCI.7,478
		W 001	UHV	LED,FEM	MELMED,AJ+CARROLL,JJ 1970 SURFACE SCI 19,243
		CU 001	HV, UHV	TEM	OLSEN,GH+JESSER,WA 1971 ACTA MET.19,1009
		NI 001,CU 001	UHV	---	OLSEN,GH+JESSER,WA 1971 ACTA MET.19,1299
		AG 111	HV	TEM	OLSEN,GH+SNYMAN,HC 1973 ACTA MET.21,769
		AG 111	HV	TEM	SNYMAN,HC+OLSEN,GH 1973 J.APPL.PHYS.44,889
		AU 001	HV	TEM	SPAIN,RJ+PUCHALSKA,IB 1964 J.APPL.PHYS.35,824
		AU 001,111	UHV	TEM	WASSERMAN,EF+JABLONSKI,HP 1970 SURFACE SCI.22,69
		CU 110	ED	RED	WRIGHT,JG 1971 PHIL.MAG.24,217
	2	PBTE 001	HV	TED	BLACKBURN,WJS+FERRIER,RP 1968 J.APPL.PHYS.39,1163
		PBS 001	HV	RED	MIYAKE,S+KUBO,M 1947(A) J.PHYS.SOC.JAPAN 2,15
	3	NACL 001	HV	TEM	BOYD,EL 1960 IBM J.RES.DEVELOP.4,116
		NACL 001,110	HV	RED	BRUCK 1936.SEE AL 3
		NACL 001,110,111	HV	RED,TED	COLLINS,LE+HEAVENS,OS 1957 PROC.PHYS.SOC.LONDON B70,265
		NACL 001	HV	TEM	ESCAIG,J+SELLA,C 1970*7ICEM 2,419
		LIF 001,111	HV	TED	GILLET ET AL 1965.SEE 1
		LIF 001,NACL 001	HV	RED,TED	HAASE,O 1956 Z.NATURFORSCH.11A,46/ 1959.SEE 1
		NACL 001,110,KBR	HV	TED	HEAVENS,OS 1964*ICTFBB, 383
		NACL 001	HV	TEM	HEAVENS,OS+BROWN,MM+HINTON,V 1959 VACUUM 9,17
		NACL 001,KCL 001, KBR 001,KI 001	HV	OM,TEM	HEAVENS,OS+MILLER,RF+ZAFAR,MS 1966 ACTA CRYST.20,288
		NACL 001	HV	TEM	HOTHERSALL,DC 1969 PHIL.MAG.20,433
		NACL 001	HV	TED	KIRENSKII,LV+PYN≠KO,VG+EDEL≠MAN,IS 1965 SOVIET PHYS.CRYST.9,572
		LIF 001,NACL 001, KCL 001	HV	RED,TED	KIRENSKII,LV+PYN≠KO,VG+SIVKOV,NI+PYN≠KO,GP+SUKHANOVA,RV+OVSYANNIKOV,MA 1966 PHYS.STAT.SOL.17,243
		NACL 001	UHV	XD	LARSON,DC+CRISTOPHER,JE+COLEMAN,RV+ISIN,A 1969 J.VAC.SCI.TECHN.6,670
		NACL 001	HV, UHV	TED	MATTHEWS,JW 1965 APPL.PHYS.LETT.7,255
		NACL 001	UHV	TEM	MATTHEWS.JW 1966 J.VAC.SCI.TECHN.3,133
		NACL 001	HV	TEM	PUCHALSKA,IB+PRUTTON,M 1972 THINSOLIDFILMS 13,S9
		NACL 001,110,111, CSCL 001,110,111	UHV	---	REALE,C 1972.SEE K 3
		NACL 001	HV	RED	SATO,H+ASTRUE,RW 1962 J.APPL.PHYS.33,2956
		NACL 001	HV	TED	SELLA+TRILLAT 1964.SEE AL 3
		NACL 001	HV	TED	SHINOZAKI,S+SATO,H 1965 J.APPL.PHYS.36,2320
		NACL 001	HV	TEM	SHINOZAKI,S+SATO,H 1967*1ICCG, 515
		NACL 001	UHV	TED	SHINOZAKI,S+SATO,H 1969 J.VAC.SCI.TECH.6,534
		NACL 001	HV	TEM	SHINOZAKI,S+SATO,H+HONJO,G 1966*6ICEM 1,501
		NACL 001,KCL 001, KBR 001,KI 001	HV	TED	SHIRAI,S 1937 PROC.PHYS.MATH.SOC.JAPAN 19,937
		NACL 001	HV	TED	SHIRAI,S 1938 PROC.PHYS-MATH.SOC.JAPAN 20,854
		NACL 001,110,111, KCL 001,KBR 001	HV	TEM	TSUKAHARA,S+KAWAKATSU,H 1966 JPHYSSOCJAPAN 21,313
	4	MGO 001	HV	RED	GONDO,Y 1962 J.PHYS.SOC.JAPAN 17,1129
		MICA 00.1	UHV	XD	LARSON ET AL 1969.SEE 3
		MG 001	HV	RED	SATO+ASTRUE. 1962.SEE 3
		MGO 001	HV	RED	SATO,H+TOTH,R+ASTRUE,RW 1962 J.APPL.PHYS.33(SUPPL),1113/ 1964*ICTFBB,395
IRON-CHROMIUM ALLOYS	3	NACL 001	HV	TEM	JAUNET+SELLA 1965.SEE AL 3
IRON OXIDES					
FEO	4	MGO 001	CVD	OM,RED, XD	CECH,RE+ALESSANDRINI,EI 1959 TRANS.AMER.SOC.METALS 51,150
		MGO 001	CVD	OM,XD	ROBINSON,LB+WHITE,WB+ROY,R 1966 J.MAT.SCI.1,336
FE2O3	4	MGO 001	CVD	OM,XD	TAKEI,H+CHIBA,S 1966 J.PHYS.SOC.JAPAN 21,1255
ORTHOFERRITES					
TBFEO3		TBFEO3 001	LPE	OM,XD	SHICK,LK+NIELSEN,JW 1971 J.APPL.PHYS.42,1554
		YALO3 001, SAPPHIRE 10.2	SP	RED,EP, XD	SOSNIAK,J 1971 J.APPL.PHYS.42,1802
SM-TBFEO3		SM-TBFEO3 001	LPE	OM,XD	SHICK+NIELSEN 1971.SEE TBFEO3
YBFEO3		YBFEO3 001	LPE	OM,XD	SHICK+NIELSEN 1971.SEE TBFEO3

Z= 26 OVERGROWTH	SUBSTRATE	PREP	OBSERV	REFERENCE
SPINEL FERRITES				
GENERAL	MGO,MGAL2O4, SAPPHIRE	CVD	RED,XD, XM	ARCHER,JL+PULLIAM,GR+WARREN,RG+MEE,JE 1967*1ICCG, 337
		CVD		MEE,JE+PULLIAM,GR+ARCHER,JL+BESSER,PJ 1969 IEEE TRANS.MAGN,VOL.MAG-5,P.717
LI0.5FE2.5O4	MGO 001,111	LPE	OM,XD	GAMBINO,RJ 1967 J.APPL.PHYS.38,1129
	MGO 001,110,111	CVD	---	MEE ET AL 1969.SEE 2.REF.THIS GROUP
MGFE2O4	MGO 001	CG	XD	HANAK+JOHNSON 1968.SEE MGO 4
	MGO	CVD+ CG		MEE,JE+KOLB,L (UNPUBLISHED WORK). SEE MEE ET AL 1967 UNDER Y3FE5O12
	MGO 001,110,111, MGAL2O4 001,110, 111, SAPPHIRE	CVD	XD,XM	PULLIAM,GR 1967 J.APPL.PHYS.38,1120
MG-MNFE2O4	MGO 001	CVD	---	TELESNIN,RV+NESTRELAY,TI+KOSHKIN,LI+SHISHKOV,AG+ ECONOMOV,NA+ANTONOV,LI 1971 PHYS.STAT.SOL.A4,805
MNFE2O4	MGO 001,110,111	CVD	---	HANAK+JOHNSON 1968.SEE MGO4
	MGO 001	CVD		PULLIAM,GR+MEE,JE+ARCHER,JL+WARREN,RG 1965 PROC. NAT.AEROSPACE ELECTRON.CONF(DAYTON,OHIO1965),241
	MGO 001,110,111	CVD	OM	MARSHALL,DJ 1971 J.CRYST.GROWTH 9,305
	MGO 001,110,111, MGAL2O4 001,110, 111, SAPPHIRE	CVD	XD,XM	PULLIAM 1967.SEE MGFE2O4
MN-COFE2O4	MGO 001,110,111, MGAL2O4 001,110, 111, SAPPHIRE	CVD	XD,XM	PULLIAM 1967.SEE MGFE2O4
MN-NIFE2O4	MGAL2O4 001	CVD	---	MEE 1969.SEE 2 REF. THIS GROUP
	MGO 001,110,111, MGAL2O4 001,110, 111, SAPPHIRE	CVD	XD,XM	PULLIAM 1967.SEE MGFE2O4
	MGO 001,110,111, MGAL2O4 001	CVD	---	PULLIAM ET AL 1965.SEE MNFE2O4
	MGO 001	CVD	---	PULLIAM,GR+WARREN,RG+HOLMES,RE+ARCHER,JL 1967 J.APPL.PHYS.38,2192
FE3O4	MGO 001,110,111, MGAL2O4 001,110, 111, SAPPHIRE	CVD	XD,XM	PULLIAM 1967.SEE MGFE2O4
	MGO 001	CVD	OM,TEM, XD	ROBINSON ET AL 1966.SEE FEO 4
	MGO 001	CVD	OM,XD	TAKEI,H+TAKASU,S 1964 JAPAN.J.APPL.PHYS.3,175
COFE2O4	MGO 001,111	LPE	OM,XD	GAMBINO 1967.SEE LI0.5FE2.5O4
	MGO 001	CVD	---	MEE 1969.SEE 2. REF. THIS GROUP
	MGO 001,110,111, MGAL2O4 001,110, 111, SAPPHIRE	CVD	XD,XM	PULLIAM 1967.SEE MGFE2O4
	MGO 001	CVD	OM,XD	TAKEI+TAKASU 1964.SEE FE3O4
NIFE2O4	MGO 001,111	LPE	OM,XD	GAMBINO 1967.SEE LI0.5FE2.5O4
	MGO 001,110,111	CVD	OM	MARSHALL 1971.SEE MNFE2O4
	MGO 110,111	CVD	OM,XD	NAGASAWA,K+BANDO,Y+TAKADA,T 1968 JAPAN.J.APPL.PHYS.7,174
	MGO 001,110,111, MGAL2O4 001,110, 111, SAPPHIRE	CVD	XD,XM	PULLIAM 1967.SEE MGFE2O4
	MGO 001	CVD	---	PULLIAM ET AL 1965.SEE MNFE2O4
	MGO 001	CVD, SP	OM,XD	SCHRODER,H+GLAUCHE,E 1968 J.APPL.PHYS.39,1155
	MGO 001	CVD	OM,XD	TAKEI+TAKASU 1964.SEE FE3O4
	MGO 001	SP	OM,TED, XD	WESTWOOD,WD+EASTWOOD,HK+POULSEN,RG+SADLER,AG 1967 J.AMER.CERAMIC SOC.50,119
NI-ZNFE2O4	MGO 001,111	LPE	OM,XD	GAMBINO 1967.SEE LI0.5FE2.5O4
	MGO 001,110,111	CVD	OM	MARSHALL 1971.SEE MNFE2O4
GARNETS				
GENERAL	GD3GA5O12,ETC.	LPE	OM	BLANK,SL+NIELSEN,JW 1972 J.CRYST.GROWTH 17,302
		CVD		MEE ET AL 1969.SEE SPINEL FERRITES (2. REF.)
	GD3GA5O12 111, ND3GA5O12 111	LPE	OM,XM	MILLER,DC 1973 J.ELECTROCHEM.SOC.120, 678
				TIEN,PK+MARTIN,RJ+BLANK,SL+WEMPLE,SH+VARNERIN,LJ 1972 APPL.PHYS.LETT.21, 207
		CVD, LPE	---	VARNERIN 1971 IEEE TRANS.MAGN.MAG-7,404
Y3FE5O12	GARNETS 001,111	CVD	OM	BESSER,PJ+MEE,JE+ELKINS,PE+HEINZ,DM 1971 MAT.RES.BULL.6,1111
	GD3GA5O12 001,111	CVD	RED,EP	BRAGINSKI,AI 1971 IEEE TRANS.MAGN.MAG-7,467

Z= 26

OVERGROWTH	SUBSTRATE	PREP	OBSERV	REFERENCE
Y3FE5012 (CONT.)	GD3GA5012 001, GD-DY3GA5012 001	CVD	EP,XD	BRAGINSKI,AI+OEFFINGER,TR+TAKEI,WJ 1972 MATER.RES.BULL.7,627
	GD3GA5012 001,110, 111,211	SG	OM	FERRAND,B+DAVAL,J+JOUBERT,JC 1972 J.CRYST.GROWTH 17,312
	GD3GA5012	LPE	OM,SEM	GILL,GP+FAIRHOLMER,RJ 1973 J.MATER.SCI.8,1115
	GD3GA5012 111	CVD	XM	GLASS,HL+HAMILTON,TN 1972 MATER.RES.BULL.7,761
	GD3GA5012 111	CVD	OM,XD	KEMPTER,K+BOEGNER,W 1972 THIN SOLID FILMS 12,35
	GD3GA5012 001,110, 111	LPE	OM	LINARES,RC 1968 J.CRYST.GROWTH 3,4,443
	Y3AL5012, GD3GA5012	CVD, LPE	XD	LINARES,RC+MCGRAW,RB+SCHROEDER,JB 1965 J.APPL.PHYS.36,2884
	GARNETS	---	---	MATTHEWS,JW+KLOKHOLM,E 1972 MATER.RES.BULL.7,213
	Y3AL5012 001,110	CVD	EP,XD	MEE,JE 1967 IEEE TRANS.MAGN.MAG-3,190
	Y3AL5012 001	CVD	RED,XD, EP	MEE,JE+ARCHER,JL+MEADE,RH+HAMILTON,TN 1967 APPL.PHYS.LETT.10,289
	Y3AL5012 111, Y3FE5012 001,110, Y3GA5012 211, GD3GA5012 001,110, 111,211	CVD	---	MEE ET AL 1969.SEE SPINEL FERRITES (2.REF.)
	GD3GA5012 111	LPE	OM,SEM, XD	TOLKSDORF,W+BARTELS,G+ESPINOSA,GP+HOLST,P MATEIKA,D+WELZ,F 1972 J.CRYST.GROWTH 17,322
	Y3AL5012 001,111, GD3GA5012 111	CVD	OM	WILKINS,CW 1973 J.CRYST.GROWTH 20,207
Y3FE-GA5012	GD3GA5012 110,111	CVD	OM	HEINZ,DM+BESSER,PJ+OWENS,JM+MEE,JE+PULLIAM,GR 1971 J.APPL.PHYS.42,1243
	GD3GA5012	---	---	MATTHEWS+KLOKHOLM 1972.SEE Y3FE5012
	GD3GA5012 110,111	CVD	OM,ED	MEE,JE+PULLIAM,GR+HEINZ,DM+OWENS,JM+BESSER,PJ 1971 APPL.PHYS.LETT.18,60
	GD3GA5012 111	LPE	OM,SEM, XD	TOLKSDORF ET AL 1972.SEE Y3FE5012
	GD3GA5012 111	CVD	OM	WILKINS 1973.SEE Y3FE5012
	DY-GD3GA5012 111	CVD	XD	WOLFE,R+NORTH,JC+BARNS,RL+ROBINSON,M+ LEVINSTEIN,HJ 1971 APPL.PHYS.LETT.19, 298
Y-LA3FE-GA5012, Y-SM3FE-GA5012	GD3GA5012 111	LPE	OM,SEM, XD	TOLKSDORF ET AL 1972.SEE Y3FE5012
Y-EU3AL-FE5012	GD3GA5012 111	LPE	---	SUEMUNE,Y+INOUE,N 1974 JAPAN.J.APPL.PHYS.13, 204
Y-EU3FE-GA5012	GD3GA5012 111	LPE	OM,XD	CRONEMEYER,DC+GIESS,EA+KLOKHOLM,E+ARGYLE,BE+ PLASKETT,TS 1972*17ACMMM,115
	GARNETS 111	LPE	XD,EP	GIESS,EA+CRONEMEYER,DC 1973 APPL.PHYS.LETT.22,601
	GD3GA5012 111	LPE	XD,EP	GIESS,EA+ARGYLE,BE+CALHOUN,BA+CRONEMEYER,DC+ KLOKHOLM,E+MC GUIRE,TR+PLASKETT,TS 1971 MATER.RES.BULL.6,1141
	GD3GA5012 111	LPE	EP	GIESS,EA+ARGYLE,BE+CRONEMEYER,DC+KLOKHOLM,E+ MCGUIRE,TR+O≠KANE,DF+PLASKETT,TS+SADAGOPAN,V 1972*17ACMMM,110
	GD3GA5012 111	LPE	EP	GIESS,EA+KUPTSIS,JP+WHITE,EAD 1972 J.CRYST.GROWTH 16,36
	GD3GA5012 111	LPE	XD	HAGEDORN,FB+BLANK,SL+BARNS,RL 1973 APPL.PHYS.LETT.22,209
	GD3GA5012 001,111	LPE	XD	KLOKHOLM,E+MATTHEWS,JW+MAYADAS,AF+ANGILELLO,J 1972*17ACMMM,105
Y-EU-TM3FE-GA5012	GD3GA5012 111	LPE	XD	BONNER,WA+GEUSIC,JE+SMITH,DH+VAN UITERT,LG+ VELLA-COLEIRO,GP 1973 MATER.RES.BULL.8,785
Y-GD3FE-GA5012	SM3GA5012 111, GD3GA5012 111, DY-GD3GA5012 111, ER3GA5012 111	CVD	XD	BESSER,PJ+MEE,JE+GLASS,HL+HEINZ,DM+AUSTERMANN,SB+ ELKINS,PE+HAMILTON,TN+WHITCOMB,EC 1972*17ACMMM,125
	GD3GA5012 111	LPE	OM,XD	CRONEMEYER ET AL 1972.SEE Y-EU3FE-GA5012
	GD3GA5012 111	SP	RED	CUOMO,JJ+SADAGOPAN,V+DE≠LUCA,J+CHAUDHARI,P+ ROSENBERG,R. 1972 APPL.PHYS.LETT.21,581
	GD3GA5012 111	LPE	EP,XD	GIESS ET AL 1971.SEE Y-EU3FE-GA5012
	GD3GA5012 001,111	LPE	XD	KLOKHOLM ET AL 1972.SEE Y-EU3FE-GA5012
	GD3GA5012	---	---	MATTHEWS+KLOKHOLM 1972.SEE Y3FE5012
	GD3GA5012 111	LPE	XM	ROBERTSON,JM+VAN HOUT,MJG+JANSSEN,MM+STACY,WT 1973 J.CRYST.GROWTH 18,294
	GD3GA5012 111	LPE	OM,SEM, XD	TOLKSDORF ET AL 1972.SEE Y3FE5012
	GD3GA5012 111	CVD	OM	WILKINS 1973.SEE Y3FE5012
Y-GD-TM3FE-GA5012	GD3GA5012 111	LPE	OM,XM	MILLER 1973. SEE GARNETS GENERAL
Y-GD-YB3FE-GA5012	GD3GA5012 111	LPE	---	SUEMUNE+INOUE 1974.SEE Y-EU3AL-FE5012
ND3GA5012	ND3GA5012 001,111	LPE	XM	ROBERTSON ET AL 1973.SEE Y-GD3FE-GA5012
SM3GA5012	SM3GA5012 001,111	LPE	XM	ROBERTSON ET AL 1973.SEE Y-GD3FE-GA5012

Z= 26-27	OVERGROWTH	SUBSTRATE	PREP	OBSERV	REFERENCE
	EU3GA5012	GD3SC2AL3012	LPE	OM	TIEN ET AL 1972. SEE GARNETS GENERAL
	EU-ER3FE-AL5012	GD3GA5012 001	LPE	XD	WOLFE ET AL 1971.SEE Y3FE-GA5012
	EU-ER3FE-GA5012	GD3GA5012 111	LPE	OM	GEUSIC,JE+LEVINSTEIN,HJ+LICHT,SJ+SHICK,LK+ BRANDLE,CD 1971 APPL.PHYS.LETT.19,93
		GD3GA5012 111	LPE	XD,XM	GLASS,HL 1972 MATER.RES.BULL.7,385
		GD3GA5012	LPE	OM	HAGEDORN,FB+TABOR,WJ+GEUSIC,JE+LEVINSTEIN,HJ+ LICHT,SJ+SHICK,LK 1971 APPL.PHYS.LETT.19,95
		GD3GA5012 111	LPE	OM	LEVINSTEIN,HJ+LICHT,SJ+LANDORF,RW+BLANK,SL 1971 APPL.PHYS.LETT.19,486
		GD3GA5012 111	LPE	OM,XM	MILLER 1973. SEE GARNETS GENERAL
		GD3GA5012 111	LPE	OM	QUON,HH+POTVIN,AJ 1972 MATER.RES.BULL.7,463
		GD3GA5012 110,111	LPE	OM,XD	SCHICK,LN+NIELSEN,JW+BOBECK,AH+KURTZIG,AJ+ MICHAELIS,PC+REEKSTIN,JP 1971 APPLPHYSLETT.18,89
	EU-YB3FE5012	GD3GA5012 111	LPE	OM,SEM	GHEZ,R+GIESS,EA 1973 MATER.RES.BULL.8,31
	GD3FE5012	Y3AL5012 111	SG	XD	COMSTOCK,RL+MOORE,EB+NEPELA,DA 1970 IEEE TRANS.MAGN.MAG-6,558
		Y3AL5012 001,110, 111	SP	XD	JOSEPHS,RM 1970 IEEE TRANS.MAGN.,MAG-6,553
		GD3GA5012, Y3AL5012	---	---	MATTHEWS+KLOKHOLM 1972.SEE Y3FE5012
		Y3AL5012 001,110	CVD	EP,XD	MEE 1967.SEE Y3FE5012
		Y3AL5012 001	CVD	RED,XD, EP	MEE ET AL 1967.SEE Y3FE5012
		Y3AL5012	SP	RED,XD	SAWATSKY,E+KAY,E 1969 J.APPL.PHYS.40,1460
		Y3AL5012 111	SP	RED,XD	SAWATSKY,E+KAY,E 1971 J.APPL.PHYS.42,367
		Y3AL5012 001,110, 111,320	CVD	OM,ER, XD	STEIN,BF 1970 J.APPL.PHYS.41,1262
		Y3AL5012 001,110, 111, GD3GA5012 110,211	CVD	OM,ER, XD	STEIN,BF 1971 J.APPL.PHYS.42,2336
		Y3AL5012, GD3GA5012, ND-GD3GA5012, LA-GD3GA5012	LPE	---	STEIN,BF+KESTIGIAN,M 1971 J.APPL.PHYS.42,1806
		GD3GA5012 111, ND3GA5012 111	CVD	OM	WILKINS 1973.SEE Y3FE5012
	GD3GA5012	GD3GA5012 001,111	LPE	XD,XM	ROBERTSON ET AL 1973.SEE Y-GD3FE-GA5012
	GD-TB3FE5012	Y3AL5012 001,111	SG	XD	COMSTOCK ET AL 1970.SEE GD3FE5012
	GD-ER3FE-GA5012	GD3GA5012	CVD	OM	ROBINSON,MCD 1973 J.CRYST.GROWTH 18,143
	TB3FE5012	ND-SM3GA5012 111, SM3GA5012 111, GD3GA5012 111	CVD	XD	BESSER ET AL 1972.SEE Y-GD3FE-GA5012
	TB-ERFE5012	SM3GA5012 001	CVD	OM	ROBINSON,MCD+BOBECK,AH+NIELSEN,JW 1971 IEEE TRANS.MAGN.MAG-7,464
		SM3GA5C12 111	CVD	XD	WOLFE ET AL 1971.SEE Y3FE-GA5012
	ER3FE-GA5012	DY-GD3GA5012 001	CVD	XD	BESSER ET AL 1972.SEE Y-GD3FE-GA5012
		GD-DY3GA5012 111	CVD	XD,XM	GLASS 1972. SEE EU-ER3FE-GA5012
27 COBALT	1	AG 111	HV	RED	BUDER,R+MASSENET,O 1969 C.R.ACAD.SCI.B,269,744
		CO 00.1	ED	OM,REM	CLIFFE,DR+FARR,JPG 1964 J.ELECTROCHEM.SOC.111,299
		CU 110	ED	RED	COCHRANE 1936.SEE CR 1
		CU 001	UHV	TEM	FEDORENCO,AI+VINCENT,R 1971 PHIL.MAG.24,55
		PT 110,AU 001	ED	TED	FINCH+SUN 1936.SEE FE 1
		CU 001	ED	RED	FINCH,GI+WILMAN,H+YANG,L 1947 DISCUSS.FARADAY SOC.A43,144
		CU 110	ED	RED	FISHER,JE 1970 THIN SOLID FILMS 5,53
		CU 111	ED	RED	FISHER,JE 1973 THIN SOLID FILMS 17,S31
		CU 110	ED	RED	FISHER,JE+GODDARD,J 1968 J.PHYS.SOC.JAPAN 25,413
		CU 111	HV	RED	GARIGUE,J+LAFOURCADE,L+NGUYEN QUAT TI+SONIER,F 1960 C.R.ACAD.SCI.250,3296
		CU 001,110	ED	OM,RED	GODDARD,J+WRIGHT,JG 1964 BRIT.J.APPL.PHYS.15,807
		CU 001,110,111	ED	---	GODDARD,J+WRIGHT,JG 1965 BRIT.J.APPL.PHYS.16,1251
		AG 111	HV	RED	GONZALEZ,C+GRUNBAUM,E 1962*5ICEM 1,DD-1
		CU 111	HV, UHV	RED,TEM	GRADMANN 1970.SEE FE 1
		CU 111	UHV	RED	GRADMANN,U+MULLER,J 1970 Z.ANGEW.PHYSIK 30,87
		AG 111	HV	RED	GRUNBAUM,E+KREMER,G 1968 J.APPL.PHYS.39,347
		AG 111	UHV	RED	GRUNBAUM,E+KREMER,G+REYMOND,C 1969 J.VAC.SCI.TECHN.6,475
		CU 001	UHV	TEM	JESSER,WA+MATTHEWS,JW 1968 PHIL.MAG.17,461
		NI 001	UHV	TEM	JESSER,WA+MATTHEWS,JW 1968ACTA MET.16,1307
		CU 110	ED	RED	LAUE,M 1937 ANN.PHYSIK 29,211
		NI-PD 001	UHV	TEM	MATTHEWS,JW 1968 PHIL.MAG.18,1149
		CU 001	UHV	TEM	MATTHEWS,JW 1970 THIN SOLID FILMS 5,369

Z= 27-28 OVERGROWTH		SUBSTRATE	PREP	OBSERV	REFERENCE
COBALT (CONT.)	1	CU 110	ED	RED	NEWMAN,RC 1956 PROC.PHYS.SOC.LOND.B 69,432
		CU 110,111	HV	RED	NGUYEN QUAT TI+COUDERC,JJ+PILOD,MF 1961*2ERCEM 1,289
		AU 111	HV	TEM	PASHLEY,DW+MENTER,JW+BASSETT,GA 1957 NATURE (LONDON) 179,752
	2	SI 111	HV	RED	WRAY,L+PRUTTON,M 1973 THIN SOLID FILMS 15,173
	3	NACL 001	HV	RED	BRUCK 1936.SEE AL 3
		NACL 001,110,111	HV	RED,TED	COLLINS+HEAVENS 1957.SEE FE 3
		NACL 001,KCL 001	HV	TEM	HONNA,T+WAYMAN,CM 1965 J.APPL.PHYS.36,2791
		NACL 001	HV	TEM	KATO,T+OGAWA,S 1970 JAP.J.APPL.PHYS.9,875
		LIF 001,NACL 001, KCL 001	HV	RED,TED	KIRENSKII ET AL 1966.SEE FE 3
		LIF 001,111	HV	TEM	MIDDLETON,BK 1972 PHYS.STATUS SOLIDI A12,67
		NACL 001	HV	TEM	OTSUKA,K+WAYMAN,CM 1967 PHYS.STAT.SOL.22,559+579
		NACL 001	UHV	OM,TEM	PATRICIAN,TJ+WAYMAN,CM 1971 PHYS.STATUS SOLIDI A6,111,449+461
		NACL 001	HV	TEM	TAKAHASHI+MARTINA 1966.SEE CR 3
		NACL 001	UHV	RED	ZIMMERMAN,R+LEIBOVICH,HA+GRUNBAUM,E 1971 PHYS.STATUS SOLIDI A5,K5
	4	MGO 001	HV	RED	DOYLE,WD 1964 J.APPL.PHYS.35,929
		MG 001	HV	RED	DOYLE,WD+FLANDERS,PJ 1964PROC.INT.CONF.MAGNETISM (NOTTINGHAM 1964), 751.INST.PHYS+PHYS.SOC.LONDON
		MICA 00.1	UHV	RED	GRUNBAUM+KREMER 1968.SEE 1
		MICA 00.1	HV	TED	KATO+OGAWA 1970.SEE 3
		MGO 001	HV	RED	SATO,H+TOTH,RS+ASTRUE,RW 1963J.APPL.PHYS.34,1062/ 1964.SEE FE 4
COBALT-NICKEL ALLOYS	1	CU 110	ED	---	LORD,DG+GODDARD,J 1970 PHYS.STATUS SOLIDI 37,665
	3	NACL 001	HV	TEM	BOYD 1960.SEE FE 3
COBALT-COPPER ALLOYS	2	MOS2 00.1	HV	TEM	MADER,S 1965 J.VAC.SCI.TECHN.2,35
		MOS2 00.1	HV, UHV	TED	MADER,S+WIDMER,H+D≠HEURLE,FM+NOWICK,AS 1963 APPL.PHYS.LETT.3,201
COBALT-PALLADIUM ALLOYS	3	NACL 001	HV	TEM	MATSUO,Y 1972 J.PHYS.SOC.JAPAN 32,972
		NACL 001	HV	TED	MATSUO,Y+HAYASHI,F 1970 J.PHYS.SOC.JAPAN 28,1375
COBALT-PHOSPHOR	1	CU 001	ELD	TEM	JONES,GA+ASPLAND,M 1972 PHYS.STAT.SOL.A11,637
COBALT SULPHIDE	3	NACL 001	HV	RED,TED	AGGARWAL,PS+GOSWAMI,A 1959 Z.NATURFORSCH.14B,419
COBALT SELENIDE	3	NACL 001,110,111	HV	RED,TED	GOSWAMI,A+NIKAM,PS 1971 J.CRYST.GROWTH 8,247
COBALT OXIDES					
COO	4	MGO 001	CVD	OM,RED, XD	CECH+ALESSANDRINI 1959.SEE FEO 4
		MGO 001,111	CVD	OM,RED, XD	GREINER,JH+BERKOWITZ,AE+WEIDENBORNER,JE 1966 J.APPL.PHYS.37,2149
		MGO 001	CVD	OM,XD	ROBINSON ET AL 1966.SEE FEO 4
		MGO 001	CVD	OM,XM	SPOONER,FJ+VERNON,MW 1969 J.MATER.SCI.4,734
CO3O4	4	MGO 001	CVD	OM,XD	ROBINSON ET AL 1966.SEE FEO 4
28 NICKEL	1	CU 001	HV	TEM	ANDERSON,JC 1961 PROC.PHYS.SOC.78,25
		CU 001	HV, UHV	TEM	ANDERSON,JC 1962*2IVC 2,930
		AU 111	HV	TEM	BASSET,GA+MENTER,JW+PASHLEY,DW 1958 PROC.ROY.SOC.A 246,345
		AU 111	ED	TEM	BASSET,GA+MENTER,JW+PASHLEY,DW 1959 DISCUSS.FARADAY SOC.28,7
		AG 111	HV	TEM	BIRAGNET ET AL 1966.SEE FE 1
		NI 111	ED	OM,REM	CLIFFE+FARR 1964.SEE CO 1
		CU 110	ED	RED	COCHRANE 1936.SEE CR 1
		AG 111	UHV	RED,LED	DOBSON,PJ+SERNA,CR 1969 J.PHYS.D=APPL.PHYS.2,1779
		PT 110,AU 001	ED	TED	FINCH+SUN 1936.SEE FE 1
		CU 001	ED	RED	FINCH ET AL 1947.SEE CO 1
		CU 110	ED	RED	FISHER+GODDARD 1968.SEE CO 1
		CU 001	ED	TEM	GAIGHER,HL+VAN WYK,GN 1973 THINSOLIDFILMS 15,163
		AG 111,AU 111	HV	RED	GLOSSOP,AB+PASHLEY,DW(UNPUBLISHED).SEE PASHLEY,DW 1965 ADVAN.PHYS.14,327
		AG 111	HV, UHV	RED,TEM	GONZALEZ,C 1967 ACTA MET.15,1373/ 1971 J.PHYS.D=APPL.PHYS.4,351
		CU 001	ED	RED	GOSWAMI,A 1956 J.SCI.IND.RES.15B,322
		FE 001	ELD	RED	GOSWAMI,A 1957 J.SCI.IND.RES.16B,186
		CU 111	HV, UHV	RED	GRADMANN,U 1964 ANN.PHYSIK 13,213/ 1966 IBID.17,91
		CU 111	UHV	LED	GRADMANN,U 1969 SURFACE SCI.13,498
		CU 111	HV, UHV	RED,TEM	GRADMANN 1969.SEE FE 1
		AG 111	UHV	RED	GRUNBAUM ET AL 1969.SEE CO 1

Z= 28 OVERGROWTH		SUBSTRATE	PREP	OBSERV	REFERENCE
NICKEL (CONT.)	1	CU 001,111	HV	RED	HAASE 1961(A).SEE FE 1
		CU 001	HV	RED	HAASE 1961(B).SEE FE 1
		CU 111	UHV	LED	HAQUE,CA+FARNSWORTH,HE 1966 SURFACE SCI.4,195
		CU 001,110,111	HV	TED	HEAVENS,OS 1962*9NVSAVS, 52
		CU 001,110,111	HV	TEM	HEAVENS,OS+MILLER,RF+MOSS,GL+ANDERSON,JC 1961 PROC.PHYS.SOC.78,33
		CU 001	ED	TEM	IVES,AG+EDINGTON,JW+ROTHWELL,GP 1970 ELECTROCHIM.ACTA 15,1797
		CU 001	ED	TEM	JONES ET AL 1965. SEE FE 1
		W	UHV	FEM,FIM	JONES,JP 1966 NATURE(LONDON) 211,479
		CU 001	HV	RED	KORITKE 1961.SEE FE 1
		CU 110,111,AU 110	HV	RED	KUBO,M+MIYAKE,S 1948 J.PHYS.SOC.JAPAN.3,114
		CU 001	HV	TEM	KUNTZE,R+CHAMBERS,A+PRUTTON,M 1969 THIN SOLID FILMS 4,47
		CU 110	ED	RED	LAUE 1937.SEE CO 1
		CU 001,110,111	ED	RED,TEM	LAWLESS,KR 1965 J.VAC.SCI.TECH.2,24
		CU 001,111	ED	RED,TEM, REM	LAWLESS,KR+GARMON,L+LEIDHEISER,JR.H 1961*2ERCEM 1,396
		CU 001,NI 001	ED	OM,XD	LEIDHEISER,JR.H+GWATHMEY,AT 1951 J.ELECTROCHEM.SOC.98,225
		PD 001	UHV	TED	MATTHEWS,JW 1972(B) THIN SOLID FILMS 12,243
		CU 001	UHV	TEM	MATTHEWS,JW+CRAWFORD,JL 1970 THINSOLIDFILMS 5,187
		CU 001	ED	TEM	NAKAHARA,S+WEIL,R 1970*7ICEM 2,435
		CU 110	ED	RED	NEWMAN 1956.SEE CO 1
		AG 111	HV	RED	NEWMAN,RC 1957 PHIL.MAG.2,750
		CU 110,111	HV, ED	RED	NGUYEN QUAT TI ET AL 1960.SEE CO 1
		CU001,110,310,531	ED	RED,TEM, REM	OGAWA,S+MIZUNO,J+WATANABE,D+FUJITA,FE 1957 J.PHYS.SOC.JAPAN 12,999
		AU 111	HV	TEM	PASHLEY ET AL 1957.SEE CO 1
		CU 001,ZN 00.1	ED	TEM	REIMER,L+FICKER,J 1961*2ERCEM 1,387
		NI 001,PD 001, AG 001	HV	TED	SHIRAI,S+FUKUDA,J+NOMURA,M 1961 J.PHYS.SOC.JAPAN 16,1989
		W	UHV	FEM	SMITH,GDW 1973 SURFACE SCI.35,304
		W	UHV	FEM,FIM	SMITH,GDW+ANDERSON,JS 1971 SURFACE SCI.24,459
		CU 001,111	ED	TEM	THOMPSON,ER+LAWLESS,KR 1966 APPL.PHYS.LETT.9,138
		CU 001,111	ED	RED,TEM, REM,XD	THOMPSON,ER+LAWLESS,KR 1969ELECTROCHIMACTA 14,269
		CU 001	ED	RED	TREHAN,YN+GOSWAMI,A 1958 TRANS.INDIAN INST.METALS 11,41
		CO 110,111,211	ED	TEM	WRIGHT,JG 1972 THIN SOLID FILMS 9,309
		CU 001,110,111, CO 10.0,00.1	ED	RED	WRIGHT,JG+GODDARD,J 1965 PHIL.MAG.11,485
	2	SI 111	UHV	LED	CHARIG,JM+SKINNER,DK 1970 SURFACE SCI.19,283
		DIAMOND 111	HV	RED	GORODETSKII,AE+TROFIMOV,VI+LUK≠YANOVICH,VM 1972 SOVIET PHYS.-SOLID ST.13,2555
		MOS2 00.1	HV	RED	KAINUMA,Y 1951 J.PHYS.SOC.JAPAN 6,135
		ZNS 110	HV	RED	KUBO+MIYAKE 1948.SEE 1
		ZNS 110	HV	RED	KURIYAMA,M+YAMANOUCHI,H+HOSOYA,S 1961 J.PHYS.SOC.JAPAN 16,701
		PBS 001	HV	RED	MIYAKE+KUBO 1947(A).SEE FE 2
		SI 111	HV	RED	WRAY+PRUTTON 1973.SEE CO 2
	3	NACL 001	HV	TEM	ADAMSKY,RF+LEBLANC,RE 1963*10NVSAVS, 453
		NACL 001	HV	TED	ALESSANDRINI,EI 1964 J.APPL.PHYS.35,1606
		NACL 001	HV	TEM	ANDERSON 1962.SEE NI 1
		NACL 001	UHV	TEM,REM	BAKER,BG+BRUCE,LA 1968 TRANS.FARADAY SOC.64,2533
		NACL 001,110,111	UHV	TEM	BALTZ,A 1964*ICTFBB,315
		NACL 001	HV	TEM	BOYD,EL 1960 IBM J.RES.DEVELOP.4,116
		NACL 001	HV	TEM	BROCKWAY,LO+MARCUS,RB+ROWE,AP 1964*ICTFBB, 231
		NACL 001,KCL 001	HV	TED	BRUCK 1936.SEE AL 3
		NACL 001	HV	TEM	CHAMBERS,A+PRUTTON,M 1967 THIN SOLID FILMS 1,235
		LIF 001	HV	RED	CHAMBERS,A+PRUTTON,M 1967/68THINSOLIDFILMS 1,393
		NACL 001	HV	---	CHIKAZUMI,S 1961 J.APPL.PHYS.32,81S
		NACL 001	HV	RED,TEM	COLLINS,LE+HEAVENS,OS 1954 PHIL.MAG.45,283
		NACL 001	HV	RED,TEM	COLLINS,LE+HEAVENS,OS 1957 PROC.PHYS.SOC.LONDON B70,265
		NACL 001	HV	OM,TEM, XD	FREEDMAN,JF 1962 IBM J.RES.DEVELOP 6,449
		NACL 001,110,111	HV, UHV	TEM,REM	GRENGA,HE+LAWLESS,KR+GARMON,LB 1971 J.APPL.PHYS.42,3629
		NACL 001,110,111	HV	TEM	HEAVENS 1962.SEE 1
		NACL 001,110,111	HV	OM,TEM	HEAVENS ET AL 1961.SEE 1
		NACL 001	HV, UHV	TEM	INO ET AL 1964.SEE AL 3
		NACL 001,KCL 001	UHV	TEM	INO+OGAWA 1966.SEE AL 3
		NACL 001	UHV	RED,LED, AES	JACKSON,DC+GALLON,TE+CHAMBERS,E 1973 SURFACE SCI.36,381
		NACL 001	HV	TEM	JAUNET+SELLA 1965.SEE AL 3
		KCL 001,KBR 001, KI 001	HV	TED	KATO 1968.SEE AL 3
		NACL 001	HV	TEM	KIRENSKII,LV+PYN≠KO,VG+PYN≠KO,GP 1966 SOVIET PHYS.-CRYST.10,768

Z= 28 OVERGROWTH		SUBSTRATE	PREP	OBSERV	REFERENCE
NICKEL (CONT.)	3	LIF 001,NACL 001, KCL 001	HV	RED,TEM	KIRENSKII ET AL 1966.SEE FE 3
		NACL 001	HV	TEM	KLEEFELD,J+HIRSCH,AA 1969 VACUUM 19,561
		NACL 001,110,111	HV	TEM	KLEEFELD,J+PRATT,B+HIRSCH,AA 1973 J.CRYST.GROWTH 19,141
		NACL 001	HV	TEM	KUNTZE ET AL 1969.SEE 1
		NACL 001	HV	RED,TED	KURIYAMA ET AL 1961.SEE 2
		NACL 001	HV	TED	LAUE 1937.SEE CO 1
		NACL 001	UHV	XD	LARSON ET AL 1969.SEE FE 3
		NACL 001	UHV	TEM	LAWLESS 1965.SEE 1
		NACL 001	HV	TED	LEIBOVICH,HA+ZIMMERMANN,RS 1968JAPPLPHYS.39,5908
		NACL 001	UHV	TEM	MATTHEWS 1966.SEE FE 3
		NACL 001	HV	TED	MENZER,G 1938 Z.KRISTALLOGR.99,410
		NACL 001	UHV	TEM	MURR,LE 1964 BRIT.J.APPL.PHYS.15,1511
		NACL 001	UHV	TEM	MURR,LE 1969 THIN SOLID FILMS 3,321
		NACL 001	HV, UHV	TEM	MURR+INMAN 1966.SEE AL 3
		NACL 001,KCL 001	UHV	TED	OGAWA ET AL 1966.SEE AL 3
		NACL 001	HV	RED,TEM	OGAWA,S+WATANABE,D+FUJITA,FE 1955 J.PHYS.SOC.JAPAN 10,429
		NACL 001	HV, UHV	TEM	POMERANTZ,M+FREEDMAN,JF+SUITS,JC 1962 J.APPL.PHYS.33,1164
		NACL 111	HV	TEM	PRATT,TE+VOOK,RW 1972*30EMSA, 512
		NACL 001	HV	TEM	PUCHALSKA+PRUTTON 1972.SEE FE 3
		NACL 001,110,111, CSCL 001,110,111	UHV	---	REALE 1972.SEE K 3
		NACL 001	HV	RED	SATO,H+TOTH,RS+ASTRUE,RW 1962JAPPLPHYS.33,1113
		NACL 001	HV	TED	SHIRAI,S 1943(A) PROC.PHYS.-MATH.SOC.JAPAN 25,168
		NACL 001	HV	TEM	SCHLOTTERER,H 1965 Z.NATURFORSCH 20A,1201
		NACL 001	HV	TED	SELLA+TRILLAT 1964.SEE AL 3
		NACL 001	HV	TEM	TAKAHASHI,N+MARTINA,H+TRILLAT,JJ 1966(A) C.R.ACAD.SCI.262B, 134
		NACL 001	HV, UHV	TEM	TAKAHASHI,N+MARTINA,H+TRILLAT,JJ 1966(B)*6ICEM 1,525
		NACL 001,110,111, KCL 001,KBR 001	HV	TEM	TSUKAHARA,KAWAKATSU 1966.SEE FE 3
		NACL 001	UHV	RED	ZIMMERMANN ET AL 1971.SEE CO 3
	4	MICA 00.1	UHV	TEM	ALLPRESS,JG+SANDERS,JV 1966 PHIL.MAG.14,937
		MICA 00.1	UHV	TEM	ALLPRESS,JG+SANDERS,JV 1967 SURFACE SCI.7,1
		MICA 00.1	UHV	TEM	ALLPRESS,JG+JAEGER,H+MERCER,PD+SANDERS,JV 1966*6ICEM 1,489
		MICA 00.1	HV	RED,TED	BIRAGNET ET AL 1966.SEE FE 1
		MGO 001	HV	---	CHIKAZUMI 1961.SEE 3
		MICA 00.1	UHV	XD	LARSON ET AL 1969.SEE FE 3
		MICA 00.1	HV	XD	REICHELT,K 1971 J.CRYST.GROWTH 11,182
		MGO 001	HV	RED	SATO ET AL 1962.SEE 3/1963.SEE CO 4/1964.SEE FE 4
NICKEL-IRON ALLOYS	1	AG 111	HV	TEM	ESCUDIER,P+BIRAGNET,F+DEVENYI,J+YELON,A 1966 PHYS.STATUS SOL.16,295
		CU	ED	TEM	FUCHS,E+POLITYCKI,A 1961 Z.ANGEW.PHYSIK 11,541
		CU 111	HV	RED	GRADMANN,U 1966*ISTFCL,485
		CU 111	HV, UHV	RED,TEM	GRADMANN 1969.SEE FE 1
		CU 111	HV	RED,TEM	GRADMANN,U+MULLER,J 1968 J.APPL.PHYS.39,1379/ 1968 PHYS.STATUS SOLIDI 27,313
		CU 001,110,111	HV	TEM	HEAVENS ET AL 1961.SEE NI 1
		CU 001	ED	TEM	JONES ET AL 1965.SEE FE 1
		CU 111,AG 111, AU 111	UHV	TED	PASCARD,H+HOFFMANN,F+SELLA,C 1972(A) THIN SOLID FILMS 12,41
		AG 111	HV, UHV	TEM	PASCARD,H+QUINTANA,C+HOFFMANN,F+SELLA,C 1972(B) J.CRYST.GROWTH 13/14,225
	3	NACL 001	HV, UHV	TEM	BALTZ,A 1963 J.APPL.PHYS.34,1575
		NACL 001	HV	TEM	BOYD 1960.SEE FE 3
		NACL 001	HV	RED,TEM	BURBANK,RD+HEIDENREICH,RD 1960 PHIL.MAG.5,373
		NACL 001,110,111	HV	TEM	HEAVENS ET AL 1961.SEE NI 1
		NACL 001	HV	TEM	LO,DS 1966 J.APPL.PHYS.37,3246
		NACL 001	HV	TEM	PUCHALSKA+PRUTTON 1972.SEE FE 3
		NACL 001	HV	TEM	SCHOENING,FRL+BALTZ,A 1962 J.APPL.PHYS.33,1442
		NACL 001,KCL 001, KBR 001	HV	TEM	TSUKAHARA,+KAWAKATSU 1966.SEE FE 3
		NACL 001	HV	OM,TEM	YELON,A+VOEGELI,O 1964*ICTFBB, 32
		NACL 001	HV	OM,TEM	YELON,A+VOEGELI,O+PUGH,EW 1965 J.APPL.PHYS.36,101
	4	MGO 001	UHV	XD	KRYDER,MH+HUMPHREY,FB 1971 J.APPL.PHYS.42,1808
		MGO 001	HV	XD	MAYADAS,AF+JANAK,JF+GANGULEE,A 1974 J.APPL.PHYS.45, 2780
		NIO 111	HV	RED	SCHLENKER,C+BUDER,R 1971 PHYS.STAT.SOL. A4,K79
		CAF2 001,111	HV	XD	VERDERBER,RR+KOSTYK,BM 1961 J.APPL.PHYS.32,696

Z= 28-29 OVERGROWTH		SUBSTRATE	PREP	OBSERV	REFERENCE
NICKEL-COPPER ALLOYS	3	NACL 001	UHV	TEM	MURR 1969.SEE NI 3
		NACL 001	HV	TEM	OLSEN,GH+BOTHA,JC 1972 J.APPL.PHYS.43,3581
		NACL 001	HV	TEM	SHINOHARA,K+HIRTH,JP 1973 THIN SOLID FILMS 16,345
		NACL 001	HV	TED	WALDEN,RH+COTELESSA,RF 1967 J.APPL.PHYS.38,1335
NICKEL-PALLADIUM ALLOYS	3	NACL 001	UHV	TEM	MATTHEWS,JW+JESSER,WA 1969 J.VAC.SCI.TECHN.6,641
NICKEL-PHOSPHOR	1	CU 001	ELD	TEM	JONES+ASPLAND 1972.SEE CO-P
NICKEL BROMIDE	4	CRBR3 00.1	VG	TEM	GRUNBAUM,E+MITCHELL,JW 1964*ICTF88,221
NICKEL OXIDES	4	MGO 001	CVD	OM,RED, XD	CECH+ALESSANDRINI 1959.SEE FEO 4
		MGO 001	CVD	TEM	KATADA,K+NAKAHIGASHI,K+SHIMOMURA,Y 1970 JAP.J.APPL.PHYS.9,1019
		MGO 001	CVD	---	LUBEZKY,I+TANNHAUSER,DS 1972 JCRYSTGROWTH 17,162
		MGO 001,110,111	CVD	OM	MARSHALL 1971.SEE SPINEL FERRITES MNFE2O4
		MGO 001	CVD	OM,XB	ROBINSON ET AL 1966.SEE FEO 4
		SAPPH. 10.0,11.0, 10.1,10.4,00.1	VG	RED	THIRSK,HR+WHITMORE,EJ 1940 TRANS.FARADAY SOC. 36,565
		MGO 001	CVD	OM,XB	VERNON,MW+SPOONER,FJ 1967 J.MATER.SCI.2,415
		MGO 001	CVD	OM,XM	VERNON,MW+SPOONER,FJ 1969 J.MATER.SCI.4,112
29 COPPER	1A	CU 001,110,111	ED	OM,XB	BARNES,SC 1959 ACTA MET.7,700
		CU	ED	OM,XB	BARNES,SC 1961 ELECTROCHIM. ACTA 5,79
		CU 001	ED	OM,REM, XD	BARNES,SC 1964 J.ELECTROCHEM.SOC.111,296
		CU 001	ED	OM,XD	BARNES,SC+STOREY,G+PICK,HJ 1960 ELECTROCHIM.ACTA 2,195
		CU 001,110,111	ED	OM,XB	BEBCZUK DE CUSMINSKY,J 1970 ELECTROCHIMACTA 15,73
		CU 111	ED	OM,XD, XM	BERTOCCI,U+BERTOCCI,C 1971 J.ELECTROCHEM.SOC..118,1287
		CU 111	ED	OM,XB, XM	BERTOCCI,U+BERTOCCI,C+LARSON,BC 1972 J.CRYST.GROWTH 13/14,427
		CU 001,110,111	ED	OM,RED	BICELLI,L+POLI,G 1966 ELECTROCHIM.ACTA 11,289
		CU 001	ED	OM	DAMJANOVIC,A+PAVNOVIC,M+BOCKRIS,J O≠M 1965 ELECTROCHIM.ACTA 10,111
		CU 001	ED	OM,REM	DAMJANOVIC,A+PAVNOVIC,M+BOCKRIS,J O≠M 1965 J.ELECTROANAL.CHEM.9,93
		CU 110,111	ED	OM	DAMJANOVIC,A+SETTY,THV+BOCKRIS,J O≠M 1966J.ELECTROCHEM.SOC.113,429
		CU	ED	OM	ECONOMOU,NA+TRIVICH,D 1961 ELECTROCHIMACTA 3,292
		CU 001	ED	OM,REM	ECONOMOU,NA+FISCHER,H+TRIVICH,D 1960 ELECTROCHIM.ACTA 2,207
		CU 001,111	ED	RED	FINCH ET AL 1947.SEE CO 1
		CU 001,110,111	ED	OM,XD	GIRON,I+OGBURN,F 1961 J.ELECTROCHEM.SOC.108,842/ SEE ALSO BARNES,SC 1962 IBID.109,546
		CU 111	UHV	RED	GRADMANN 1966.SEE NI 1
		CU 111	HV	RED	HAASE,O 1956.SEE FE 1
		CU 110	HV	RED	HAASE,O 1961(A).SEE FE 1
		CU 001,110,111	HV	RED	KRAUSE,GO 1966(A) J.APPL.PHYS.37,3691
		CU 111	HV	RED	KRAUSE,GO 1966(B) J.APPL.PHYS.37,3694
		CU 111	HV	RED	KRAUSE,G+MENZEL-KOPP,CHR+MENZEL,E 1964 PHYS.STATUS SOLIDI 6,121
		CU	HV	RED	KRAUSE,G+MENZEL-KOPP,CHR+MENZEL,E 1965 SURFACE SCI.3,421
		CU 111	HV	RED	LAFOURCADE,L+LARROQUE,P+NGUYEN QUAT TI 1959 C.R.ACAD.SCI.249,230
		CU 111	HV	RED	LAFOURCADE,L+BRUNEL,A+GARIGUE,J 1961*2ERCEM 1,189
		CU 001	ED	OM,XB	LIGHTY,PE+SHANEFIELD,D+WEISSMAN,S+SHRIER,A 1963 J.APPL.PHYS.34,2233
		CU 111	HV	RED	NGUYEN QUAT TI 1959 C.R.ACAD.SCI.249,2301
		CU 110,111	HV	RED	NGUYEN QUAT TI ET AL 1960.SEE CO 1
		CU 001,110,111	ED	OM,XB	OREM,TH 1958 J.RES.NAT.BUR.STAND.60,597
		CU	ED	OM,REM, XD	PICK,HJ+STOREY,GG+VAUGHAN,TB 1960 ELECTROCHIM.ACTA 2,165
		CU 001	ED	OM,TEM, REM,XD	SARD,R+WEIL,R 1970 ELECTROCHIM.ACTA 15,1977
		CU	ED	OM,XD	SCHULTZE,WA 1972 J.CRYST.GROWTH 13/14,421
		CU 001	ED	OM,XD	SHANEFIELD,D+LIGHTY,PE 1963JELECTROCHEMSOC110,973
		CU 001,110,111	ED	OM	SHESHADRI,BS+SETTY,THV 1974 J.CRYST.GROWTH 21,110
		CU	ED	TEM	SWANN,PR 1966 ACTA MET.14,900
		CU 001,111	ED	RED	THOMPSON,GP 1931 PROC.ROY.SOC,SER.A 133,1
		CU	ED	OM	TURNER,DR+JOHNSON,GR 1962 JELECTROCHIMSOC.109,798
		CU 111	ED	OM,RED, TED,XD	UENO,Y+TSUIKI,M 1972 DENKI KAGAKU 40,825
		CU 111	ED	TEM	UENO,Y+KIDOKORO,N+TSUIKI,M 1974 J.ELECTROCHEM.SOC.121, 202
		CU	ED	OM,XM	VAUGHAN,TB 1961 ELECTROCHIM.ACTA 4,72
		CU	ED	REM	VAUGHAN,TB+PICK,HJ 1960 ELECTROCHIM,ACTA 2,179

Z= 29	OVERGROWTH	SUBSTRATE	PREP	OBSERV	REFERENCE
COPPER	18	AU 111 + C	HV	TEM	ACKERMAN,M+VERMAAK,JS+SNYMAN,HC 1973 SURFACE SCI.34,394
		AU 111	HV	TEM	BASSET ET AL 1958.SEE NI 1
		AG 111	HV	REM	BASSETT,GA+MENTER,JW+PASHLEY,DW 1959*ICSPTF, 11
		ZN 00.1	HV	RED	COUDERC,JJ+GARIGUE,J+LAFOURCADE,L+NGUYEN QUAT TI 1959 C.R.ACAD.SCI.249,2037
		CU-IN	ED	TED	EMBURY,JD+DUFF,WR 1966 ELECTROCHIM.ACTA 11,1491
		PD 001,PT 110	ED	TED	FINCH+SUN 1936.SEE FE 1
		W	UHV	FEM	FRANKLIN,WM+LAWLESS,KR 1967*SGSI, 132
		AG 111,AU 111	HV	RED	GLOSSOP+PASHLEY (UNPUBLISHED).SEE NI 1
		FE 001	ELD	RED	GOSWAMI 1957.SEE NI 1
		AG 111	HV	RED	GRADMANN,U 1964 PHYS.KONDENS.MATERIE.3,91
		NI 111	UHV	RED	GRADMANN 1966.SEE NI 1
		AG 111	HV	RED	GRUNBAUM,E+NEWMAN,RC+PASHLEY,DW 1958PHILMAG3,1337
		AG 111	UHV	RED	GRUNBAUM ET AL 1969.SEE CO 1
		AG 111	UHV	RED,TEM,REM	HORNG,CT+VOOK,RW 1974 J.VAC.SCI.TECHN.11,140
		W	UHV	FEM	JONES,JP 1965 PROC.ROY.SOC,SER.A 284,469
		AG 111	HV	RED	KEHOE,RB+NEWMAN,RC+PASHLEY,DW 1956 BRIT.J.APPL.PHYS.7,29
		AG 111	HV	OM,RED	KEHOE,RB+NEWMAN,RC+PASHLEY,DW 1956 PHILMAG.1,783
		AU 110,NI 111	HV	RED	KUBO+MIYAKE 1948.SEE NI 1
		NI 001	HV	TEM	KUNTZE ET AL 1969.SEE NI 1
		NI	HV, ED	TEM	LAWLESS,KR 1965 J.VAC.SCI.TECHN.2,24
		NI	ED	TEM	LAWLESS,KR+GARMON,LB 1962*5ICEM 1,DD-7
		NI			LAWLESS,KR+GOODMAN,WL 1967*25EMSA,298
		NI 001	UHV	TEM	MATTHEWS,JW 1972(A) SURFACE SCI.31,241/ 1972(B).SEE NI 1
		W	UHV	FEM	MELMED,AJ 1963 J.CHEM.PHYS.38,1444/ 1965 J.APPL.PHYS.36,3585
		W 110	UHV	TED,REM, LED,ES, AES	MOSS,ARL+BLOTT,BH 1969 SURFACE SCI.17,240
		AG 111	HV	RED	NEWMAN,RC+PASHLEY,DW 1955 PHIL.MAG.46,927
		AU 111	HV	TEM	PASHLEY ET AL 1957.SEE CO 1
		TI 00.1	UHV	LED	SCHLIER,RE+FARNSWORTH,HE 1958JPHYSCHEMSOLIDS6,271
		NI 001,PD 001, AG 001	HV	TED	SHIRAI ET AL 1961.SEE NI 1
		W	UHV	FEM	SMITH 1973.SEE NI 1
		CU-ZN 211	ED	RED,REM	TAKAHASHI,N 1952 C.R.ACAD.SCI.234,1619/ 1953 J.DE CHIMIE PHYSIQUE 50,624
		W 110	UHV	LED	TAYLOR,NJ 1966 SURFACE SCI.4,161
	2	MOS2 00.1	HV	RED	KAIMUNA 1951.SEE NI 2
		PBS 001	HV	RED	MIYAKE,S+KUBO,M 1947(A+B)JPHYSSOCJAPAN 2,15+20
		ZNS 110	HV	RED	KUBO+MIYAKE 1948.SEE NI 1
	3	NACL 001	HV	TEM	ADAMSKY+LEBLANC 1963.SEE NI 3
		NACL 001	HV, UHV	RED,TED, XD	BRINE,DA+YOUNG,RA 1963 PHIL.MAG.8,651
		NACL 001	UHV	TEM	BROCKWAY,LO+ADLER,IM 1972J.ELECTROCHEMSOC.119,899
		NACL 001	HV	TEM	BROCKWAY,LO+MARCUS,RB 1963 J.APPL.PHYS.34,921
		NACL 001	HV	TEM,REM	BROCKWAY,LO+ROWE,AP 1967*SGSI, 147
		NACL 001	HV	TEM	BROCKWAY ET AL 1964.SEE NI 3
		NACL 001	HV	TED	BRUCK 1936.SEE AL 3
		NACL 001	HV	TEM	CHOPRA,KL 1965 APPL.PHYS.LETT.7,140
		NACL 001	HV	TEM	CZANDERNA,AW+BRENNAN,BW+CLARKE,JR.EG 1971 PHYS.STATUS SOLIDI A8,K75
		LIF 001,NACL 001, KCL 001,KI 001	HV	RED	GOTTSCHE 1956.SEE AL 3
		LIF 001,110	HV	XD	HALL,MJ+THOMPSON,MW 1961 BRIT.J.APPL.PHYS.12,495
		NACL 001, KI 001	HV	RED	HARSDORFF,M 1963 SOLID STATE COMMUN.1,218/ 1964 IBID.2,133
		NACL 001,KCL 001, KI 001	HV, UHV	RED,TED	HARSDORFF,M 1968(A) Z.NATURFORSCH, 23A,1059
		NACL 111	HV, UHV	RED,TEM	HORNG,CT+VOOK,RW 1973 J.VAC.SCI.TECHN.10,160
		NACL 001	HV	TEM	INO,S+WATANABE,D+OGAWA,S 1962 J.PHYS.SOC.JAPAN.17,1074/ 1964 IBID.19,881
		NACL 001	HV	TEM	JAUNET+SELLA 1965.SEE AL 3
		NAF 001,NACL 001, NABR 001,NAI 001, KF 001,KCL 001, KBR 001,KI 001, RBCL 001,RBBR 001, RBI 001	UHV	TEM	JESSER,WA+MATTHEWS,JW 1969 J.CRYST.GROWTH 5,83
		KCL 001,KBR 001, KI 001	HV	TED	KATO 1968.SEE AL 3
		NACL 001,KCL 001, KBR 001,KI 001	HV	RED	KEHOE,RB 1957 PHIL.MAG.2,455
		NACL 111	UHV	RED,TEM	KOCH,FA+HORNG,CT+VOOK,RW 1972J.VAC.SCITECHN.9,511
		NACL 001	HV	TEM	KCMODA 1966.SEE AL 3
		NACL 001	HV	TEM	KUNTZE ET AL 1969.SEE NI 1

Z= 29-30 OVERGROWTH		SUBSTRATE	PREP	OBSERV	REFERENCE
COPPER (CONT.)	3	NACL 001	HV	TEM	MATTHEWS,JW 1959 PHIL.MAG.4,1017
		NACL 001	UHV	TEM	MATTHEWS 1966.SEE FE 3
		NACL 001	UHV	TEM	MURR 1969.SEE NI 3
		NACL 001	HV	TEM	NIELSEN,NA+KEATING,KB+MILLER,WR 1967 SURFACE SCI.8,307
		NACL 001	HV	RED,TEM	OGAWA ET AL 1955.SEE NI 3
		KCL 001	UHV	TEM,LED	PALMBERG,PW+TODD,CJ+RHODIN,TN 1968 J.APPL.PHYS.39,4650
		NACL 001	HV	TEM	POSTNIKOV ET AL 1968.SEE AL 3
		NACL 001,110,111, CSCL 001,110,111	UHV	---	REALE 1972.SEE K 3
		LIF 001,NACL 001, KCL 001	HV	TEM,REM	ROWE,AP+BROCKWAY,LO 1966 J.APPL.PHYS.37,2703
		NACL 001	HV	TEM	SCHLOTTERER 1965.SEE NI 3
		NACL 001	HV	TED	SELLA+TRILLAT 1964.SEE AL 3
		NACL 001	HV	TEM	STOWELL,MJ+LAW,TJ 1969 PHIL.MAG.19,1257
		NACL 001	HV	XD	SHANK,TF+LAWLESS,KR 1965 J.APPL.PHYS.36,2089
		NACL 001	UHV	TEM	TAKAHASHI ET AL 1966(A).SEE NI 3
		NACL 001	HV	TEM	VOOK,RW+HORNG,CT 1973 THIN SOLID FILMS 18,295
		NACL 001,111	HV	TEM	YELON,A+HOFFMAN,RW 1960 J.APPL.PHYS.31,1672
		NACL 001	HV	XD	
	4	MGO 001	HV, UHV	RED,TED, XD	BRINE+YOUNG 1963.SEE 3
		MICA 00.1	HV	TEM	CHOPRA 1965.SEE 3
		MICA 00.1	HV	XD	HALL+THOMPSON 1961.SEE 3
		SAPPH. 00.1	HV	XD	KATZ,G 1968 APPL.PHYS.LETT.12,161
		SAPPH. 00.1	HV	SEM,XD	KATZ,G 1970 J.MATER.SCI.5,736
		MGO 001	UHV	LED,AES	PALMBERG,PW+RHODIN,TN 1968(A) J.PHYS.CHEM.49,134
		MICA 00.1	HV	TEM	POSTNIKOV ET AL 1968.SEE AL 3
		MICA 00.1	UHV	XD	REICHELT,K 1973 SURFACE SCI.36.725
COPPER-PALLADIUM ALLOYS	3	NACL 001	HV	TED	WATANABE,D+HIRABAYASHI,M+OGAWA,S 1955 ACTA CRYST.8,510
		NACL 001	HV	TED	WATANABE,D+OGAWA,S 1956 J.PHYS.SOC.JAPAN 11,226
COPPER-TUNGSTEN ALLOYS	1	W 110	UHV	LED	TAYLOR 1966.SEE CU 1B
COPPER-GOLD ALLOYS	1	AU 111	HV	TEM	ACKERMAN ET AL 1973.SEE CU 1B
		AG 111	HV	TEM	BASSETT+PASHLEY 1958-59.SEE AU 1
		AG 001	HV	RED,TEM	GLOSSOP,AB+PASHLEY,DW 1959 PROC.ROY.SOC.A250,132
		AU 111	HV	RED	GLOSSOP+PASHLEY (UNPUBLISHED).SEE NI 1
		AG 001	HV	TEM	HUNT,AM+PASHLEY,DW 1962 J.PHYS.RADIUM 23,846
		CU 001	UHV	LED,AES	PALMBERG+RHODIN 1968(A).SEE CU 4/ 1968(C) J.APPL.PHYS.39,2425
		AG 001	HV	TEM	PASHLEY,DW+PRESLAND,AEB 1958-59 J.INST.METALS 87, 419/ 1959*ICSPTF, 199
	3	NACL 001	HV	TED	ALESSANDRINI,EI 1959 ACTA CRYST.12,471
		NACL 001	HV	TEM	MIHAMA,K 1970*7ICEM 2,405
		NACL 001	HV	TED	OGAWA,S+WATANABE,D 1952 ACTA CRYST.5,848/ 1954 IBID.7,377/ 1954 J.PHYS.SOC.JAPAN 9,475
		NACL 001	HV	TEM	OGAWA,S+WATANABE,D+WATANABE,H+KOMODA,T 1958 ACTA CRYST.11,872
		NACL 001	HV	TEM	PHILLIPS,VA 1960 J.APPL.PHYS.31,697
		NACL 001	HV	TED	RAETHER,H 1950. SEE AL-CU 3/ 1951 ACTA CRYST.4,70/ 1952 Z.ANGEW.PHYS.4,53
		NACL 001	HV	TED	RAETHER,H 1952 Z.NATURFORSCH.7A,210
		NACL 001	HV	TEM	SATO,H 1964*ICTFBB, 341
		NACL 001	HV	TEM	SATO,H+TOTH,RS 1961 PHYS.REV.124,1833
		NACL 001	HV	TEM	SATO,H+TOTH,RS 1966 J.APPL.PHYS.37,3367
		NACL 001	SP	TEM	TONG,HC+WAYMAN,CM 1972 J.CRYST.GROWTH 15,211
		NACL 001	HV	TED	TOTH,RS+SATO,H 1962 J.APPL.PHYS.33,3250/ 1964 IBID.35,698
		NACL 001	HV	TEM	WATANABE,D+FISHER,PMJ 1965 JPHYSSOC.JAP.20,2170
		NACL 001	HV	TEM	YAMAGUCHI,S+WATANABE,D+OGAWA,S 1961 J.PHYS.SOC.JAP.17,1030
COPPER-GOLD-PALLADIUM	3	NACL 001	HV	TED	NAGASAWA,A 1965 J.PHYS.SOC.JAPAN 20,1520/ 1966 IBID.21,955
		NACL 001	HV	TED	SATO+TOTH 1961.SEE CU-AU 3
COPPER-GOLD-ZINC ALLOYS	3	NACL 001	HV	TEM	OGAWA,S+WATANABE,D+WATANABE,H+KOMODA,T 1959 J.PHYS.SOC.JAPAN 14,936
		NACL 001	HV	TEM	SATO+TOTH 1961.SEE CU-AU 3
COPPER SELENIDE	3-4	NACL 001,110,111, MICA 00.1	HV	RED,TED	GOSWAMI,A+NIKAM,PS 1970 INDIAN J.PURE APPL.PHYS.8,710
COPPER IODIDE	3-4	NACL 001,110,111, KCL 001,MICA 00.1	HV	RED,TED	BADACHNAPE,SB+GOSWAMI,A 1966*ISTFCL, 704
30 ZINC	1	CU 001,110,111	ED	RED	GOSWAMI,A 1969 INDIAN J.PURE.APPL.PHYS.7,232
		CU 111	HV	RED	HAASE 1956.SEE FE 1
		CU,ZN	ED	OM,XD	KEEN,JM+FARR,JPG 1962 J.ELECTROCHEM.SOC.109,668

Z= 30 OVERGROWTH		SUBSTRATE	PREP	OBSERV	REFERENCE
ZINC	2	GAAS 001,111	UHV	RED,LED	ARTHUR,JR 1973 SURFACE SCI.38,394
		ZNS 110	HV	RED	SEGMULLER 1956.SEE SI 2B
ZINC SULPHIDE	1	MO,W	HV	---	GALLI,G+HOLMES,RE 1968 ELECTROCHEM.TECHNOL.6,358
	2	GAAS 001	HV	RED,XD	BEHRNDT,ME+MORENO,SC 1971 J.VAC.SCI.TECHN.8,494
		GAP 111	CVD	OM,XD	BERTOTI,I+FARKAS-JAHNKE,M+LENDVAY,E+NEMETH,T 1969 J.MATER.SCI.4,699
		ZNS	VG	RED	BLANCONNIER,P+HENOC,P 1972 J.CRYST.GROWTH 17,218
		ZNS 10.0,11.0, CDS 10.0,11.0	VG	TEM,REM	CAVENEY,RJ 1968 J.CRYST.GROWTH 2,85
		CDS 10.0,00.1, GAAS 111	CVD	RED,XD	CUSANO,DA 1967 PHYSICS AND CHEMISTRY 2-6 COMP. (M.AVEN+JS.PRENER,EDS.),709.NORTH-HOLLAND PUBL.
		GE 111	HV	OM,XD	DEASLEY,PJ+OWEN,SJT+WEBB,PW 1970 JMATERSCI.5,1054
		SI 111	HV	OM,TEM, XD	JONES,PL+LITTING,CNW+MASON,DE+WILLIAMS,VA 1968 J.PHYS.D=APPL.PHYS.1,283
		SI 111	CVD	OM,RED, TED,XD	LILLEY,P+JONES,PL+LITTING,CNW 1970JMATERSCI.5,891
		SI 111	CVD	OM,RED, TED	LILLEY,P+JONES,PL+LITTING,CNW 1972 J.CRYST.GROWTH 13/14,371
		SI 001,110,111	UHV	RED,TED, SEM,XD	RAWLINS,TGR 1970 J.MATER.SCI.5,881
		SI 111	UHV, SP	RED	RAWLINS,TGR+JONES,CR 1971 J.MATER.SCI.6,1041
		GAP	CVD	OM	SIRTL 1963.SEE SI 2A
		GAP 111	CVD	OM,XD	SLEGER,KJ+MILNES,AG 1972 INT.J.ELECTRON.33,565
		GAAS 001	CVD	OM,RED, XD	VOHL,P+BUCHAN,WR+GENTHE,JE 1971 J.ELECTROCHEM.SOC.118,1842
		SI 111	CVD	TEM	WILKES,P 1969 J.MATER.SCI.4,91
		GAP 001,111, GAAS 001	CVD	OM,RED, XD	YIM,WM+STOFKO,EJ 1972 J.ELECTROCHEM.SOC.119,381
	3	NACL 001,110,111	HV	TED	AGGARWAL,PS+GOSWAMI,A 1963 INDIAN J.PURE APPL.PHYS.1,366
		NACL 001+C,SIOX	HV	TEM	BARNA,A+BARNA,PB+POCZA,JF 1969THINSOLIDFILMS4,R32
		NACL 001	SP	TED	BUNTON,GV+DAY,SCM 1972 THIN SOLID FILMS 10,11
		NACL 001	HV, UHV	TEM	HOLT,DB+WOODCOCK,JM 1970 J.MATER.SCI.5,275
		NACL 001	HV, UHV	TEM	UNVALA,BA+WOODCOCK,JM+HOLT,DB 1968 J.PHYS.D=APPL.PHYS.1,11
		NACL 001	HV, UHV	OM,TEM	WOODCOCK,M+HOLT,DB 1969 J.PHYS.D=APPL.PHYS.2,775
	4	SAPPHIRE 00.1, SPINEL111,BE010.1	CVD	OM,RED, XD	MANASEVIT,HM+SIMPSON,WI 1971 J.ELECTROCHEM.SOC.118,644
		MICA 00.1	HV	RED	MURAVJEVA,KK+KALINKIN,IP+ALESKOVSKY,VB+ BOGOMOLOV,NS. 1970 THIN SOLID FILMS 5,7
ZINC SELENIDE	2	GE 111	HV	TED	AUNG SAN 1973 J.APPL.PHYS.44,523/1974 IBID.45,968 SEE ALSO HOLT,DB 1974 IBID.45, 966
		GAAS 111	CVD	OM,XD	BACZEWSKI,A 1965 J.ELECTROCHEM.SOC.112,577
		ZNSE	VG	RED	BLANCONNIER+HENOC 1972.SEE ZNS 2
		GAAS 111	CVD	OM,XD	BOUGNOT,G+ETIENNE,D+CHEVRIER,J+BOHE,C 1971 MATER.RES.BULL.6,145
		GE 001,110,111	HV	XD	CALOW,IT+OWEN,SJT+WEBB,PW 1968 PHYS.STATUS SOLIDI 28,295
		GE 001,110,111	HV, UHV	OM,XD	CALOW,IT+KIRK,DL+OWEN,SJT 1972 THIN SOLID FILMS 10,409
		GAAS 111	CVD	OM,XD	CHEVRIER,J+ETIENNE,D+CAMASSEL,J+AUVERGNE,D+PONS, JC+MATHIEU,H+BOUGNOT,G 1972MATER.RES.BULL.7,1485
		GAAS 001	VG	RED,XD	GENTHE,JE+ALDRICH,RE 1971 THIN SOLID FILMS 8,149
		GE 111,GAAS 111, ZNSE 111	CVD	OM,XD	HOVEL,HJ+MILNES,AG 1969 J.ELECTROCHEM.SOC.116,843
		GAAS 110,111	CVD	---	MACH,R+LUDWIG,W+EICHHORN,G+ARNOLD,H 1970 PHYS.STATUS SOLIDI A2,701
		GE 001,110,111	HV	RED,XD	NAKAMURA,S+FUKAI,M 1967 JAPAN.J.APPL.PHYS.6, 1473
		GE 001	HV	---	NEWBURY,DM+KIRK,DL 1973 THIN SOLID FILMS 17,S25
		GAAS 001,110,111, 112,115	CVD	OM,XD	PARKER,SG 1971 J.CRYST.GROWTH 9,177
		GAAS 001,110,111, 112,115, GE110,111,112,115	CVD	OM,XM	PARKER,SG+PINNEL,JE+SWINK,LN 1971 J.PHYS.CHEM.SOLIDS 32,139
		SI 001,111, GE 001,111	HV	RED,SEM, XD	RAWLINS 1970.SEE ZNS 2
		CDSE 00.1	HV	RED	SERGEEVA,LA+KALINKIN,IP+ALESKOVSKII,VB 1965 SOVIET PHYS-CRYST.10,178
		GAP 111	CVD	OM,XD	SLEGER+MILNES 1972.SEE ZNS 2
		GAAS 001	CVD	OM,RED, XD	VOHL ET AL 1971. SEE ZNS 2
		GE	CVD, VG	RED	YAMATO,T 1965 JAPAN.J.APPL.PHYS.4,541
		GAAS 001,111	CVD	OM,RED, XD	YIM+STOFKO 1972.SEE ZNS 2

Z= 30-31 OVERGROWTH		SUBSTRATE	PREP	OBSERV	REFERENCE
ZINC SELENIDE	3	NACL 001,110,111	HV	RED,TED	DHERE,NG+GOSWAMI,A 1969 THIN SOLID FILMS 3,439
		NACL 001,111	HV	RED	SERGEEVA ET AL 1965.SEE 2
	4	CAF2 111,MICA 00.1	HV	RED,TED	BATAILLER,G+POINDESSAULT,R+GENDRON,F+AUZARY,C 1972 THIN SOLID FILMS 12,143
		MICA 00.1	HV	RED,TED	DHERE+GOSWAMI 1969.SEE 3
		SAPPH. 10.2	HV	OM,RED, XD	GOODMAN,AM 1969 J.ELECTROCHEM.SOC.116,364
		SAPPH. 00.1, SPINEL 111,BEO10.1	CVD	OM,RED, XD	MANASEVIT+SIMPSON 1971.SEE ZNS 4
		MICA 00.1	HV	RED	MURAVJEVA ET AL 1970.SEE ZNS 4
		MICA 00.1	HV	TED	POINDESSAULT,R+AUZARY,C+LEBOURG,F 1971 C.R.ACAD.SCI.272,691
		MICA 00.1	HV	OM,RED	SERGEYEVA,LA+KAZANNIKOVA,TP+KHARLAMOV,IA+ ALESKOVSKY,VB 1972 THIN SOLID FILMS 11,105
		SAPPHIRE 00.1	CVD	OM,XD	YIM+STOFKO 1972.SEE ZNS 2
ZINC TELLURIDE	2	CDSE 00.1	CVD	OM,RED, XD	ARNOLD,H+KAUFMANN,T+MACH,R 1970 PHYS.STAT.SOLIDI A1, K5
		ZNSE,ZNTE	VG	RED	BLANCONNIER+HENOC 1972.SEE ZNS 2
		GE 111	HV	RED	HOLT,DB+MUFTI,AR 1973 SOLID-ST.ELECTRON.16, 1213
		INAS 111	VG	OM,RED	MORIIZUMI,T+TAKAHASHI,K 1970 JAPANJAPPLPHYS.9,849
		GE 001,111	HV	OM,TEM, SEM	MUFTI,AR+HOLT,DB 1972 J.MATER.SCI.7,694
		ZNSE 110,111	HV	RED,KD	NAKAMURA+FUKAI 1967. SEE ZNSE 2
		GAP 111,GAAS 111, INAS 111	HV	RED	OTA,T+KANAMORI,A+TAKAHASHI,K 1973 PHYS.STATUS SOLIDI A17, K5
		INAS 111	LPE	OM,RED, EP	TAMURA,T+MORIIZUMI,T+TAKAHASHI,K 1971 JAPAN.J.APPL.PHYS.10,813/ 1972 IBID.11,1024
	3	NACL 001,111	HV, UHV	TEM	HOLT,DB 1969 J.MATER.SCI.4,935
		NACL 001,110,111	HV	RED,TED	DHERE+GOSWAMI 1969.SEE ZNSE 3
	4	MICA 00.1	HV	RED,TED	DHERE+GOSWAMI 1969.SEE ZNSE 3
		CAF2 111,BAF2 111	HV	TEM	HOLT,DB 1966 BRIT.J.APPL.PHYS.17,1395
		CAF2 111,BAF2 111	HV, UHV	TEM	HOLT 1969.SEE 3
		SAPPH. 00.1	CVD	OM,RED, XD	MANASEVIT+SIMPSON 1971.SEE ZNS 4
		MICA 00.1	HV	RED	MURAVJEVA ET AL 1970.SEE ZNS 4
ZINC SELENIDE-TELLURIDE	2	GE 111	HV	RED,KD	NAKAMURA+FUKAI 1967. SEE ZNSE 2
ZINC SULPHIDE- GALLIUM PHOSPHIDE	2	GAAS 111 + ZNS	CVD	OM,XD	YIM,WM+DISMUKES,JP+KRESSEL,H 1970 RCAREV.31,662
ZINC SELENIDE- GALLIUM PHOSPHIDE, GALLIUM ARSENIDE	2	GAAS 001 + ZNSE	CVD	OM,XD	YIM ET AL 1970.SEE ZNS-GAP 2
ZINC OXIDES	2	ZNO 00.1	CVD	OM,RED	RABADANOV,RA+SEMILETOV,SA+MAGOMEDOV,ZA 1970 SOVIET PHYS.-SOLID STATE 12,1124
		CDS 00.1	SP	RED,XD	ROZGONI,GA+POLITO,WJ 1969 J.VAC.SCI.TECHN.6,115
		GAAS 001,110,111, GE 001,110,111	CVD	OM,RED	SEMILETOV,SA+RABADANOV,RA 1972 SOVIET.PHYS.-CRYST.17,380
	4	MICA 00.1, SAPPH. 11.0,10.1, 10.2,00.1	CVD	OM,RED	RABADANOV ET AL 1970.SEE 2
		SAPPHIRE	SP	RED,XD	ROZGONI+POLITO 1969.SEE 2
		MICA 00.1	CVD	RED	RYABOVA,LA+SAVITSKAYA,YA.S 1968 THIN SOLID FILMS 2,141
31 GALLIUM	1	W	MG	FIM	NISHIKAWA,O+UTSUMI,T 1973 J.APPL.PHYS.44,945
GALLIUM NITRIDE	2	SIC 00.1	CVD	OM,RED	CHU,TL 1971 J.ELECTROCHEM.SOC.118,1200
		GAAS 111	VG	RED,REM	KOSICKI,BB+KAHNG,D 1969 J.VAC.SCI.TECHN.6,593
		SIC	CVD	OM	MANASEVIT ET AL 1971.SEE ALN 2
		SI 111, GAP 111	CVD	OM,XD	MORIMOTO,Y+UCHIHO,K+USHIO,S 1973 J.ELECTROCHEM.SOC.120, 1783
		SIC 00.1	CVD	OM,XD	WICKENDEN,DK+FAULKNER,KR+BRANDER,RW+ISHERWOOD,BJ 1971 J.CRYST.GROWTH 9,158
	4	SAPPH. 00.1	CVD	OM,RED, XD	ILEGEMS,M 1972 J.CRYST.GROWTH 13/14,360
		SAPPH. 00.1	VG	RED,REM	KOSICKI+KAHNG 1969.SEE 2
		SAPPHIRE	CVD	---	LAGERSTEDT,O+MONEMAR,B 1974 J.APPL.PHYS.45,2266
		SAPPH. 00.1	CVD	XD	MARUSKA,HP+TIETJEN,JJ 1969 APPL.PHYS.LETT.15,327
		SAPPH. 10.2,10.4, 00.1,SPINEL 111	CVD	OM	MANASEVITS ET AL 1971.SEE ALN 2
		SAPPHIRE 00.1	CVD	OM,XD	MORIMOTO ET AL 1973. SEE 2
		SAPPH. 10.0,11.0, 00.1	CVD	OM,XD	WICKENDEN ET AL 1971.SEE 2

Z= 31 OVERGROWTH		SUBSTRATE	PREP	OBSERV	REFERENCE
GALLIUM PHOSPHIDE	2A	GAP	UHV	OM,RED	ARTHUR,JR+LEPORE,JJ 1969 J.VAC.SCI.TECHN.6,545
		GAP	LPE	OM	BERGH,AA+SAUL,RH+PAOLA,CR 1973 J.ELECTROCHEM.SOC.120, 1558
		GAP 111	CVD	OM	BLOM,GM+BHARGAVA,RN 1972 J.CRYST.GROWTH 17,38
		GAP 111	LPE	XM	BROWN,AS+SPRINGTHORPE,AJ 1971 PHYSSTATSOL. A7,495
		GAP 111	CVD, LPE	TEM	CHASE,BD+HOLT,DB 1972 J.MATER.SCI.7,265
		GAP 111	LPE	SEM	CHASE,BD+HOLT,DB 1973 PHYS.STAT.SOLIDI A19, 467
		GAP	CVD	OM,XD	FLICKER,H+GOLDSTEIN,B 1964 J.APPL.PHYS.35,2959
		GAP	LPE	OM,XD	KOWALCHIK,M+JORDAN,AS+READ,MH 1972 J.ELECTROCHEM.SOC.119,756
		GAP 111	LPE	OM,XM	LADANY,I+MCFARLANE III,SH+BASS,SJ 1969 J.APPL.PHYS.40,4984
		GAP 111	LPE	OM	LIEN,SY+BESTEL,JL 1973 J.ELECTROCHEM.SOC.120,1571
		GAP 111	LPE	OM	LORENZ,MR+PILKUHN,M 1966 J.APPL.PHYS.37,4094
		GAP	LPE	---	LORIMOR,OG+HACKETT,JR.WH+BACHRACH,RZ 1973 J.ELECTROCHEM.SOC.120, 1424
		GAP 001,110,111	CVD	OM,XD	MOTTRAM,A+PEAKER,AR+SUDLOW,PD 1971 J.ELECTROCHEM.SOC.118,318
		GAP 111, OFF 111	LPE	OM	ROZGONYI,GA+IIZUKA,T 1973J.ELECTROCHEMSOC.120,673
		GAP 111	LPE	SEM,XM	ROZGONYI,GA+SAUL,RH 1972 J.APPL.PHYS.43,1186
		GAP 111	CVD, LPE, SG	OM	SAUL,RH 1968 J.ELECTROCHEM.SOC.115,1184
		GAP 111	LPE	OM	SAUL,RH 1971 J.ELECTROCHEM.SOC.118,793
		GAP	LPE	OM	SAUL,RH+ROCCASECCA,DD 1973 J.APPL.PHYS.44,1983/ 1973 J.ELECTROCHEM.SOC.120, 1128
		GAP 111	LPE	---	SHIH,KK+WOODALL,JM+BLUM,SE+FOSTER,LM 1968 J.APPL.PHYS.39, 2962
		GAP 001,111	LPE	OM	SUDLOW,PD+MOTTRAM,A+PEAKER,AR 1972JMATERSCI.7,168
		GAP 111	CVD	OM	TAYLOR,RC+WOODS,JF+LORENZ,RM 1968JAPPLPHYS39,5404
	2B	GAAS	CVD	OM,RED	ALFEROV,ZHI+KOROL≠KOV,VI+MIKHAILOVA-MIKHEEVA,IP+ ROMANENKO,VN+TUCHKEVICH,VM 1965 SOVIET PHYS.-SOLID STATE 6,1865
		GAAS 111	CVD	LPE,TEM	CHASE+HOLT 1972.SEE 2A
		GAAS	CVD	TEM	CHASE,BD+HOLT,DB+UNVALA,BA 1972 J.ELECTROCHEM.SOC.119,310
		SI 001,110,111	ED	OM,XD	CUOMO,JJ+GAMBINO,RJ 1968 JELECTROCHEM.SOC.115,755
		GAAS 001,110,111	CVD	---	D≠YAKONOV,LI+MASLOV,VN+SAKHAROV,BA 1966 SOVIET PHYS.DOKLADY 10,650
		GAAS 001,110,111, 211,511, GAP-AS, GE	CVD	OM,XD	FLICKER+GOLDSTEIN 1964.SEE 2A
		GAAS 001,110,111	CVD	OM,XD	FROSCH,CJ 1964 J.ELECTROCHEM.SOC.111,180
		GAAS 001,110,111, 211	CVD	OM	GROVES,WO 1967*1ICCG,669
		GAAS	CVD		HOLONYAK,JR.N+JILLSON,DC+BEVACQUA,SF 1962*AIMEMSM 15,49
		GAAS 111	CVD	---	HOSS,PA+MURRAY,LA+RIVERA,JA 1968 J.ELECTROCHEM.SOC.115,553
		SI 001,110,111	VG	OM,XD	IGARASHI,O 1970 J.APPL.PHYS.41,3190
		SI 001,110,111	VG, CVD	OM,XD	IGARASHI,O 1972 J.ELECTROCHEM.SOC.119,1430
		GAAS 110	CVD	OM,RED, XD	ING,JR.SW+MINDEN,HT 1962 JELECTROCHEMSOC.109,995
		GAAS 111	CVD	OM,XD	KAMATH,GS+BOWMAN,D 1967 J.ELECTROCHEMSOC.114,192
		GAAS 111	CVD	OM	LUTHER,LC+ROCCASECCA,DD 1968 J.ELECTROCHEM.SOC.115,850
		GAAS,GE 111	CVD	OM,RED	MANASEVIT,HM+SIMPSON,WI 1969 J.ELECTROCHEM.SOC.116,1725
		SI 001	CVD	XD,XM	MCFARLANE III,SH+WANG,CC 1972 J.APPL.PHYS.43,1724
		GAAS 110,111	CVD	XD,EP	MICHEL,J+FAHMY,D 1967 J.MATER.SCI.2,299
		GAAS	CVD	OM,RED, XD	MOEST,RR+SHUPP,BR 1962 J.ELECTROCHEM.SOC.109,1061
		GAAS 001,110,111	CVD	OM,XD	MOTTRAM ET AL 1971.SEE 2A
		GAAS 001	CVD	OM	NICOLL,FH 1963 J.ELECTROCHEM.SOC.110,1165
		SI 110,111	CVD	OM,XD, XM	NOAK,J+WOHLING,W 1970 PHYS.STATUS SOLIDI A3,K229
		GAAS 001,111	CVD	OM,XM	OGIRIMA,M 1972 JAPAN.J.APPL.PHYS.11,281
		GAAS 001,111	CVD	OM	OLDHAM,WG 1965 J.APPL.PHYS.36,2887
		GE 001,110,111	HV	OM,RED, XD	RICHARDS ET AL 1963,1964.SEE ALSB 2
		GAAS 001,110,221	CVD	---	RICHMAN,D+TIETJEN,JJ 1967 TRANS.AIME 239,418
		SI	LPE	OM,EP, XD	ROSZTOCZY,FE+STEIN,WW 1972JELECTROCHEMSOC119,1119
		GAAS	CVD, LPE, SG	OM	SAUL 1968.SEE 2A
		GAAS	CVD	---	SAUL,RH 1969 J.APPL.PHYS.40,3273
		GAAS 111	CVD	TEM	TAMURA,M 1973 J.APPL.PHYS.44,1913
		GAAS 111	CVD	OM	TAYLOR ET AL 1968.SEE 2A
		SI 111	CVD	RED,EP	THOMAS,RW 1969 J.ELECTROCHEM.SOC.116,1449

Z= 31 OVERGROWTH		SUBSTRATE	PREP	OBSERV	REFERENCE
GALLIUM PHOSPHIDE(CONT)	2B	SI 001	CVD	RED,SEM, XD,XM	WANG,CC+MCFARLANE III,SH 1972 J.CRYST.GROWTH 13/14,262
		GAAS 111,GE	CVD, SG	OM,XD	WEINSTEIN,M+BELL,RO+MENNA,AA 1964 J.ELECTROCHEM.SOC.111,674
	4	CAF2 111	UHV	RED,REM	CHO,AY 1970 J.APPL.PHYS.41,782
		CAF2 111	UHV	RED	CHO,AY+CHEN,YS 1970 SOLID-ST.COMMUN.8,377
		SAPPH. 00.1	CVD	OM,RED	MANASEVIT+SIMPSON 1969.SEE 2B
		SAPPH. 00.1,10.2	CVD	XD,XM	MCFARLANE+WANG 1972.SEE 2B
		SAPPH. 00.1,10.2, SPINEL 111	CVD	RED,SEM, XD,XM	WANG+MCFARLANE 1972.SEE 2B
GALLIUM ARSENIDE	1	W 111	CVD	OM,XD	AMICK,JA 1963 RCA REV.24,555/ 1964*ICTF88,283
		W 111,NB 111	CVD	OM,XD	KNAPPETT,JE+OWEN,SJT 1967 J.MATER.SCI.2,295
		AL 110	UHV	RED	LUDEKE ET AL 1973.SEE AL 2
		W 111	CVD	SEM	STEVENSON,GA+TUCK,B+OWEN,SJT 1971 JMATERSCI.6,413
	2A	GAAS 001	CVD, LPE	OM	ABRAHAMS,MS+BUIOCCHI,CJ 1966 J.APPL.PHYS.37,1973
		GAAS 001,111	CVD	OM,TEM, REM	ABRAHAMS,MS+BUIOCCHI,CJ 1967 J.PHYS.CHEM.SOLIDS 28,927
		GAAS	CVD	TEM	ABRAHAMS,MS+BUIOCCHI,CJ+TIETJEN,JJ 1967 J.APPL.PHYS.38,760
		GAAS 001	CVD	TEM	ABRAHAMS,MS+BUIOCCHI,CJ+WILLIAMS,BF 1971 APPL.PHYS.LETT.18,220
		GAAS 001,111	LPE	OM	AHN,BH+SHURTZ,RR+TRUSSELL,CW 1971 J.APPL.PHYS.42, 4512
		GAAS 111	CVD	OM	ALEXANDROV,LN+SIDOROV,YUG 1971 J.VAC.SCI.TECHN.8,571
		GAAS 111	CVD	RED,REM	ALEXANDROV,LN+SIDOROV,YUG+KRIVOROTOV,EA 1969 THIN SOLID FILMS 3,395
		GAAS 001,111,ETC.	CVD	RED,REM	ALEXANDROV,LN+KRIVOROTOV,EA+SIDOROV,YUG 1971 PHYS.STATUS SOLIDI A4,339/ 1970*7ICEM 2,427
		GAAS 001, OFF 001	CVD	REM	ALEKSANDROV,LN+ZALETIN,VM+KRIVOROTOV,EA+ SIDOROV,YU.G 1973 PHYS.STAT.SOLIDI A15,367
		GAAS OFF 001	CVD	OM	AOKI,T+YAMAGUCHI,M 1972 JAPAN.J.APPL.PHYS.11,1775
		GAAS 111	UHV	RED	ARTHUR,JR.JR 1968 J.APPL.PHYS.39,4032
		GAAS 001,110,111	UHV	OM,RED, TED	ARTHUR+LEPORE 1969.SEE GAP 2A
		GAAS 111,001	CVD	OM,XD	BLAKESLEE,AE 1969 TRANS.AIME 245,577
		GAAS 110		OM,SEM	BLUM,FA+LAWLEY,KL+SCOTT,WC+HOLTON,WC 1974 APPL.PHYS.LETT.24,430
		GAAS 001,110,111	CVD	XD	BOBB,LC+HOLLOWAY,H+MAXWELL,KH+ZIMMERMAN,E 1966 J. APPLPHYS.37,3909/ 1966J.PHYS.CHEM.SOLIDS 27,1679
		GAAS	CVD	---	BOUCHER,A+HOLLAN,L 1970 J.ELECTROCHEM.SOC.117,932
		GAAS	LPE	XD,XM	CASTET,L+MAYET,L+MESNARG,G 1970 C.R.ACAD.SCI.,SER.B.,271,908
		GAAS 001,111,113	CVD	OM	CHANE,JP+HOLLAN,L+SCHILLER,C 1972 J.CRYST.GROWTH 13/14,325
		GAAS 001	UHV	RED,REM	CHANG,LL+ESAKI,L+HOWARD,WE+LUDEKE,R+SCHUL,G 1973 J.VAC.SCI.TECHN.10,655
		GAAS 111	UHV	RED	CHO,AY 1969 SURFACE SCI.17,494
		GAAS 111	UHV	RED,REM	CHO,AY 1970 J.APPL.PHYS.41,2780
		GAAS 001	UHV	RED,REM, SEM	CHO,AY 1971 J.VAC.SCI.TECHN.8,S31
		GAAS 001	CVD+ UHV	RED	CHO,AY. 1971 J.APPL.PHYS.42,2074
					CHO+CASEY 1974. SEE AL-GAAS 2
		GAAS 001	UHV	OM	CHO,AY+HAYASHI,I 1971 SOLID-ST.ELECTRON.14,125
		GAAS 111	UHV	RED	CHO,AY+PANISH,MB+HAYASHI,I 1970 PROC.3RD.INT.SYMP
		GAAS 111	UHV	RED	GALLIUM ARSENIDE+REL.COMP.AACHEN,GERMANY1970,18
		GAAS 001	CVD	---	CONRAD,RW+REYNOLDS,RA+JEFFCOAT,MW 1967 SOLID-ST.ELECTRON.10,507
		GAAS 001,111	LPE	OM	CROSSLEY,I+SMALL,MB 1972 J.CRYST.GROWTH 15,275
		GAAS 001,111	VG	OM,RED, REM	DAVEY,JE+PANKEY,T 1968 J.APPL.PHYS.39,1941
		GAAS 001	CVD	OM	DERSIN,HJ+SIRTL,E 1966 Z.NATURFORSCH.21A,332
		GAAS 001,111,211, 311	CVD	OM	DILORENZO,JV 1972 J.CRYST.GROWTH 17,189
		GAAS 001,111, 211,311	CVD	OM	DILORENZO,JV+MACHALA,AE 1971 J.ELECTROCHEM.SOC.118, 1516
		GAAS 001	UHV	RED	DOVE,DB+LUDEKE,R+CHANG,LL 1973 JAPPLPHYS.44,1897
		GAAS	CVD	OM	EFFER,D 1965 J.ELECTROCHEM.SOC.112,1020
		GAAS 111	LPE	OM,XM	ETTENBERG,M 1974 J.APPL.PHYS.45,901
		GAAS 001,111	CVD	OM	EWING,RE+GREENE,PE 1964 J.ELECTROCHEMSOC.111,1266
		GAAS	CVD	---	FAIRMAN,RD+SOLOMON,R 1973 J.ELECTROCHEM.SOC.120,541
		GAAS 001	UHV	OM,SEM, EP	FOXON,CT+HARVEY,JA+JOYCE,BA 1973 J.PHYS.CHEM.SOLIDS 34,1693
		GAAS	CVD	OM,XD	FROSCH 1964.SEE GAP 2B
		GAAS 001,111	CVD	RED,XD	GOLDSMITH,N+OSHINSKY,W 1963 RCA REV.24,546
		GAAS	CVD	OM	HOLLAN,L+SCHILLER,C 1972 J.CRYST.GROWTH 13/14,319
		GAAS 001	CVD	XM	HOLLAN,L+HALLAIS,J+SCHILLER,C 1971 J.CRYST.GROWTH 9,165

Z= 31 OVERGROWTH	SUBSTRATE	PREP	OBSERV	REFERENCE
GALLIUM ARSENIDE(CONT.)2A	GAAS 111	CVD	OM,XM	HOLLOWAY,H+BOBB,LC 1967(A) J.APPL.PHYS.38,2711
	GAAS 001	CVD	OM,XM	HOLLOWAY,H+BOBB,LC 1967(B) J.APPL.PHYS.38,2893
	GAAS 110	CVD	OM,XD	HOLONYAK ET AL 1962.SEE GAP 2B
	GAAS 111	CVD	---	HOSS ET AL 1968.SEE GAP 2B
	GAAS OFF 001	CVD	OM	IHARA,M+DAZAI,K+RYUZAN,O 1974 J.APPL.PHYS.45,528
	GAAS	CVD	OM,RED, XD	ING+MINDEN 1962.SEE GAP 2B
	GAAS 001	CVD	---	ITO,S+SHINOHARA,T+SEKI,Y 1973 J.ELECTROCHEM.SOC.120, 1419
	GAAS 111	CVD	OM	JOYCE,BD+MULLIN,JB 1966 SOLID-ST.COMMUN.4,463
	GAAS 111	CVD	OM	JOYCE,BD+MULLIN,JB 1967 SOLID-ST.COMMUN.5,727
	GAAS 111	CVD	OM,RED, XD,EP	KASANO,H+IIDA,S 1967 JAPAN.J.APPL.PHYS.6,1038
	GAAS OFF 001	CVD	OM	KENNEDY,JK+POTTER,WD 1973 J.CRYST.GROWTH 19,85
	GAAS 001	CVD, LPE	OM,XM	KISHINO,S+IIDA,S 1972 J.ELECTROCHEM.SOC.119,1113
	GAAS 110	CVD	---	KNIGHT,JR+EFFER,D+EVANS,PR 1965 SOLID-ST.ELECTRON.8,178
	GAAS 110,111	CVD	OM	KOVDA,AV+SEMILETOV,SA 1967SOVIETPHYSCRYST.12,468
	GAAS 001,110,111	CVD	XD	LINDEKE,K+SACK,W+NICKL,JJ 1970 J.ELECTROCHEM.SOC.117,1316
	GAAS 001,110,111, 211,311	LPE	OM	LONGO,JT+HARRIS,JR.JS+GERTNER,ER+CHU,JC 1972 J.CRYSTAL GROWTH 15, 107
	GAAS 111	CVD	OM,RED	MAGOMEDOV,KHA+SHEFTAL,NN 1965SOVIETPHYSCRYST9,756
	GAAS 111	CVD	OM	MAGOMEDOV,KHA+MAGOMEDOV,NN 1967 SOVIET PHYS.CRYST.12,286
	GAAS	CVD	OM,RED, EP,XD	MANASEVIT+SIMPSON 1969.SEE GAP 2B
	GAAS 001	LPE	OM	MARUYAMA,S 1972 JAPAN.J.APPL.PHYS.11,424
	GAAS 001	CVD	OM,EP	MIKI,H+ITO,M+ODA,T 1972 JAPAN.J.APPL.PHYS.11,623
	GAAS 001	CVD	OM,REM	MINDEN,HT 1971 J.CRYST.GROWTH 8,37
	GAAS 001,110,111	CVD	OM	MOEST,RR 1966 J.ELECTROCHEM.SOC.113,141
	GAAS	CVD	OM,RED, XD	MOEST+SHUPP 1962.SEE GAP 2B
	GAAS 001	LPE	OM	MOON,RL+KINOSHITA,J 1974 J.CRYST.GROWTH 21,149
	GAAS 001	LPE	OM,XD	NELSON,H 1963 RCA REV.24,603
	GAAS 110,111	CVD	OM	NEWMAN,RL+GOLDSMITH,N 1961JELECTROCHEMSOC108,1127
	GAAS 111	CVD	OM,XD	OKADA,T 1963 JAPAN.J.APPL.PHYS.2,206
	GAAS	LPE	---	PILKUHN,MH+RUPRECHT,H 1967 J.APPL.PHYS.38,5
	GAAS 001,110,111, 112,113	CVD	OM,XD	PIZZARELLO,FA 1963 J.ELECTROCHEM.SOC.110,1059
	GAAS 001	CVD	RED,TEM	RAI-CHOUDHURY,P 1969 J.ELECTROCHEM.SOC.116,1745
	GAAS 001	CVD	OM,TEM, SEM	RAI-CHOUDHURY+SCHROEDER 1971.SEE SI 2A
	GAAS 111	HV	OM,RED, XD	RICHARDS ET AL 1963,1964.SEE ALSB 2
	GAAS 001,110,111	CVD	OM	RUBENSTEIN,M+MYERS,E 1966 JELECTROCHEMSOC.113,365
	GAAS 001,110,111	CVD	OM,REM, XD	SHAW,DW 1966 J.ELECTROCHEM.SOC.113,904/ 1968 IBID.115,777
	GAAS 001,111,112, 113,115	CVD	OM	SHAW,DW 1968 J.ELECTROCHEM.SOC.115,405
	GAAS 001	CVD	---	SHAW,DW 1970 J.ELECTROCHEM.SOC.117,683
	GAAS 001	CVD	OM,SEM	SHAW,DW 1972 J.CRYST.GROWTH 12,249
	GAAS 111	CVD	OM	SHEFTAL,NN+MAGOMEDOV,KHA 1967*1ICCG, 533
	GAAS 111 + SIO2	CVD	OM,XD	TAUSCH,JR.FW+LAPIERRE,III AG 1965 J.ELECTROCHEM.SOC.112,706
	GAAS 001,110,111	CVD	OM	TAYLOR,RC 1967 J.ELECTROCHEM.SOC.114,410
	GAAS 001	LPE	OM	VILMS,J+GARRETT,JP 1972 SOLID-ST.ELECTRON.15,443
	GAAS 001	CVD	OM	WANG,P+PINK,F+SCIOLA,J 1967 J.ELECTROCHEM.SOC.114,879
	GAAS	CVD	OM	WEINSTEIN ET AL 1964.SEE GAP 2B
	GAAS 001,110,111	CVD	OM	WILLIAMS,FV 1964 J.ELECTROCHEM.SOC.111,886
	GAAS 001	CVD	ICH	WOOD,DR+MORGAN,DV 1973 PHYS.STAT.SOLIDI A17,K143
	GAAS 001	LPE	SEM	YANG,L+BALLANTYNE,JM 1974 APPL.PHYS.LETT.25,67
	GAAS 001,111	CVD, LPE	OM	YOUNG,ML+ROWLAND,MC 1973 PHYS.STAT.SOLIDI A16,603
	GAAS 001,111	CVD	OM	ZIMMERLI,U+STEINEMANN,A 1967 SOLID-STATE COMMUN.5,447
2B	GAP 001	CVD	TEM	ABRAHAMS ET AL 1972.SEE 2A
	GE 111,SI 111	CVD	RED,TEM, REM,EP	ALEXANDROV ET AL 1970,1971.SEE 2A
	GE 111	CVD	OM,XD	AMICK 1963,1964.SEE 1
	ZNSE 110,111	LPE	OM,XD	BALCH,JW+ANDERSON,WW 1972 J.CRYST.GROWTH 15,204
	GE001,111,311,511	CVD	OM,XD	BOBB,LC+HOLLOWAY,H+MAXWELL,KH+ZIMMERMAN,E 1966 J.APPL.PHYS.37,3909 + 4687
	GE	CVD	OM	CHASHCHINOV,YUM+MOKIEVSKII,VA 1966 SOVIET PHYS.-SOLID STATE 8,1201
	GE 001,111	VG	RED	DAVEY+PANKEY 1968.SEE 2A
	GAP 111	LPE	OM,XM	ETTENBERG 1974. SEE 2A
	GAP	CVD	OM,XD	FROSCH 1964.SEE GAP 2B
	GE 001,110,111	SP	TEM	FRANCOMBE,H 1963*10NVSAVS, 316
	GE 111	CVD	OM	GABOR,T 1964 J.ELECTROCHEM.SOC.111,817 + 825
	GE 111	CVD	OM,XD	GOTTLIEB,GE+CORBOY,JF 1963 RCA REV.24,585

Z= 31 OVERGROWTH		SUBSTRATE	PREP	OBSERV	REFERENCE
GALLIUM ARSENIDE(CONT.)	2B	GE 001	CVD	OM,XM	HOLLOWAY+BOBB 1967(B).SEE 2A
		GE 311	CVD	XM	HOLLOWAY,H+BOBB,LC 1968 J.APPL.PHYS.39,2467
		GE001,110,111,113	CVD	RED	HOLLOWAY,H+HOLLMAN,K+JOSEPH,AS 1965 PHIL.MAG.11,263
		GE	CVD		HOLONYAK ET AL 1962.SEE GAP 2B
		GE 111	CVD	OM,RED, XD	ING+MINDEN 1962.SEE GAP 2B
		GE001,110,111,211, INAS 111	CVD	OM,RED, SEM, EP,XD	KASANO+IIDA 1967.SEE 2A
		ZNSE 110,111	CVD	---	MACH ET AL 1970.SEE ZNSE 2
		ALAS 111	CVD	OM,XD	MANASEVIT 1971.SEE ALAS 4
		GE 111	CVD	OM,RED, EP,XD	MANASEVIT+SIMPSON 1969.SEE GAP 2B
		GE 111	CVD	OM,XM	MEIERAN,ES 1967 J.ELECTROCHEM.SOC.114,292
		GE	CVD	OM,RED, XD	MOEST+SHUPP 1962.SEE GAP 2B
		GE 001,110,111	SP	OM,RED, REM	MOLNAR,B+FLOOD,JJ+FRANCOMBE,MH 1964 J.APPL.PHYS.35,3554
		GE 001,110,111	HV	RED	MULLER,EK 1964 J.APPL.PHYS.35,580
		GE 111	CVD	OM	NICOLL 1963.SEE GAP 2B
		GE 111	CVD	OM	OKADA 1963.SEE 2A
		GE 111	CVD	OM,XD	OKADA,T+KANO,T+SASAKI,Y 1961 J.PHYS.SOC.JAPAN 16,2591
		GE 001,110,111	HV	OM,RED, XD	RICHARDS ET AL 1963,1964.SEE ALSB 2
		GE 001,111	CVD	OM,XD	ROBINSON,PH 1963 RCA REV.24,574
		GE 111	CVD	XD	SCHULZE,RG 1966 J.APPL.PHYS.37,4295
		GE 111	CVD	OM,XD	SIRTL 1963.SEE SI 2A
		GE 001,110,111	CVD	---	TAKABAYASHI 1962.SEE GE 2A
		GE, GAP	CVD, SG	OM,XD	WEINSTEIN ET AL 1964.SEE GAP 2B
		GAP 001,111, AL-GAAS 001,111	CVD, LPE	OM	YOUNG+ROWLAND 1973.SEE 2A
		GE 111	SP	TED	YURASOVA,VE+PROCHOROV,YUA+SHELJAKIN,LB+ NEVZOROVA,LN 1966*6ICEM 1,509
	3	NACL 001	SP	TEM	BUNTON+DAY 1972.SEE ZNS 3
		NACL 001	SP	TEM	EVANS,T+NOREIKA,AJ 1966 PHIL.MAG.13,717
		NACL 001	HV	TEM	O≠NEAL,JE+LEONHARD,FW 1968 J.CRYST.GROWTH 2,80
		NACL 001	HV	TED	STEINBERG,RF+SCRUGGS,DM 1966 JAPPL.PHYS.37,4586
	4	SAPPHIRE	UHV	OM,RED	ARTHUR+LEPORE 1969.SEE GAP 2A
		CAF2 111	UHV	RED	CHO+CHEN 1970.SEE GAP 4
		SAPPHIRE 00.1, SPINEL 001	CVD	OM,RED, XD	GUTIERREZ,WA+POMMERRENIG,HD+JASPER,MA+ MANTZOURANIS,AP 1970 SOLID-ST.ELECTRON.13, 1199
		SAPPH. 10.2,20.1, 00.1	SP	XD	HYDER,SB 1971 J.VAC.SCI.TECHN.8,228
		SPINEL 110,111, SAPPH. 00.1	CVD+ LPE	OM,SEM, XD	LADANY,I+WANG,CC 1972 J.APPL.PHYS.43,236
		SAPPH. 00.1, SPINEL 001,111, BE010.0,10.1,00.1, THO2 001	CVD	OM,RED, XD	MANASEVIT,HM 1968 APPL.PHYS.LETT.12,156
		SAPPH. 00.1	CVD	RED,REM	MANASEVIT,HM 1972 J.CRYST.GROWTH 13/14,306
		SAPPH. 00.1, SPINEL001,110,111, BEO 10.0,10.1,00.1, THO2 001	CVD	OM,RED, REM,EP, XD	MANASEVIT+SIMPSON 1969.SEE GAP 2B
		SAPPH. 00.1, SPINEL 110	CVD	---	MANASEVIT,HM+THORSEN,AC 1972JELECTROCHEMSOC119,99
		SPINEL 110,111	CVD	XD,XM	MCFARLANE+WANG 1972.SEE GAP 2B
		SAPPH. 00.1	CVD	RED,TEM	RAI-CHOUDHURY 1969.SEE 2A
		CAF2 111	HV	OM,RED, XD	RICHARDS ET AL 1963,1964.SEE ALSB 2
		SPINEL001,110,111, OFF 001,110,111	CVD	OM,RED, SEM,XD	WANG,CC+DOUGHERTY,FC+ZANZUCCHI,PJ+ MCFARLANE III,SH 1974 J.ELECTROCHEM.SOC.121, 571
GALLIUM ANTIMONIDE	2	GASB 111	CVD	OM,XD	ARIZUMI,T+KAKEHI,M+SHIMOKAWA,R 1971 J.CRYST.GROWTH 9,151
		GAAS 001	CVD	XD	CLOUGH,RB+TIETJEN,JJ 1969 TRANS.AIME 245,583
		INAS 111	VG+ CG	TED,REM	COHEN-SOLAL ET AL 1966.SEE CD-HGTE 2
		GE 111	CVD	OM,XD	IDO,T+KAKEH,I+ARIZUMI,T 1971 JAPJAPPLPHYS.10,1388
		GASB 111	CVD	OM,XD	KAKEHI,M+SHIMOKAWA,R+ARIZUMI,T 1970 JAPAN.J.APPL.PHYS.9,1039
		GASB 001,111	LPE	---	MIKI,H+SEGAWA,K+FUJIBAYASHI,K 1974 JAPAN.J.APPL.PHYS.13, 203
		GE 001,110,111	HV	OM,RED, XD	RICHARDS ET AL 1963,1964.SEE ALSB 2

Z= 31 OVERGROWTH		SUBSTRATE	PREP	OBSERV	REFERENCE
GALLIUM PHOSPHIDE-ARSENIDE	2	GAAS 001	CVD	TEM	ABRAHAMS,MS+TIETJEN,JJ 1969JPHYSCHEMSOLIDS30,2491
		GAAS 001	CVD	TEM	ABRAHAMS,MS+WEISBERG,LR+BUIOCCHI,CJ+BLANC,J 1969 J.MATER.SCI.4,223
		GAAS	CVD	TEM	ABRAHAMS,MS+WEISBERG,LR+TIETJEN,JJ 1969 J.APPL.PHYS.40,3754
		GAAS 001,110,111, GAP	UHV	OM,RED	ARTHUR+LEPORE 1969.SEE GAP 2A
			CVD	OM,XD	BAN,VS+GOSSENBERGER,HF+TIETJEN,JJ 1972 J.APPL.PHYS.43,2471
		GAAS 001	CVD	OM,XD	BELOUET,CHR 1972 J.CRYST.GROWTH 13/14,342
		GAAS 001,113	CVD	OM,REM, SEM,XD	BLAKESLEE,AE 1971 J.ELECTROCHEM.SOC.118,1459
		GAAS	CVD	REM	BLAKESLEE,AE+ALIOTTA,CF 1970IBMJRESDEVELOP.14,686
		GE 001,111,311	CVD	OM,XD	BURMEISTER,JR.RA+REGEHR,RW 1969 TRANS.AIME 245,565
		GAAS 001	CVD	OM,EP, XD	BURMEISTER,JR.RA+PIGHINI,GP+GREENE,PE 1969 TRANS.AIME 245,587
		GAAS 001	CVD	EP,XD, XM	EWING,RE+SMITH,DK 1968 J.APPL.PHYS.39,5943
		GAAS 111	CVD	OM,XD	FINCH,WF+MEHAL,EW 1964 J.ELECTROCHEM.SOC.111,814
		GAAS 001,111	CVD	OM,XD	GOTTLIEB,GE 1965 J.ELECTROCHEM.SOC.112,192
		GAAS 001	CVD	OM	GREENING,DA+HERZOG,AH 1968 J.APPL.PHYS.39,2783
		GAP-AS	CVD		HOLONYAK ET AL 1962.SEE GAP 2B
		GAAS 111	CVD	---	HOSS ET AL 1968.SEE GAP 2B
		GAAS 001,111	CVD	XM	HOWARD,JK+DOBROTT,RD 1966J.ELECTROCHEMSOC.113,567
		GAAS 111	CVD	---	HULL,EM 1964 J.ELECTROCHEM.SOC.111,1295
		GAAS 001	CVD	OM,XD	INOUE,M+ASAHI,K 1972 JAPAN.J.APPL.PHYS.11,919
		GAAS 001	CVD	OM,EP, XD,XM	KISHINO,S+OGIRIMA,M+KURATA,K 1972 J.ELECTROCHEM.SOC.119,617
		GAAS 001,110,111, 112, GAP	CVD	XD	KU,S 1963 J.ELECTROCHEM.SOC.110,991
		GAAS,GE 111	CVD	OM,RED	MANASEVIT+SIMPSON 1969.SEE GAP 2B
		GE 111	HV	RED	MULLER,EK+RICHARDS,JL 1964 J.APPL.PHYS.35,1233
		GAAS 001	CVD	OM,XD	NAGAI,H 1972 J.APPL.PHYS.43,4254
		GAAS 001	CVD	---	PUROHIT,RK 1968 J.MATER.SCI.3,330
		GE 001,110,111	HV	RED,XD	RICHARDS ET AL 1964.SEE ALSB 2
		GAAS 111	CVD	OM	ROSENBERG,R+KOZLOWSKI,M+MCALEER,WJ+POLLAK,PI 1965 J.ELECTROCHEM.SOC.112,459
		GAAS 001	CVD	OM	SAITOH,T+MINAGAWA,S 1973 J.ELECTROCHEMSOC.120,656
		GAP 111	MG	OM	SHIH,KK 1970 J.ELECTROCHEM.SOC.117,387
		GAAS 001	CVD	OM,XM	STRINGFELLOW,GB+GREENE,PE 1969 J.APPL.PHYS.40,502
		GAAS 001,110,111, GE 111	CVD	OM,XD	TIETJEN,JJ+AMICK,JA 1966 J.ELECTROCHEMSOC.113,724
		GAAS 111	CVD	OM,SEM	VON PHILIPSBORN,H 1970 J.CRYST.GROWTH 7,246
		GAAS 001	CVD	OM	WILLIAMS,FV 1967 TRANS.AIME 239,702
		GAAS	CVD	OM	WU,TY 1974 J.CRYST.GROWTH 21,85
	4	SAPPH. 00.1, SPINEL 111	CVD	OM,RED	MANASEVIT+SIMPSON 1969.SEE GAP 2B
GALLIUM PHOSPHIDE-GALLIUM ANTIMONIDE	2	GE 001	HV	RED	MULLER+RICHARDS 1964.SEE GAP-AS 2
GALLIUM ARSENIDE-ANTIMONIDE	2	GAAS 001	CVD	XD	CLOUGH+TIETJEN 1969.SEE GASB 2
		GAAS 001	CVD	OM,RED	MANASEVIT+SIMPSON 1969.SEE GAP 2B
		GAAS 001, AL-GAAS-SB 001	LPE	EP	SUGIYAMA+SAITO 1972. SEE AL-GAAS-SB 2
	4	SAPPH. 00.1	CVD	OM,RED	ARTHUR+LEPORE 1969.SEE GAP 2A
		SAPPH. 00.1	CVD	OM,RED	MANASEVIT+SIMPSON 1969.SEE GAP 2B
GALLIUM-INDIUM PHOSPHIDE	2	GAAS 001	LPE	OM	CAMPBELL,JC+HITCHENS,WR+HOLONYAK,JR.N+LEE,MH+LUDOWISE,MJ+COLEMAN,JJ 1974APPL.PHYS.LETT.24,327
		GAAS 111	LPE	EP	HAKKI,BW 1971 J.ELECTROCHEM.SOC.118, 1469
		GAP,GAAS,INP,INAS	CVD	EP,XD	JOYCE,BD+FAIRHURST,KM+CLARKE,RC+BORN,PJ 1971 J.CRYST.GROWTH 11,243
		GAP-AS 001	LPE	OM	MACKSEY,HM+LEE,MH+HOLONYAK,JR.N+HITCHENS,WR+DUPUIS,RD+CAMPBELL,JC 1973 J.APPL.PHYS.44,5035
		GAAS	CVD	---	NUESE,CJ+RICHMAN,D+CLOUGH,RB 1971 MET.TRANS.2,789
		GAP OFF 001	CVD	OM,XD, EP	SIGAI,AG+NUESE,CJ+ENSTROM,RE+ZAMEROWSKI,T 1973 J.ELECTROCHEM.SOC.120, 947
		GAAS	LPE	OM,XD	STRINGFELLOW,GB 1972 J.APPL.PHYS.43,3455
		GAAS	LPE		STRINGFELLOW,GB+LINDQUIST,PF+BURMEISTER,RA 1972 J.ELECTRON.MATER.1, 437
GALLIUM-INDIUM ARSENIDE	2	GAAS 001,111, INAS 111	LPE	OM	ANTYPAS,GA 1970 J.ELECTROCHEM.SOC.117,1393
		GAAS	CVD	EP	BAILEY,LG+ROGERS,GG 1971 J.ELECTROCHEM.118,834
		GAAS 001	CVD	OM,XD	CONRAD,RW+HOYT,PL+MARTIN,DD 1967 J.ELECTROCHEM.SOC.114,164
		INAS 111	MG	XD	EWING,RE (UNPUBLISHED). SEE STRINGFELLOW,GB+GREENE,PE 1969 J.PHYS.CHEM.SOLIDS 30,1779

Z=31-32 OVERGROWTH		SUBSTRATE	PREP	OBSERV	REFERENCE
GALLIUM-INDIUM ARSENIDE (CONT.)	2	GAAS 001,111	CVD	XM	HOWARD+DOBROTT 1966.SEE GAP-AS 2
		GE 111	UHV	RED,TEM	KRAUSE,GO 1969 J.VAC.SCI.TECHN.6,582
		GAAS 111	CVD	EP	MANASEVIT,HM+SIMPSON,WI 1973 J.ELECTROCHEM.SOC.120,135
		GE 111	HV	RED	MULLER+RICHARDS 1964.SEE GAP-AS 2
		GAAS 001,110	CVD	OM,XD	NAGAI 1972.SEE GAP-AS 2
		GAAS 111,INAS 111	LPE	OM	WU,TY+PEARSON,GL 1972 J.PHYS.CHEM.SOLIDS 33,409
	4	CAF2 111	UHV	RED,TEM	KRAUSE 1969.SEE 2
		SAPPHIRE 00.1	CVD	EP	MANASEVIT+SIMPSON 1973.SEE 2
GALLIUM-INDIUM ANTIMONIDE	2	GASB,INSB	MG	OM,XD	MROCZKOWSKI,RS+WITT,AF+GATOS,HC 1968 J.ELECTROCHEM.SOC.115,750
GALLIUM-INDIUM PHOSPHIDE-ARSENIDE	2	INP 111	LPE	OM,XD, EP	ANTYPAS,GA+MOON,RL 1973 J.ELECTROCHEM.SOC.120, 1574
		INP 111	LPE	OM	MOON+ANTYPAS 1973.SEE AL-GAAS 2
GALLIUM ANTIMONIDE-INDIUM ARSENIDE	2	GE 001	HV	RED	MULLER+RICHARDS 1964.SEE GAP-AS 2
32 GERMANIUM	1	W	UHV	FEM,FIM	JANSSEN,AP+JONES,JP 1974 SURFACE SCI.41,257
		STEEL,IRON	HV	OM,RED	KUROV,GA+VASIL≠EV,VD+KOSAGANOVA,MG 1963 SOVIET PHYS-CRYST.7,625
	2A	GE 111	UHV	RED	ADAMSKY,RF 1969 J.APPL.PHYS.40,4301
		GE 111	HV, UHV	RED	ADAMSKY,RF+BEHRNDT,KH+BROGAN,WT 1969 J.VAC.SCI.TECHN.6,542
		GE	HV	RED	ALEKSANDROV,LN+DAGMAN,EI+PETROSYAN,VI 1968*4IVC, 505
		GE 111	CVD	RED,REM	ALEXANDROV ET AL 1969.SEE GAAS 2
		GE 001,110,111	CVD	OM	ARIZUMI,T+AKASAKI,I 1963 JAPAN.J.APPL.PHYS.2,143
		GE	HV, UHV, SP	---	BEHRNDT,KH 1966 J.APPL.PHYS.37,3841
		GE 110	CVD	OM	BERKENBLIT,M+REISMAN,A+LIGHT,TB 1968 J.ELECTROCHEM.SOC.115,966
		GE 001,111 + AL	SPE	SEM,ECH	CANALI ET AL 1974. SEE SI 2A
		GE	LPE	XD,XM	CASTET ET AL 1970.SEE GAAS 2A
		GE 111	CVD	OM,XD	CAVE+CZORNY 1963.SEE SI 2A
		GE 111	HV	OM	COURVOISIER,JC+JANSEN,L+HAIDINGER,W 1962*9NVSAVS, 14
		GE	HV	OM	COURVOISIER,JC+HAIDINGER,W+JOCHEMS,PJW+TUMMERS,LJ 1963 SOLID-ST.ELECTRON.6,265
		GE 111,110	HV	RED	DAVEY,JE 1962 J.APPL.PHYS.33,1015
		GE 111	HV	RED,XD	DAVEY,JE 1965 J.VAC.SCI.TECHN.2,12
		GE 001,110,111	CVD	OM	DAVIS,M+LEVER,RF 1956 J.APPL.PHYS.27,835
		GE 111	CVD	OM,XD	DEM≠YANOV,EA+KOLESNIKOV,VN+SLEPTSOV,GV 1965 SOVIET PHYS-CRYST.9,761
		GE	HV	RED	DONOVAN,TM+ASHLEY,EJ 1964 J.OPT.SOC.AMER.54,1141
		GE 111	CVD	OM	GIVARGIZOV,EI 1963 SOVIET PHYS-SOLID STATE 5,840
		GE 001,110,111	CVD	OM	GIVARGIZOV 1964.SEE SI 2A
		GE 001,110,111	CVD	OM	GIVARGIZOV,EI+SHEFTAL,NN 1967*1ICCG, 277
		GE 110	CVD	OM,SEM, EP	GREEN,JM 1970 METALLURG.TRANS.1,647
		GE 001,111	HV	TEM	HAASE,O 1962*AIMEMSM 15,159
		GE 111	UHV	OM	HAIDINGER,W+COURVOISIER,JC 1963LEVIDE 104,141
		GE 111	SP	OM,XD	HAQ,KE 1965 J.ELECTROCHEM.SOC.112,500
		GE 111	CVD	OM	INGHAM,JR.HS+MCDADE,PJ 1960 IBM JRESDEVELOP.4,302
		GE 111	CVD	OM	INOUE,M 1972 JAPAN.J.APPL.PHYS.11,1232
		GE 111	UHV	OM,XD	JOHANNESSEN,JS 1972 PHYS.STATUS SOLIDI A10,569
		GE 111	SP	RED	KHAN,IH 1973 J.APPL.PHYS.44,14
		GE 111	LPE	OM	KIJIMA,K+MIYAMOTO,N+NISHIZAWA,J 1971 J.APPL.PHYS.42,486
		GE 001,111	SP	RED	KRIKORIAN,E 1964*ICTFBB, 113
		GE 001,111	SP	RED	KRIKORIAN,E+SNEED,RJ 1963*10NVSAVS, 368
		GE 001,111	UHV, SP	RED	KRIKORIAN,E+SNEED,RJ 1966 J.APPL.PHYS.37,3665
		GE 111	CVD	TEM	KUPER,AB+CHRISTENSEN,H 1963 J.ELECTROCHEM.SOC.110,988
		GE 111	HV	---	KUROV,GA 1961 SOVIET PHYS.SOLID STATE 3,1207/ 1962 IBID.4,412
		GE 111	CVD	OM	KUROV,GA 1963 SOVIET PHYS-CRYST.7,774
		GE	HV	---	KUROV,GA 1964 SOVIET PHYS-SOLID STATE 5,2226
		GE 111	CVD	OM	KUROV,GA+FILATOVA,IV 1965 SOVIET PHYS.CRYST.9,489
		GE	HV	OM,RED	KUROV,GA+SEMILETOV,SA+PINSKER,ZG 1956 SOVIET PHYS.DOKLADY 1,604
		GE 111,ETC.	HV	OM,RED, REM	KUROV,GA+SEMILETOV,SA+PINSKER,ZG 1957 SOVIET PHYS.-CRYST.2,53
		GE 111	LPE	XD	LAUGIER,A+GAVAND,M+MAYET,L+CASTET,L 1970 THIN SOLID FILMS 6,217

Z= 32 OVERGROWTH		SUBSTRATE	PREP	OBSERV	REFERENCE
GERMANIUM (CONT.)	2A	GE 111	SP	RED,REM	LAYTON,CK+CROSS,KB 1967 THIN SOLID FILMS 1,169
		GE001,110,111,112	CVD	OM	LEVER,RF+JONA,F 1963 J.APPL.PHYS.34,3139
		GE	CVD		LI 1966. SEE SI 2A
		GE	CVD	OM,TEM	LIGHT 1962.SEE SI 2A
		GE 111	SPE	EP,ICH	MARELLO,V+CAYWOOD,JM+MAYER,JW+NICOLET,MA 1972 PHYS.STAT.SOL.A13,531
		GE 001,110,111	CVD	OM,XD	MARINACE,JC 1960 IBM J.RES.DEVELOP.4,248
		GE	CVD	---	MARINACE,JC 1960 IBM J.RES.DEVELOP.4,280
		GE 111	HV	OM,RED, TEM	MEZENTSEVA,NL+PETRIN,AI+KUROV,GA 1965 SOVIET PHYS-SOLID STATE 6,1599
		GE 111	LPE	OM,XD	NELSON 1963.SEE GAAS 2
		GE 111	UHV	OM,XD	REICHELT,H+MUELLER,GFP 1962*2IVC 2,956
		GE 111	CVD	OM	REISMAN,A+BERKENBLIT,M 1965JELECTROCHEMSOC112,315
		GE	SP		REIZMAN,F+BASSECHES,H 1962*AIMEMSM 15,169
		GE 001,110,111	CVD	OM,XD	RIBEN,AR+FEUCHT,DL+OLDHAM,WG 1966 J.ELECTROCHEM.SOC.113,245
		GE 111	CVD	OM,XD	ROTH,EA+GOSSENBERGER,H+AMICK,JA 1963 RCA REV.24,499
		GE	CVD	OM,RED, XD	RUTH,RP+MARINACE,JC+DUNLAP,JR.WC 1960 J.APPL.PHYS.31,995
		GE	HV	RED	SEMILETOV,SA 1956 SOVIET PHYS.-CRYST.1,430
		GE 001,110,111	CVD	OM	SHEFTAL,NN+GIVARGIZOV,EI+SPITSYN,BV+KEVORKOV,AM 1966 GROWTH OF CRYSTALS(AV.SHUBNIKOV+NN.SHEFTAL, EDS.),10.CONSULTANTS BUREAU,NEW YORK
		GE 001,110,111	CVD	OM	SHEFTAL,NN+GIVARGIZOV,EI 1965 SOVIET PHYS.-CRYST.9,576
		GE001,110,111,211	CVD	OM	SILVESTRI,VJ 1969 J.ELECTROCHEM.SOC.116,81
		GE	CVD	OM,XD	TAKABAYASHI,M 1962 JAPAN.J.APPL.PHYS.1,22
		GE 001	UHV	RED,REM	THOMAS,MT+SHIMAOKA,G+DILLON,JR.JA 1967 SURFACE SCI.6,261
		GE 111	HV	OM	TIKHONOVA,AA+VASIL≠EV,VD+KUROV,GA 1964 SOVIET PHYS.-CRYST.8,750
		GE 111	HV	RED	VIA,GG+THUN,RE 1962*2IVC 2,950
		GE 001,110,111	HV	OM,RED	WEINREICH,O+DERMIT,G+TUFTS,C 1961 J.APPL.PHYS.32,1170
		GE 001,111	SP	RED,XD	WOLSKY,SP+PIWKOWSKI,TR+WALLIS,G 1965 J.VAC.SCI.TECHN.2,97
		GE 111	HV, UHV	OM,RED	WOODMAN,TP 1968 THIN SOLID FILMS 2,173
		GE 111	CVD	OM,RED, XD	ZEVEKE ET AL 1968.SEE SI 2A
	2B	GAAS 110	CVD	OM,XM	BERKENBLIT ET AL 1968.SEE 2A
		SI 111	HV	TEM	BRIANTE,JD+CORBETT,JM+BOSWELL,FW 1972 THIN SOLID FILMS 14,305
		GAAS 111	LPE	XD,XM	CASTET ET AL 1970.SEE GAAS 2A
		GAAS	CVD	OM	CHASHCHINOV+MOKIEVSKII 1966.SEE GAAS 2B
		SI 111	UHV	TEM	CULLIS,AG+BOOKER,GR 1970*7ICEM 2,423
		SI 111	UHV	OM,TEM, REM,SEM	CULLIS+BOOKER 1971.SEE SI 2A
		SI 001	UHV	XD,XM	DATSENKO,LI+GUREEV,AN+KOROTKEVICH,NF+SOLDATENKO, NN+TKHORIK,YUA 1971 THIN SOLID FILMS 7, 117
		GAAS	HV	RED	DAVEY,JE 1966 APPL.PHYS.LETT.8,164
		SI 111	SG	OM,XD	DONNEL,JP+MILNES,AG 1966 JELECTROCHEMSOC.113,297
		GAAS	HV	RED	DONAVAN+ASHLEY 1964.SEE 2A
		SI 001,111	CVD	OM,SEM	DUMIN 1971.SEE SI 2A
		SI 111	UHV	RED,TEM	ITO,I+TAKAHASHI,K 1968 JAPAN.J.APPL.PHYS.7,821
		SI 111	CVD	OM,XD	KANERVA,HKJ+STUBB,H+STUBB,T 1969 Z.NATURFORSCH.24A,1343
		GAAS 001,111	HV	RED	KLIMENKO,AP+KLOCHKOV,VP+SOLDATENKO,NN+TORCHUN,NM+ TKHORIK,YUA 1968 SOVIET PHYS-CRYST.13,303
		GAAS 001,111	HV	RED	KLIMENKO,AP+POLUDIN,VI+SVECHNIKOV,SV+TORCHUN,NM+ TKHORIK,YUA 1973 THIN SOLID FILMS 16,205
		GAAS 111	CVD	XM	KRAUSE,GO 1968 J.APPL.PHYS.39,2469
		GA-INAS 111	UHV	RED,TEM	KRAUSE 1969.SEE GA-INAS 2
		SI 111	UHV	RED,SEM	KRAUSE,GO 1970 PHYS.STAT.SOLIDI A3,907
		GAAS 001	CVD	OM,XM	KRAUSE,GO+TEAGUE,EC 1967 APPL.PHYS.LETT.10,251
		GAAS 001,111	LPE	OM,EP, XD	LAUGIER ET AL 1970.SEE 2A
		GAAS 110	HV	RED	LEVER,RF+HUMINSKI,EJ 1966 J.APPL.PHYS.37,3638
		GAAS 110	CVD	OM,XM	LIGHT,TB+BERKENBLIT,M+REISMAN,A 1968 J.ELECTROCHEM.SOC.115,969
		GAAS	CVD	---	MARINACE 1960.SEE 2A
		GAAS 110	CVD	OM,XM	MATTHEWS 1970.SEE SI 2A
		GAAS 111	LPE	OM,XD	MAYET,L+LAUGIER,A 1971 J.CRYST.GROWTH 8,73
		GAAS 111	CVD	OM	OKADA 1963.SEE GAAS 2A
		SI 111,GAAS 111	CVD	---	OLDHAM,WG+MILNES,AG 1964 SOLID-ST.ELECTRON.7,153
		GAAS	CVD	OM,XD	PAPAZIAN,SA+REISMAN,A 1968JELECTROCHEMSOC.115,961
		SI 111	UHV	OM,XD, XM	PFEIFER,J+VARGA,L+SZENTPALI,B 1972 THIN SOLID FILMS 11,59
		SI 001,111, GAAS 001,110,111	HV	RED,REM	POLUDIN,VI+SVECHNIKOV,SI+TORCHUN,NM+TKHORIK,YUA+ SHVARTS,YUM 1973 THIN SOLID FILMS 16,297
		SI 111	UHV	OM,XD	REICHELT+MUELLER 1962.SEE 2A

Z= 32 OVERGROWTH		SUBSTRATE	PREP	OBSERV	REFERENCE
GERMANIUM (CONT.)	2B	SI 111	CVD	OM,XD	RIBEN ET AL 1966.SEE 2A
		GAAS 110,111			
		GAAS	LPE	OM,EP	ROSZTOCZY+STEIN 1972.SEE GAP 2B
		GAAS 111	HV	RED	RYBKA,V+DUDROVA,E+SEVCIK,Z+KREJCI,P 1971 THIN SOLID FILMS 8,R7
		GAAS 111	SP		RYBKA,V+KREJCI,P+DUDROVA,E+SEVCIK,Z 1971*ICPCSH 3,239
		GAAS 111	HV	RED	RYU,I+TAKAHASHI,K 1965 JAPAN.J.APPL.PHYS.4,850
		ZNS 110	HV	RED	SEGMULLER 1956.SEE SI 2B
		ZNSIP2 112,101, 011-ZNSIAS2 112, 101,011,	CVD	OM,XD	SPRING-THORPE,AJ+HARVEY,RJ+PAMPLIN,BR 1969 J.CRYST.GROWTH 6,104
		ZNGEP2 112,100			
		GAP	CVD	OM,XD	VAN RUYVEN,LJ+DEKKER,W 1962 PHYSICA 28,307
		GAAS 111	CVD	XD,XM	VARGA,L+SZENTPALI,B+BERTOTI,I 1970 PHYS.STATUS SOLIDI A2,K135
		SI 001,111	HV	OM	VASILEVSKAYA,VN+SOLDATENKO,NN+TKHORIK,YUA 1971 THIN SOLID FILMS 7,127
	3	NACL 001	HV	RED	ALEKSANDROV ET AL 1968.SEE 2A
		NACL 001	HV	OM,RED	COLLINS,LE+HEAVENS,OS 1952 PROC.PHYS.SOC.(LONDON) B65,825
		NACL 001	HV	TEM	JAUNET+SELLA 1965.SEE AL 3
		NACL 001	HV	TEM	MADER,S 1971 J.VAC.SCI.TECHN.8,247
		NAF 001	HV	OM	MIETZ,I 1965 NATURWISSENSCHAFTEN 52,537
		NACL 001	HV	TED,XD	SCAGGS,CW+JONES,JR 1964 J.APPL.PHYS.35,3013
		NACL 001	HV	TEM	SELLA+TRILLAT 1964.SEE AL 3
		NACL 001	UHV	RED,TEM	SHIMAOKA,G+CHANG,SC 1970*28EMSA, 512
		NACL 001	UHV	RED,TEM	SHIMAOKA,G+CHANG,SC 1971 J.VAC.SCI.TECHN.8,243
		NAF 001, NACL 001	HV	OM,TEM	SLOOPE,BW+TILLER,CO 1962 J.APPL.PHYS.33,3458
	4	CAF2 111	UHV	TEM	CATLIN,A+HUMPHRIS,RR 1966*ISTFCL, 175
		CAF2 111	UHV	TEM	CATLIN,A+BELLEMORE,JR.AJ+HUMPHRIS,RR 1964 J.APPL.PHYS.35,251
		SPINEL 001,111	CVD	OM,XD	DUMIN,DJ 1967 J.ELECTROCHEM.SOC.114,749
		SAPPH. 11.2 + SI, SPINEL 001,111+SI	CVD	---	DUMIN,DJ 1970 J.ELECTROCHEM.SOC.117,95/ SEE ALSO ALEKSANDROV,LN 1972PHYS.STAT.SOL.A11,17
		CAF2 111	HV	OM,RED	GRANT,PM+PAUL,W 1966 J.APPL.PHYS.37,3110
		CAF2 111	UHV	TEM	HUMPHRIS,RR+CATLIN,A 1965 SOLID-ST.ELECTRON.8,957
		CAF2 111	UHV	RED,TEM	KRAUSE 1969.SEE GA-INAS 2
		CAF2 111	SP	RED	KRIKORIAN 1964.SEE 2A
		CAF2 111	SP	RED	KRIKORIAN+SNEED 1963.SEE 2A
		CAF2 111	UHV, SP	RED	KRIKORIAN+SNEED 1966.SEE 2A
		MICA 00.1	UHV	RED,XD	LOMAS,RA+HAMPSHIRE,MJ+TOMLINSON,RD+KNOTT,KF 1973 PHYS.STATUS SOLIDI A16,385
		CAF2 111	HV	RED	MARUCCHI,J 1964 C.R.ACAD.SCI.258,3846
		CAF2 111	HV	RED,TED	MARUCCHI,J+NIFFENTOFF,N 1959 C.R.ACAD.SCI.249,435
		CAF2 111	HV	TEM	PUNDSACK,AL 1963 J.APPL.PHYS.34,2306
		CAF2 111	HV	TED	RUMSH,MA+LYUBITS,K+KONOROV,PP 1965 SOVIET PHYS.-CRYST.9,678
		CAF2 111,BAF2 111	HV	RED	SCHALLA,RL+TIDESWELL,NW+COFFIN,FD 1964*ICTFBB,301
		CAF2 110	HV	RED	SCHALLA,RL+THALLER,LH+POTTER,JR.AE 1962 J.APPL.PHYS.33,2554
		CALCITE 10.1, MICA 00.1	HV	RED	SEMILETOV 1965.SEE 2A
		CAF2 111,MGO 001	HV	OM,RED, TEM,REM, XD	SLOOPE,BW+TILLER,CO 1962.SEE 3/ 1963 JAPAN.J.APPL.PHYS.2,308/ 1963*10NVSAVS, 339
		CAF2 111	UHV	TED,XD	SLOOPE,BW+TILLER,CO 1965 J.APPL.PHYS.36,3174
		CAF2 111	HV, UHV	OM,TEM, REM	SLOOPE,BW+TILLER,CO 1966 J.APPL.PHYS.37,887
		SAPPH. 00.1	CVD	OM,XD	TRAMPOSCH,RF 1966 APPL.PHYS.LETT.9,83/ 1969 J.ELECTROCHEM.SOC.116,654
		CAF2 001,111	HV	RED	VIA+THUN 1962.SEE 2A
		SAPPH. 00.1	CVD	OM,XD	ZEVEKE,TA+MEDVEDEVA,TT+SHEFTAL,NN 1970 SOVIET PHYS.-CRYST.14,649
GERMANIUM-PALLADIUM ALL.	2	GE 111	CG	OM,RED, EP	HUTCHINS+SHEPELA 1973. SEE SI-PD 2
GERMANIUM-GALLIUM PHOSPHIDE	2	GE 001,110,111	HV	RED,XD	RICHARDS ET AL 1964.SEE ALSB 2
GERMANIUM SELENIDE	3	NACL 001,110,111	HV	RED,TED	GOSWAMI,A+NIKAM,PS 1970 INDIANJPUREAPPLPHYS.8,798
GERMANIUM TELLURIDE	2	COTE,PBTE	VG+ CG	---	COHEN-SOLAL ET AL 1966.SEE CD-HGTE 2
		SNTE 001	UHV	TEM	STOEMENOS,J+VINCENT,R 1972 PHYS.STAT.SOL.A11,545
	3	NACL 001	HV	TED	CHOPRA,KL+BAHL,SK 1969 J.APPL.PHYS.40,4171
		NACL 001,110,111	HV	RED,TED	GOSWAMI+NIKAM 1970.SEE GESE 3

Z= 32-37 OVERGROWTH		SUBSTRATE	PREP	OBSERV	REFERENCE
GERMANIUM TELLURIDE(CONT)	3	KCL 001	HV	OM,RED	MIKO LAICHUK,AG+KOGUT,AN 1970 SOVIET PHYS.-CRYST.15,294
		KCL 001	UHV	OM,TEM, XD	STOEMENOS+VINCENT 1972.SEE 2
		KCL 001	UHV	TEM	STOEMENOS,J+KOKKOU,S+ECONOMOU,NA 1972 PHYS.STATUS SOLIDI A13,265
	4	MICA 00.1	HV	TED	CHOPRA+BAHL 1969.SEE 3
		MICA 00.1	HV	RED,TED	GOSWAMI+NIKAM 1970.SEE GESE 3
		MICA 00.1	HV	OM,RED	MIKO LAICHUK+KOGUT 1970.SEE 3
GERMANIUM-TIN TELLURIDE	3	KCL 001	UHV	OM,TEM, XD	STOEMENOS+VINCENT 1972.SEE GETE 2
GERMANIUM IODIDES	2	GRAPHITE 00.1	VG	LED	LANDER,JJ 1967*SGSI, 25
		GRAPHITE 00.1	VG	LED	LANDER,JJ+MORRISON,J 1966 SURFACE SCI.4,103/ 1967 IBID.6,1
GERMANIUM OXIDES	4	SIO2	SG	OM,XD, EP	ROY,R+THEOKRITOFF,S 1972 J.CRYST.GROWTH 12,69
33 ARSENIC OXIDES	2	GRAPHITE 00.1	VG	LED	LANDER 1967.SEE GERMANIUM IODIDES 2
		GRAPHITE 00.1	VG	LED	LANDER+MORRISON 1966,1967.SEE GERMANIUM IODIDES 2
34 SELENIUM	2	TE 11.0	HV	OM	BARBOT,J+JOVENIAUX,M+THUILLIER,JM 1973 PHYS.STAT.SOLIDI A19, K1
		TE 00.1	ED	RED,SEM	DE BECDELIEVRE,AM+BOUDEULLE,M+BARBIER,MJ 1971 MATER.RES.BULL.6,225
		TE 10.0,00.1	HV	TEM,XD	GRIFFITHS,CH+SANG,H 1967 APPL.PHYS.LETT.11,118
		TE 10.0	MG	OM,XD, EP	LEMERCIER,C+THUILLIER,JM 1966 MATER.RESBULL.1,109
		TE 10.0,00.1	HV	RED	SAKAI,Y+FUKUDA,H 1968 JAPAN.J.APPL.PHYS.7,303
		TE 10.0,00.1	HV	OM	SHIOSAKI,T+ITO,M+KAWABATA,A+TANAKA,T 1969 JAPAN.J.APPL.PHYS.8,407
		TE	HV	REM	SHIOSAKI,T+KAWABATA,A+TANAKA,T 1970 JAPAN.J.APPL.PHYS.9,631
	3	NACL 001	HV	TED	BAMMES,P+OTTO,A+PETRI,E 1972 PHYSSTAT.SOL.A10,365
		KBR 001,110	HV	TEM	HEAVENS,OS+GRIFFITHS,CH 1965 ACTA CRYST. 18,532
		KBR 001,KI 001	HV	TEM	NABITOVICH,ID 1970 SOVIETPHYS.SOLIDSTATE 11,1726/ 1971 IBID.12,1899
		KI 001	HV	TEM	NABITOVICH,ID+BUDZHAK,YAS+OSIPOVA,VV+STETSIV,YAL 1971 SOVIET PHYS.-SOLID STATE 13,2447
SELENIUM-TELLURIUM ALL.	2	TE 10.0	HV	TEM	KEEZER,RC+GRIFFITHS,CH+VERNON,JP 1968 J.CRYST.GROWTH 3,4,755
		TE 00.1	HV	OM,EP, XD	SHIOSAKI,T+TOMISAWA,O+KAWABATA,A 1971 JAPAN.J.APPL.PHYS.10,1287
35 BROMINE	2	GRAPHITE 00.1	VG	LED	LANDER 1967. SEE GERMANIUM IODIDES 2
		GRAPHITE 00.1	VG	LED	LANDER+MORRISON 1967.SEE GERMANIUM IODIDES 2
36 KRIPTON	2	GRAPHITE 00.1	VG	TEM	BALL+VENABLES 1970.SEE AR 2
		GRAPHITE 00.1, MOS2 00.1	HV	TEM	KRAMER,HM+VENABLES,JA 1972 J.CRYST.GROWTH 17,329
37 RUBIDIUM	1	W 001	UHV	LED,AES	THOMAS+HAAS 1973.SEE K 1
RUBIDIUM FLUORIDE	2-3	PBS 001,NACL 001, KCL 001	SG	OM	ROYER 1928.SEE LICL 3
RUBIDIUM CHLORIDE	1	BI 111	SG	OM	JOHNSON 1950(A), 1951.SEE NACL 1
	2	PBS 001	SG	OM	SLOAT+MENZIES 1931.SEE LICL 3
	3	NACL 001,NABR001, KCN 001,KCL 001, KBR 001,KI 001, RBCN 001,RBBR 001	SG	OM	BARKER 1906,1907,1908.SEE NACL 3
		KBR 001	HV	RED,REM	LUDEMANN 1954.SEE LIF 3
		KCL 001,KI 001	SG	OM	ROYER 1928.SEE LICL 3
		NACL 001,KCL 001	SG	OM	SLOAT+MENZIES 1931.SEE LICL 3
	4	MICA 00.1	SG	OM	LISGARTEN 1954.SEE NH4CL 4
		MICA 00.1	SG	OM	ROYER 1928.SEE LICL 3
		CACO3 100	SG	OM	ROYER 1937.SEE NABR 4
		MICA 00.1	SG	OM,REM	UPRETI+WALTON 1966.SEE NH4I 4
RUBIDIUM BROMIDE	1	AG 001	HV, SG	RED	SCHULZ 1951(A) JCHEMPHYS.19,504/1952(C).SEE LIF 1
	2	PBS 001	SG	OM	SLOAT+MENZIES 1931.SEE LICL 3

Z= 37-42 OVERGROWTH		SUBSTRATE	PREP	OBSERV	REFERENCE
RUBIDIUM BROMIDE	3	NABR 001,KCL 001, KBR 001,KI 001, RBCN 001,RBI 001	SG	OM	BARKER 1906,1907,1908.SEE NACL 3
		NH4I 001,KBR 001, KI 001,RBI 001	SG	OM	ROYER 1928.SEE LICL 3
		NACL 001,KCL 001	SG	OM	SLOAT+MENZIES 1931.SEE LICL 3
	4	MICA 00.1	SG	OM	LISGARTEN 1954.SEE NH4CL 4
		MICA 00.1	SG	OM	ROYER 1928.SEE LICL 3
		CACO3 100	SG	OM	ROYER 1937.SEE NABR 4
RUBIDIUM IODIDE	3	NH4I 001,NABR001, KBR 001,KI 001, RBCN 001,RBBR 001	SG	OM	BARKER 1906,1907,1908.SEE NACL 3
		KBR 001,KI 001, RBBR 001	SG	OM	ROYER 1928.SEE LICL 3
		LIF 001,NACL 001, KCL 001,KBR 001	HV	RED	SCHULZ 1952(A).SEE LIF 3
		NACL 001,KCL 001	SG	OM	SLOAT+MENZIES 1931.SEE LICL 3
	4	MICA 00.1	SG	OM	LISGARTEN 1954.SEE NH4CL 4
		MICA 00.1	SG	OM	ROYER 1928.SEE LICL 3
		MICA 00.1	HV, SG	RED	SCHULZ 1951(B).SEE LIF 4
		NANO3 100, CACO3 100	HV, SG	RED	SCHULZ 1952(B).SEE LIF 4
		MICA 00.1	SG	OM,REM	UPRETI+WALTON 1966.SEE NH4I 4
		MICA 00.1	MG	OM	WEST 1945.SEE NAI 4
RUBIDIUM CYANIDE	3	NACL 001,KCN 001, KCL 001,KBR 001, KI 001,RBCL 001, RBBR 001	SG	OM	BARKER 1906,1907,1908.SEE NACL 3
38 STRONTIUM TITANATE, STRONTIUM STANNATE	3	LIF 001,NAF 001	HV	RED	MULLER ET AL 1963.SEE CATIO3 3
39 YTTRIUM	1	W	UHV	FEM	MELMED,AJ 1965 J.LESS-COMMON METALS 8,320
40 ZIRCONIUM	1	W	UHV	FEM	COLLINS,RA+BLOTT,BH 1968 SURFACE SCI.10,349
		W 001	UHV	RED,LED	HILL,GE+MARKLUND,I+MARTINSON,J+HOPKINS,BJ 1971 SURFACE SCI.24,435
	3	NACL 001	SP	RED,TED, XD	CHOPRA,KL+RANDLETT,MR+DUFF,RH 1967 PHIL.MAG.16,261
		NACL 001	HV	TEM	DENOUX,M 1966*ISTFCL, 170
	4	CAF2 111	HV	TEM	DENOUX 1966.SEE 3
41 NIOBIUM	4	SAPPHIRE 00.1	SP	XD	CUOMO,JJ+ANGILELLO,J 1973J.ELECTROCHEMSOC.120,125
		MGO 001	HV	TEM	HUTCHINSON,TE 1965 J.APPL.PHYS.36,270
		MGO 001	HV	TEM	HUTCHINSON,TE+OLSEN,KH 1967 J.APPL.PHYS.38,4933
		SAPPHIRE 00.1	SP, UHV	RED,XD	MAYADAS,AF+LAIBOWITZ,RB+CUOMO,JJ 1972 J.APPL.PHYS.43,1287
		MGO 001	HV	TEM	OLSEN,KH+HUTCHINSON,TE 1967*25EMSA, 350
		SAPPH. 10.2,00.1	UHV	OM,RED, REM	O≠NEAL,JE+WYATT,RL 1971 THIN FILMS 2,71
		SAPPH. 10.2,00.1	UHV	RED,REM	O≠NEAL ET AL 1972.SEE VANADIUM 4
		MGO	SP	TEM	SOSNIAK,J 1968 J.APPL.PHYS.39,4157
NIOBIUM-TIN ALLOYS	1	NB 110	UHV	LED	JACKSON,AG+HOOKER,MP 1969*4IMATC, 73
NIOBIUM NITRIDES	4	MGO 001	CVD	TED	OYA,G+ONODERA,Y 1971 JAPAN.J.APPL.PHYS.10,1485
42 MOLYBDENUM	1	MO	CVD	OM,SEM, XD,EP	GILLARDEAU,J+FARON,R+BARGUES,M+HASSON,R+ DEJACHY,G+DURAND,JP 1971 J.CRYST.GROWTH 9,255
	3	NACL 001	SP	RED,TED, XD	CHOPRA,KL+RANDLETT,MR+DUFF,RH 1967PHILMAG.16,261
		NACL 001	HV	TED	SHIRAI,S 1941 PROC.PHYS.MATH.SOC.JAPAN 23,12+914
	4	SAPPH. 10.2,00.1, SPINEL 111, MGO 001,110, BEO 1.00,10.1	CVD	XD	MANASEVIT,HM+MORRITZ,FL+FORBES,DH 1968 TRANS.AIME 242,75
		SAPPH. 10.2,00.1	UHV	RED,REM	O≠NEAL,JE+RATH,BB 1973*31EMSA, 40
MOLYBDENUM-TUNGSTEN ALL.	1	MO	CVD	OM,SEM, XD,EP	GILLARDEAU ET AL 1971.SEE MO 1
MOLYBDENUM-CARBIDES	2	GRAPHITE	CG	TEM	KIKUCHI,M+NAGAKURA,S+OKETANI,S 1966*6ICEM 1,497

Z= 45-47 OVERGROWTH		SUBSTRATE	PREP	OBSERV	REFERENCE
45 RHODIUM	1	AU 111	HV	TEM	BASSETT ET AL 1958.SEE NI 1
		AG 111	HV	TEM	BASSETT ET AL 1959.SEE CU 1B
		AG	HV	TEM	BRAME,DR+EVANS,T 1958 PHIL.MAG.3,971
		IR	UHV	FIM	GRAHAM,WR+REED,DA+HUTCHINSON,F 1972 J.APPL.PHYS.43,2951
46 PALLADIUM	1	AU 111	HV	TEM	BASSET ET AL 1958.SEE NI 1
		AU 111	UHV	TEM	CHERNS,D (UNPUBLISHED).SEE STOWELL,MJ 1972 THIN SOLID FILMS 12,341
		IR	UHV	FIM	GRAHAM ET AL 1972.SEE RH 1
		CU 111	HV	RED	HAASE, 1956.SEE FE 1
		CU 001	HV	RED	HAASE 1961(A+B).SEE FE 1
		AU 111	HV	TEM	PASHLEY ET AL 1957.SEE CO 1
		AU 111	HV	TEM	POPPA,H 1964 Z.NATURFORSCH.19A,835
		AU 001	HV	TEM	SCHOBER,T 1969 J.APPL.PHYS.40,4658
		NI001,PD001,AG001	HV	TED	SHIRAI ET AL 1961.SEE NI 1
		AU 111	HV	TEM	YAGI,K+KOBAYASHI,K+TAKAYANAGI,K+HONJO,G 1970*7ICEM 2,439
		AU 111	HV	TEM	YAGI,K+TAKAYANAGI,K+KOBAYASHI,K+HONJO.G 1971 J.CRYST.GROWTH 9,84
	3	NACL 001,KCL 001	HV	TED	BRUCK 1936.SEE AL 3
		NACL 001,KBR 001	HV	TED,RED	FORDHAM,S+KHALSA,RG 1939 J.CHEM.SOC.406
		LIF 001,NACL 001, KCL 001,KI 001	HV	RED	GOTTSCHE 1956.SEE AL 3
		LIF 001	HV	XD	HALL+THOMPSON 1961.SEE CU 3
		NACL 001,KCL 001	UHV	TEM	INO+OGAWA 1966.SEE AL 3
		KCL 001,KBR 001, KI 001	HV	TED	KATO 1968.SEE AL 3
		NACL 001	UHV	TEM	MATTHEWS 1966.SEE FE 3
		NACL 001	HV	TEM	MURR,LE 1967*2CTF, 73/ 1970*28EMSA, 456
		NACL 001	HV, UHV	TEM	MURR,LE 1971 THIN SOLID FILMS 7,101
		NACL 001,KCL 001	UHV	TEM	OGAWA+INO 1969.SEE AU 3
		NACL 001	HV	RED,TEM	OGAWA ET AL 1955.SEE NI 3
		NACL 001,KCL 001	UHV	TED	OGAWA ET AL 1966.SEE AL 3
	4	MICA 00.1	UHV	TEM	ALLPRESS+SANDERS 1966.SEE NI 4
		MICA 00.1	UHV	TEM	ALLPRESS+SANDERS 1967.SEE NI 4
		MICA 00.1	UHV	TEM	ALLPRESS ET AL 1966.SEE NI 4
		MICA 00.1	HV	XD	HALL+THOMPSON 1961.SEE CU 3
		MGO 001	HV, UHV	TED	HONJO,G 1964-65 UNPUBLISHED. SEE SATO ET AL 1969 IN AG 4
		MGO 001	UHV	LED	PALMBERG+RHODIN 1968(A).SEE CU 4
		MICA00.1,CAF2 111, CACO3 100	HV	RED	RUDIGER,O 1937 ANN.PHYS.(LEIPZIG).30,505
PALLADIUM-MANGANESE ALL.	3	NACL 001	HV	TEM	GJONNES,J+OLSEN,A 1973 PHYS.STATUS SOLIDI A17,71
		NACL 001	HV	TED	SATO,H+TOTH,RS 1964 SOLID-STATE COMMUN.2,249/ 1965 PHYS.REV.139,A1581
PALLADIUM-GOLD ALLOYS	1	AG	HV	TEM	BRAME+EVANS 1958.SEE RH 1
	3	KCL 001	HV	TED	KAWASAKI,Y+INO,S+OGAWA,S 1971 J.PHYS.SOC.JAPAN 30,1758
		NACL 001	HV	TED	MATSUO,Y+NAGASAWA,A+KAKINOKI,J 1966 J.PHYS.SOC.JAPAN 21,2633
		NACL 001	HV	TEM	MATTHEWS,JW 1964*ICTFBB, 165
		NACL 001	HV	TEM	MATTHEWS,JW+CRAWFORD,DL 1965 PHIL.MAG.11,977
		NACL 001	HV	TED	NAGASAWA,A 1964 J.PHYS.SOC.JAPAN 19,2344
		NACL 001	HV	TED	NAGASAWA,A+MATSUO,Y+KAKINOKI,J 1965 J.PHYS.SOC.JAPAN 20,1881
	4	MICA 00.1	HV	TEM	MATTHEWS 1964.SEE 3
PALLADIUM-LEAD ALLOYS	3	NACL 001	HV	TED	SHIRAI ET AL 1961.SEE NI 1
47 SILVER	1	CU 001,111	HV	LED	BAUER,E 1967 SURFACE SCI.7,351
		CU,W	UHV		BAUER,E+POPPA,H 1972 THIN SOLID FILMS 12,167
		AG 001,AU 001	HV	TED	BRUCK 1936.SEE AL 3
		CU 110,111	ED	RED	COCHRANE 1936.SEE CR 1
		W	UHV	FEM	DUELL,MJ+MOSS,RL 1964 BRIT.J.APPL.PHYS.15,157
		AU 001	UHV	LED	FARNSWORTH,HE 1933 PHYS.REV.43,900/ 1936 IBID.49,605
		NI 111	UHV	LED	FEINSTEIN,LG+BLANC,E+DUFAYARD,D 1970 SURFACE SCI.19,269
		CU 001,110,FE 110	ED	RED	FINCH ET AL 1947.SEE CO 1
		NI 111	HV	RED	GONZALEZ 1971.SEE NI 1
		CU 001,110,111	ELD	RED	GOSWAMI 1957.SEE NI 1
		CU 111	UHV	RED	GRADMANN 1964(A).SEE NI 1/ 1964(B).SEE CU 1B
		CU 111	HV	RED	HAASE,O 1956.SEE FE 1
		CU 001,110	HV	RED	HAASE,O 1961(A).SEE FE 1
		CU 111	UHV	RED,TEM	HORNG,CT+VOOK,RW 1973*31EMSA, 112/ 1973.SEE CU 3
		NI 001	UHV	TEM,AES	JACKSON ET AL 1973.SEE NI 3

Z= 47 OVERGROWTH

SILVER (CONT.)

	SUBSTRATE	PREP	OBSERV	REFERENCE
1	W	UHV	FEM	JONES,JP 1972 SURFACE SCI.32,29
	CU 111	HV	RED	KRAUSE 1966(B). SEE CU 1A
	CU	HV	RED	KRAUSE ET AL 1965.SEE CU 1A
	AG 001	HV	TED	LASSEN+BRUCK 1935.SEE 3
	W 110	UHV	LED	LO,CM+HUDSON,JB 1972 THIN SOLID FILMS 12,261
	W	UHV	FEM	MELMED,AJ+MCCARTHY,RF 1965 J.CHEM.PHYS.42,1466
	AG 111	HV	RED	NEWMAN 1957.SEE NI 1
	AU 001	UHV	LED	PALMBERG+RHODIN 1967.SEE 3
	CU 001,PD 001	UHV	LED,AES	PALMBERG,PW+RHODIN,TN 1968(A+B) J.CHEM.PHYS.49,134 + 147
	AU 001	UHV	LED,AES	PALMBERG+RHODIN 1968(C).SEE CU-AU 1
	PT 111,AU 111	HV	TEM	POPPA 1964.SEE PD 1
	AU 001	UHV	LED	SCHLIER+FARNSWORTH 1958.SEE CU 1B
	AU 001	HV	TEM	SCHOBER 1969.SEE PD 1
	AU 001	HV	TEM	SCHOBER,T+BALLUFI,RW 1968 PHYS.STAT.SOL.27,195
	AG 001,110,111	ED	OM,RED	SETTY,THV+WILMAN,H 1955 TRANS.FARADAY SOC.51,984
	AG 001,110,111	ED	OM,RED	SETTY,THV+WILMAN,H 1960 ELECTROCHIM.ACTA 11,297
	NI001,PD001,AG001	HV	TED	SHIRAI ET AL 1961. SEE NI 1
	AU 111	UHV	TEM	SNYMAN,HC+BOSWELL,FW 1970*7ICEM 2,429
	AU 111	UHV	TEM	SNYMAN,HC+BOSWELL,FW 1973 J.APPL.PHYS.44,3347
	AU 111	UHV	TEM	SNYMAN,HC+BOSWELL,FW 1973 SURFACE SCI.36,74
	AU 111	UHV	TEM	SNYMAN,HC+ENGELBRECHT,JA 1973 ACTA MET.21,479
	AU 111	HV	TEM	SNYMAN,HC+BOSWELL,FW+CORBETT,JW 1970 J.APPL.PHYS.41,816
	AG 001	ED	OM	SUZUKI,T 1972 J.CRYST.GROWTH 16,80
	AG 001	ED	OM	SUZUKI,T 1973 J.CRYST.GROWTH 20,202
	AG 001	ED	RED	THOMSON 1931.SEE CU 1A
2	MOS2 00.1	HV	TEM	BASSET,GA 1961*2ERCEM 1,270
	MOS2 00.1	HV	TEM	BASSET,GA 1964*ISCES, 599
	SI,PBS,PBSE,PBTE	HV, UHV		BAUER+POPPA 1972.SEE 1
	MOS2 00.1	UHV	TEM	CORBETT,JM+BOSWELL,FW 1969 J.APPL.PHYS.40,2663
	PBS 001	UHV	RED	GREEN,AK+DANCY,C+BAUER,E 1971 JVAC.SCITECHN.8,165
	GAAS 110	UHV	LED	HANAWA,T+SAGAWA,M+NAGATA,S 1971 JAPAN.J.APPL.PHYS.10,1111
	SI 111	HV, UHV	---	HARSDORFF,M 1968(B) Z.NATURFORSCH.23A,1253
	MOS2 00.1	HV	TEM	HONJO,G+YAGI,K 1969 J.VAC.SCI.TECHN.6,576
	MOS2 00.1	HV	RED	KAINUMA 1951.SEE NI 2
	MOS2 00.1	HV	TEM	KAYIMA,Y+UYEDA,R 1961 ACTA CRYST.14,70
	SI 111	SP	TEM	MAA,JS+HUTCHINSON,TE 1973*31EMSA, 122
	SI 111	SP	TEM	MAA,JS+LEE,JI+HUTCHINSON,TE 1974 J.VAC.SCI.TECHN.11,136
	MOS2 00.1	HV	TEM	MATTHEWS 1972(A).SEE CU 1
	MOS2 00.1 + AU	HV	TEM	MIHAMA,K+ITO,K 1966*6ICEM 1,491
	PBS 001	HV	RED	MIYAKE+KUBO 1947(B).SEE CU 2
	MOS2 00.1	HV	TEM	PASHLEY,DW+STOWELL,MJ 1962*5ICEM 1,GG-1
	MOS2 00.1	HV	TEM	PASHLEY,DW+STOWELL,MJ+JACOBS,MH+LAW TJ 1964 PHIL.MAG.10,127
	MOS2 00.1	HV	TEM	PILKINGTON,G+HIRTH,JP 1972 SURFACE SCI.29,363
	MOS2 00.1	HV	TEM	POPPA 1964.SEE PD 1
	SI 001, PBSE 001	UHV	TEM	POPPA,H+MOORHEAD,RD+HEINEMANN,K 1974 J.VAC.SCI.TECHN.11,132
	SI 111	UHV	LED	SPIEGEL,K 1967 SURFACE SCI.7,125
	FES2 001,ZNS 110, MOS2 00.1,PBS 001	HV	RED	UYEDA,R 1940 PROC.PHYS-MATH.SOC.JAPAN 22,1023
	ZNS 110,MOS2 00.1	HV	RED	UYEDA,R 1942 PROC.PHYS-MATH.SOC.JAPAN 24,809
	GE 110	UHV	LED	WUBBENHORST,R+HARTING,K+NIEDERMAYER,R 1969 J.VAC.SCI.TECHN.6,865
3	NACL 001,111	HV	TEM	ADAMSKY,RF 1966*NATOUTF, 243
	NACL 001	HV	TEM	ADAMSKY+LEBLANC 1963.SEE NI 3
	NACL 001,111	HV	TEM	ADAMSKY,RF+LEBLANC,RE 1965 J.VAC.SCI.TECHN.2,79
	NACL 001	HV	TED	ALEKSANDROV ET AL 1968.SEE GE 2A
	NACL 001	HV	REM	ALLPRESS,JG+SANDERS,JV 1964 PHIL.MAG.9,645
	LIF 001,NACL 001, KCL 001	UHV		BAUER+POPPA 1972.SEE 1
	NACL 001,KCL 001	HV, UHV	TEM	BETHGE,H 1969 J.VAC.SCI.TECHN.6,460
	NACL 001	HV	TEM,REM	BETHGE,H+KROHN,M 1966*ISTFCL, 157
	NACL 001	HV, UHV	RED,TED, XD	BRINE+YOUNG 1963.SEE CU 3
	NACL 110,111	HV	RED	BRU,L+GHARPUREY,MK 1951 PROC.PHYS.SOC.(LONDON) A64,283/ 1951 AN.SOC.ESP.FIS.QUIM.A 47,101
	NACL 001,KCL 001	HV	TED	BRUCK 1936.SEE AL 3
	NACL 001	SP	TEM	CAMPBELL,DS+STIRLAND,DJ 1964 PHIL.MAG.9,703
	NACL 001	HV	TEM	CHOPRA 1965.SEE CU 3/ 1966(A) J.APPL.PHYS.37,2249
	NACL 001	HV, SP	TEM	CHOPRA,KL 1966(B) J.APPL.PHYS.37,3405/ 1966(C)*6ICEM 1,519
	NACL 001 + C,SIO, SIO2, BATIO3	HV	RED,TED	CHOPRA,KL 1969 J.APPL.PHYS.40,906
	NACL 001	SP	TEM	CHOPRA,KL+RANDLETT,R 1966 APPL.PHYS.LETT.8,241

Z= 47 OVERGROWTH	SUBSTRATE	PREP	OBSERV	REFERENCE
SILVER (CONT.) 3	NACL 001 + C	HV	TEM	DISTLER,GI+LEBEDEVA,VN+MOSKVIN,VV 1971 J.CRYST.GROWTH 9,98
	NACL 001,110,111	HV	---	ENGEL,OG 1952 J.CHEM.PHYS.20,1174
	KCL 001	UHV	TEM,LED, AES	GALLON,TE+HIGGINBOTHAM,IG+PRUTTON,M+TOKUTAKA,H 1968 THIN SOLID FILMS 2,369
	NACL 001	HV	RED,TED	GOCHE,O+WILMAN,H 1939 PROC.PHYS.SOC(LONDON)51,625
	NACL 001	HV	RED	GOSWAMI,A 1954 J.SCI.IND.RES.13B,677
	NACL 001	HV	RED,TED	GOSWAMI,A 1958 J.SCI.IND.RES.17B,324
	NACL 001	HV	RED,TED	GOTTSCHE,H 1953 Z.PHYSIK 134,517
	LIF 001,NACL 001, KCL 001,KI 001	HV	RED	GOTTSCHE 1956.SEE AL 3
	LIF 001	UHV	RED,TEM	GREEN ET AL 1970.SEE AL 3
	LIF 001	HV	XD	HALL+THOMPSON 1961.SEE CU 3
	NACL 001	HV	RED	HARSDORFF 1963,1964.SEE CU 3
	NACL 001,KCL 001, KI 001	HV, UHV	RED,TED	HARSDORFF 1968(A).SEE CU 3
	NACL 001	HV, UHV	---	HARSDORFF 1968(B).SEE 2
	NACL 001	HV	TED	HARSDORFF,M+RAETHER,H 1964 ZNATURFORSCH.19A,1497
	NACL 001 + C,SIO2	HV	TEM	HAYEK,K+SCHWABE,V 1970 SURFACE SCI.19,329
	NACL 001,KCL 001	UHV	---	HENNING,CAO+VERMAAK,JS 1970 PHIL.MAG.22,281
	NACL 001	HV, UHV	TEM	HYDER,SB+WILKOV,MA 1967 J.APPL.PHYS.38,2386
	NACL 001,KCL 001	UHV	TEM	INO+OGAWA 1966.SEE AL 3
	NACL 001	HV	TEM	INO,S+WATANABE,D+OGAWA,S 1962,1964.SEE CU 3
	NACL 001	HV	TEM,REM,	JAEGER,H 1967 J.CATAL. 9,237
	NACL 001	UHV	OM,TEM, REM	JAEGER,H 1971 ACTA MET.19,621 AND 637
	NACL 001	HV	TED	JAHRREISS+ISKEN 1966.SEE AL 3
	NACL 001	HV	TEM	JAUNET+SELLA 1965.SEE AL 3
	NACL 001	HV	TEM	JAUNET,J+SELLA,C+TRILLAT,JJ 1964 C.R.ACAD.SCI.258,135
	LIF 001,NAF 001, NACL001,NABR 001, NAI 001,KF 001, KCL 001,KBR 001, KI 001,RBCL 001, RBBR 001,RBI 001	UHV	TEM	JESSER+MATTHEWS 1969.SEE CU 3
	NACL 001	HV	OM	JOHNSON 1950(A).SEE NACL 1
	KCL 001,KBR 001, KI 001	HV	TED	KATO 1968.SEE AL 3
	NACL 001,KBR 001	HV	RED	KEHOE 1957.SEE CU 3
	NACL 001	HV	RED,TED	KIRCHNER,F+CRAMER,H 1938 ANN.PHYS.(LEIPZIG)33,138
	NACL 001	HV	RED	KIRCHNER,F+LASSEN,H 1935 ANN.PHYS.(LEIPZIG)24,113
	NACL 001	HV	RED	KIRCHNER,F+RUDIGER,O 1937 ANN.PHYS(LEIPZIG)30,609
	NACL 111	UHV	RED,REM	KOCH+VOOK 1971.SEE NACL 4
	NACL 111	HV	RED,TEM	KOCH,FA+VOOK,RW 1971 J.APPL.PHYS.42,4510
	NACL 111	UHV	RED,TEM	KOCH ET AL 1972.SEE CU 3
	KCL 001	UHV	RED,TEM	KUNZ ET AL 1966.SEE AL 3
	NACL 001	HV	TED	LASSEN,H 1934 PHYS.ZEITSCHR.35,172
	NACL 001	HV	RED,TED	LASSEN,H+BRUCK,L 1935 ANN.PHYS.(LEIPZIG).22,65/ 1935 IBID.23,18
	NACL 001	HV	TED	LAUE 1937.SEE CO 1
	NACL 001	SP	TED	LAYTON,CK+CAMPBELL,DS 1966 J.MATER.SCI.1,367
	NAF 001	UHV	TEM	LORD,G 1971 THIN SOLID FILMS 7,R39
	NACL 001	HV	TEM	MATTHEWS 1959.SEE CU 3
	NACL 001	UHV	TEM	MATTHEWS 1966.SEE FE 3
	NACL 001	HV	TED	MENZER 1938.SEE NI 3
	NACL 111	HV	TEM	MISSIROLI 1972.SEE NACL 4
	NACL 001	HV	XD	MIWA 1954.SEE AL 3
	NACL 001	UHV	TEM	MURR 1964.SEE NI 3
	NACL 001	HV, UHV	TEM	MURR+INMAN 1966.SEE AL 3
	NACL 001,110,111	HV	TEM	NAGAKURA,S+WAKASHIMA,K+FUKACHAMI,M 1969 J.VAC.SCI.TECHN.6,611
	NACL 001	HV	TEM	NAGASAWA,A+OGAWA,S 1960 J.PHYS.SOC.JAP.15,1421
	NACL 001	HV	TEM,XD	NOGGLE,TS 1972 NUCL.INSTR,METH. 102,539
	NACL 001,KCL 001	UHV	TEM	OGAWA+INO 1969.SEE AU 3
	NACL 001	HV	RED,TEM	OGAWA ET AL 1955.SEE NI 3
	NACL 001,KCL 001	UHV	TEM	OGAWA ET AL 1966.SEE AL 3
	KCL 001	UHV	LED	PALMBERG,PW+RHODIN,TN 1967 PHYS.REV.161,586
	KCL 001	UHV	TEM,LED	PALMBERG,PW+RHODIN,TN+TODD,CJ 1967(A) APPL.PHYS.LETT.10,122
	KCL 001	UHV	TEM,LED	PALMBERG ET AL 1968.SEE CU 3
	NACL 001	HV	RED,TED	PANDE,A 1958 J.SCI.IND.RES.17B,1
	NACL 001	HV	TEM	PATEL,AR+MOHANA,S 1973 THIN SOLID FILMS 16,369
	NACL 001	HV	TEM	PHILLIPS,VA 1960 PHILMAG.5,571/ 1961*2ERCEM 1,413
	NACL	HV	TEM	POSTNIKOV ET AL 1968.SEE AL 3
	NACL 001	UHV	TEM	QUINTANA+SACEDON 1973.SEE AU 3
	NACL 001	HV	RED	RAETHER,H 1946 OPTIK 1,296/ 1951 ERG.EXAKT.NATURW. 24,121
	NACL 001	HV	TEM	REIMER,L 1959 OPTIK 16,30
	KCL 001	UHV	TEM	RHODIN,TN+PALMBERG,PW+TODD,CJ 1969*CMPSS, 499

Z= 47 OVERGROWTH		SUBSTRATE	PREP	OBSERV	REFERENCE
SILVER (CONT.)	3	NACL001,NABR 001, NAF 001,KCL 001, KBR 001	UHV	TEM	ROOS,JR+VERMAAK,JS 1972 J.CRYST.GROWTH 13/14,217
		NACL 001	HV	---	ROYER,L 1935 ANN.PHYS.(LEIPZIG) 23,16
		NACL 001	HV	RED	RUDIGER 1937.SEE PD 4
		NACL 001	HV	TEM	SCHLOTTERER,H 1962*5ICEM 1,DD-5
		NACL 001	HV	TEM	SCHLOTTERER 1965.SEE NI 3
		NACL 001	HV	TEM	SCHOBER 1969.SEE PD 1
		NACL 001	HV	TED	SELLA+TRILLAT 1964.SEE AL 3
		NACL 001	HV	TEM	SHARMA,SK+BAHL,OP 1971 JAPAN.J.APPL.PHYS.10,375
		NACL 001	HV	TEM	SHARMA,SK+KUSHWAHA,RPS 1969 SURFACE SCI.18,449
		NACL001,NABR 001, KCL 001,KBR 001, KI 001	HV	TED	SHIRAI,S 1943(B) PROC.PHYS.-MATH.SOC.JAPAN 25,633
		NACL 001	HV	OM,TEM	SLOOPE,BW+TILLER,CD 1961 J.APPL.PHYS.32,1331
		NACL 001	HV	TEM	TAKAHASHI ET AL 1966.SEE NI 3
		KCL 111	HV	RED	THIRSK,HR 1950 PROC.PHYS.SOC.(LONDON) B63,833
		NACL 001	HV, UHV	TEM	TOTH,RS+CICOTTE,LJ 1968 THIN SOLID FILMS 2,111
		C REPLICA NACL001	HV	TED	VERMOUT,P+DEKEYSER,W 1959 PHYSICA 25,53
		NACL	HV	TED,REM	WAKASHIMA,K+FUKAMACHI,M+NAGAKURA,S 1969 JAPAN.J.APPL.PHYS.8,1167
		NACL 001	UHV	TEM	WALTON,T+RHODIN,TN+ROLLINS,RW 1963 J.CHEM.PHYS.38,2698
		NACL	HV	TED	WATANABE,Y 1956 J.PHYS.SOC.JAPAN 11,740
	4	MICA 00.1	HV	TEM	ADAMSKY 1966.SEE 3
		MICA 00.1	HV	TEM	ADAMSKY+LEBLANC 1965.SEE 3
		MICA 00.1	HV	REM	ALLPRESS+SANDERS 1964.SEE 3
		MICA 00.1	HV, UHV	TEM	ALLPRESS+SANDERS 1966.SEE NI 4
		MICA 00.1	UHV	TEM	ALLPRESS ET AL 1966.SEE NI 4
		MICA 00.1	HV	RED,TEM, REM	BAGG,J+JAEGER,H+SANDERS,JV 1963 J.CATAL.2,449
		MGO 001	UHV		BAUER+POPPA 1972.SEE 1
		AGBR 111	HV	RED	BERRY,CHR 1949 ACTA CRYST.2,393
		MGO 001	HV, UHV	RED,TED, XD	BRINE+YOUNG 1963.SEE CU 3
		MICA 00.1	HV	TEM	CHOPRA 1965.SEE CU 3
		MICA 00.1	HV, SP	TEM	CHOPRA 1966(B+C).SEE 3
		MICA 00.1 + C, SIO2,BATIO3	HV	RED,TED	CHOPRA 1969.SEE 3
		MICA 00.1	SP	TEM	CHOPRA+RANDLETT 1966.SEE 3
		MGO 001	UHV	RED	DENNIS 1972.SEE AU 1
		MICA 00.1	HV	TEM	DICKSON,EW+PASHLEY,DW 1962 PHIL.MAG.7,1315
		MICA 00.1	HV	REM	DICKSON ET AL 1965.SEE AU 1
		MGO 001	UHV	RED,TEM	GREEN ET AL 1970.SEE AL 3
		MICA 00.1	HV	XD	HALL+THOMPSON 1961.SEE CU 3
		MGO 001,111	HV	TEM	HONJO+YAGI 1969.SEE 2
		MGO 001	HV	TEM	HONJO,G+SHINOZAKI,S+SATO,S 1966 APPLPHYSLETT.9,23
		MICA 00.1	HV	TEM,REM	JAEGER 1967.SEE 3
		MICA 00.1	UHV	OM,TEM, REM	JAEGER 1971.SEE 3
		MICA 00.1	HV, UHV	TEM,REM	JAEGER,H+MERCER,PD+SHERWOOD,RG 1967 SURFACE SCI.6,309/ 1968 IBID.11,265/ 1969 IBID.13,349 + 502
		MICA 00.1	UHV	RED	LARSON,DC+BOIKO,BT 1964 APPL.PHYS.LETT.5,155
		MICA 00.1	HV	TEM	MATTHEWS,JW 1961*2ERCEM 1,276
		MICA 00.1	HV	TEM	MATTHEWS,JW 1962 PHIL.MAG.7,915
		MICA 00.1	HV	TEM	MATTHEWS 1972(A).SEE CU 1
		MICA 00.1	UHV	LED	MULLER,K 1964 Z.NATURFORSCH.19A,1234
		MGO 001	UHV	LED	PALMBERG+RHODIN 1967.SEE 3
		MGO 001	UHV	TEM,LED, EAS	PALMBERG,PW+RHODIN,TN+TODD,CJ 1967(B) APPL.PHYS.LETT.11,33/ 1968*4IVC, 515
		MICA 00.1,MGO 001	HV	RED,TED	PANDE 1958. SEE 3
		MICA 00.1	HV	TED,REM	PASCARD ET AL 1972(A).SEE NIFE 1
		MICA 00.1	HV, UHV	TEM,REM	PASCARD ET AL 1972(B).SEE NI-FE 1
		MICA 00.1	HV	RED	PASHLEY,DW 1959(A) PHIL.MAG.4,316
		MICA 00.1	HV	TEM	POPPA,H 1962*9NVSAVS, 21
		MICA 00.1	HV	TEM	POPPA 1964.SEE PD 1
		MICA 00.1	HV	TEM	POSTNIKOV ET AL 1968.SEE AL 3
		MGO	UHV	TEM	REICHELT 1973.SEE LIF 3
		MICA 00.1	HV	XD,PCH	REICHELT,K+LUTZ,HO 1971 J.CRYST.GROWTH 10,103
		MGO 001	UHV	TEM,LED	RHODIN ET AL 1969. SEE 3
		MICA00.1,CAF2 111, CACO3 100	HV	RED	RUDIGER 1937.SEE PD 4
		MGO 001	HV	TEM	SATO,H+SHINOZAKI,S 1971 J.VAC.SCI.TECHN.8,159
		MGO 001	HV	TEM	SATO,H+SHINOZAKI,S+CICOTTE,LJ 1969 J.VAC.SCI.TECHN.6,62
		MGO 001,111	HV	TEM	SHINOZAKI,S+SATO,H 1972 J.APPL.PHYS.43,701
		MGO 001	HV	TEM	SHINOZAKI,S+HONJO,G+SATO,H 1966*6ICEM 1,505
		MICA 00.1	HV	TEM	TAKAHASHI ET AL 1966.SEE NI 3

Z= 47-48	OVERGROWTH		SUBSTRATE	PREP	OBSERV	REFERENCE
	SILVER (CONT.)	4	MICA00.1,MGO 001, CACO3 100	HV	RED	THIRSK 1950.SEE 3
			MGO 001,MICA 00.1	HV	RED	TULL,VFG 1951 PROC.ROY.SOC.SER.A 206,219
			MGO 001	HV	RED	UYEDA,R 1940,1942.SEE 2
			MICA 00.1	HV	TEM	ZIGNANI,F+MISSIROLI,GF+DESALVO,A+PETRALIA,S 1968 NUOVO CIMENTO 55B,539
	SILVER-MAGNESIUM ALLOYS	3	NACL 001	HV	TED	FUJIWARA,K+HIRABAYASHI,M+WATANABE,D+OGAWA,S 1958 J.PHYS.SOC.JAP.13,167
	SILVER-GOLD ALLOYS	3	NACL 001	UHV	TEM	MURPHY,RJ 1970*7ICEM 2,403
			NACL 001	HV	TED	RAETHER 1952.SEE CU-AU 3
	SILVER SELENIDE	3-4	NACL 001,110,111, MICA 00.1	HV	RED,TED	DHERE,NG+GOSWAMI,A 1970 THIN SOLID FILMS 5,137
	SILVER TELLURIDE	3	NACL 001,110,111	HV	RED,TED	DHERE+GOSWAMI 1970.SEE SILVER SELENIDE 3-4
	SILVER-BISMUTH TELLURIDE	3	NACL 001	HV	TED	PINSKER,ZG+IMAMOV,RM 1964 SOVIET PHYS.CRYST.9,277
	SILVER CHLORIDE	2	PBS 001	SG	OM	SLOAT+MENZIES 1931.SEE LICL 3
		3	NACL 001	HV	TEM	BRADY,LE+CASTLE,JW+HAMILTON,JF 1968 APPL.PHYS.LETT.13,76
			PVC REPL.NACL 001	HV	TEM	DISTLER,GI+TOKMAKOVA,EI 1970 THINSOLIDFILMS 6,203
			LIF 001	HV	TEM	DISTLER,GI+VLASOV,VP 1970 SOVIET PHYS.-SOLID STATE 11,1798
			NACL 001	HV	TEM	DISTLER,GI+VLASOV,VP+GERASIMOV,YM+SAROVSKY,EG 1970*7ICEM 2,465
			NACL 001,110, KBR 001	HV	RED	PASHLEY,DW 1952 PROC.PHYS.SOC.(LONDON) A65,33
			NACL 001,KCL 001	SG	OM	SLOAT+MENZIES 1931.SEE LICL 3
		4	MICA 00.1	HV	TEM	BRADY ET AL 1968.SEE 3
			TRIGLYC.SULF. 010	HV	TEM	DISTLER,GI+VLASOV,VP 1969 THIN SOLID FILMS 3,333/ 1970 SOVIET PHYS.-CRYST.14,747
			TRIGLYC.SULF. 010	HV	TEM	DISTLER,GI+KONSTANTINOVA,VP+VLASOV,VP 1969 SOVIET PHYS.-CRYST.14,70
			TRIGLYCINE SULF.	HV	TEM	DISTLER ET AL 1970.SEE 3
			MICA 00.1	HV	RED	PASHLEY 1952.SEE 3
			TRIGLYCINE SULF.	HV	TEM	VLASOV,VP+GERASIMOV,YUM+DISTLER,GI 1970 SOVIET PHYS.-CRYST.15,289
	SILVER BROMIDE	2	PBS 001	SG	OM	SLOAT+MENZIES 1931.SEE LICL 3
		3	NACL 001	HV	TEM	BRADY ET AL 1968.SEE AGCL 3
			NACL 001,KBR 001	HV	RED	PASHLEY 1952.SEE AGCL 3
			NACL 001,KCL 001	SG	OM	SLOAT+MENZIES 1931.SEE LICL 3
			NACL 001	HV	---	TRAUTWEILER,F+BRADY,LE+CASTLE,JW+HAMILTON,JF 1969*4IMATC, 83
		4	MICA 00.1	HV	TEM	BRADY ET AL 1968.SEE AGCL 3
			MICA 00.1	HV	RED	PASHLEY 1952.SEE AGCL 3
			MICA 00.1	HV	---	TRAUTWEILER ET AL 1969.SEE 3
			TRIGLYCINE SULF.	HV	TEM	VLASOV ET AL 1970.SEE AGCL 4
	SILVER IODIDE	1	AG 111,AU 111	HV	RED	OSTWALD,R+WEIL,KG 1969 J.VAC.SCI.TECHN.6,684
		3	NACL 001	HV	TEM	BRADY ET AL 1968.SEE AGCL 3
		4	MICA 00.1	HV	TEM	BRADY ET AL 1968.SEE AGCL 3
			MICA 00.1	HV	TEM	COCHRANE,G 1970 J.CRYST.GROWTH 7,109
			MICA 00.1	HV	RED	PASHLEY 1952.SEE AGCL 3
			TRIGLYCINE SULF.	HV	TEM	VLASOV ET AL 1970.SEE AGCL 4
	SILVER CYANIDE	3	NACL 001,KCL 001	SG	OM	SLOAT+MENZIES 1931.SEE LICL 3
48	CADMIUM	1	CD 10.0,11.0,00.1	ED	OM,XD	BICELLI+POLI 1966.SEE CU 1A
			CU 001,110,111	ED	RED	GOSWAMI 1969.SEE ZN 1
		2	MOS2 00.1	HV	TEM	JACOBS,MH+LAW,TJ (UNPUBLISHED).SEE PASHLEY,DW 1965 ADV.PHYS.14,327
	CADMIUM-TIN PHOSPHIDE	2	INP 001	LPE	OM,SEM	SHAY,JL+BACHMAN,KJ+BUEHLER,E 1974JAPPLPHYS45,1302
	CADMIUM SULPHIDE	1	AG 111	HV	RED,TEM	CHOPRA,KL+KHAN,IH 1967 SURFACE SCI.6,33
			AG 111	HV	OM,RED	MAGOMEDOV,ZA+SEMILETOV,SA 1967 SOVIET PHYS.-CRYST.12,470
		2	GE 001,110,111	HV	TEM	ABDALLA,MI+HOLT,DB+WILCOX,DM 1973J.MATERSCI.8,590
			GE 111,GAAS 111	CVD	OM,RED	AITHKOZHIN,SA+KOTELYANSKII,IM+BOKII,GB+ DVORYANKIN,VF+DVORYANKINA,GG+PANTELEEV,VV 1969 SOVIET PHYS.-CRYST.14,309

Z=48 OVERGROWTH	SUBSTRATE	PREP	OBSERV	REFERENCE
CADMIUM SULPHIDE(CONT.) 2	ZNTE 111	VG	OM,RED, XD	AVEN,M+GARWACKI,W 1963 J.ELECTROCHEM.SOC.110,401
	CDS	VG	RED	BLANCONNIER+HENOC 1972.SEE ZNS 2
	GAP 111,GAAS 111	CVD	OM,XD	CARDONA,M+WEINSTEIN,M+WOLFF,GA 1965 PHYS.REV.140,A633
	ZNS 11.0, CDS10.0,11.0,00.1	VG	TEM,REM	CAVENEY 1968. SEE ZNS 2
	CDS	VG	XM	CHIKAWA,J 1965 APPL.PHYS.LETT.7,193
	GAAS 001,110,111	CVD	OM,XD	CURTIS,BJ+BRUNNER,H 1970 J.CRYST.GROWTH 6,269
	CDS 00.1, INSB 001,110,111	HV	RED	HOLLOWAY,H+WILKES,E 1968 J.APPL.PHYS.39,5807
	GE 111	HV	TEM	HOLT,DB+WILCOX,DM 1971 J.CRYST.GROWTH 9,193
	GAAS 111	VG	OM,REM, XD	IGARASHI,O 1969 JAPAN.J.APPL.PHYS.8,642
	CDTE 110,GAP 110, GAAS 001,110,112, 221-INAS 001,110	CVD	XD	IGARASHI,O 1971 J.APPL.PHYS.42,4035
	GAP 111	CVD	OM,LED	KOTELYANSKII,IM+MITYAGIN,AYU+ORLOV,VP 1971 J.CRYST.GROWTH 10,191
	GE 111,GAAS 110	HV	OM,RED	MAGOMEDOV+SEMILETOV 1967.SEE 1
	CDS 00.1, GAAS 110,111	CVD	OM,XD	MOULIN,M+HUBER,A+DUGUE,M 1972 JCRYSTGROWTH 17,212
	ZNTE 111	CVD	OM,RED	OTA,T+KOBAYASHI,K+TAKAHASHI,K 1974 J.APPL.PHYS.45,1750
	GE 111	CVD	OM,XD	RATCHEVA-STAMBOLIEVA,T+TCHISTYAKOV,YUD+ DJOGLEV,DH+BAKARDJIEVA,VS 1972 PHYS.STATUS SOLIDI A10,209
	ZNS 00.1,11.0	VG	XD	SEILER,H+REIMERS,P+INDRADEV 1969 MATER.RES.BULL.4,119
	MOS2 00.1	HV	RED,TEM	SHIMAOKA,G 1971 THIN SOLID FILMS 7,405
	CDS00.1,ZNS 00.1, GAAS 111,GE 111	CVD	OM,RED, REM,PCH	STREHLOW,WH 1970 J.APPL.PHYS.41,1810
	GE 110,111	CVD	XD	VAN DIJK,H+GOORISSEN,J 1967*1ICCG, 531
	GAP 111,GAAS 111	CVD	OM,REM, EP,XD	WEINSTEIN,M+WOLFF,GA 1967*1ICCG, 537
	GAP 111,GAAS 111	CVD	OM,XD	WEINSTEIN,M+WOLFF,GA+DAS,BN 1965APPLPHYSLETT.6,73
	GAAS 111	CVD	OM,RED, XD	YIM+STOFKO 1972.SEE ZNS 2
3	NACL 001,110,111	HV	TED	AGGARWAL+GOSWAMI 1963.SEE ZNS 3
	NACL 001	HV	RED,TEM	CHOPRA+KHAN 1967.SEE 1
	NACL 001	SP	TEM	DISTLER,GI+GERASIMOV,YM+KOBZAREVA,SA+MOSKVIN,VV+ SHENYAVSKAYA,LA 1968*4ERCEM 1,517
	NACL 001,KCL 001, KBR 001	CVD	OM,XD	HEYRAUD,JC+CAPELLA,L 1968 J.CRYST.GROWTH 2,405
	NACL 001,110,111	HV	TEM	HOLT+WILCOX 1971.SEE 2
	NACL 001	HV	TED	HOLT,DB+WILCOX,DM 1972 THIN SOLID FILMS 10,141
	NACL 001	HV	RED,TED	SEMILETOV,SA 1956 SOVIET PHYS.-CRYST.1,236
	NACL 001	HV	RED,TEM	SHIMAOKA 1971.SEE 2
	NACL 001,110,111	HV	TEM	WILCOX,DM+HOLT,DB 1969 J.MATER.SCI.4,672
	NACL 001	SP	RED,TEM	YURASOVA,VE+EFREMENKOVA,VM+SYSOJEV,AA 1968*4ERCEM 1,391
	NACL 001	SP	TED	YURASOVA,VE+LEVYKINA,LN+EFREMENKOVA,VM 1966 SOVIET PHYS.SOLID STATE 7,2332
4	MICA00.1,CAF2 111	CVD	OM,RED	AITKHOZHIN ET AL 1969.SEE 2
	MICA 00.1	HV	RED,TEM	CHOPRA+KHAN 1967.SEE 1
	MICA 00.1	HV	OM,XD	ESCOFFERY,CA 1964 J.APPL.PHYS.35,2273
	MICA00.1,CAF2 111	CVD	OM,XD	HEYRAUD+CAPELLA 1968.SEE 3
	BAF2 111	HV	TEM	HOLT+WILCOX 1971.SEE 2
	MICA 00.1	HV	OM,RED	MAGOMEDOV+SEMILETOV 1967.SEE 1
	SAPPHIRE 00.1	CVD	OM,RED, XD	MANASEVIT+SIMPSON 1971.SEE ZNS 4
	MICA 00.1	HV	RED	MURAVJEVA ET AL 1970.SEE ZNS 4
	MICA 00.1,SAPPH.	HV	RED	MURAVJEVA,KK+KALINKIN,IP+ALESKOVSKY,VB+ANIKIN,IN 1972 THIN SOLID FILMS 10,355
	CAF2 111	CVD	OM,XD	RATCHEVA-STAMBOLIEVA ET AL 1972.SEE 2
	MICA 00.1	HV	RED,TED	SEMILETOV 1956.SEE 3
	MICA 00.1	HV	RED	SERGEEVA,LA+KALINKIN,IP+NECHIPORENKO,AP+ ALESKOVSKII,VB 1971*ICPCSH 3,255
	MICA 00.1	HV	OM,RED	SERGEYEVA ET AL 1972.SEE ZNSE 4
	MICA00.1,CAF2 111	HV	RED,TEM	SHIMAOKA 1971.SEE 3
	MICA 00.1	HV	RED,TEM	SHIMAOKA,G+WALCH,AS 1967*25EMSA, 346
	MICA 00.1	HV	RED,REM, XD	SIMOV,S 1973 THIN SOLID FILMS 15,79
	MICA00.1,SRF2 111	CVD	OM,RED, REM,PCH	STREHLOW 1970.SEE 2
	TRIGLYCINE SULF.	HV	OM	VLASOV ET AL 1970.SEE AGCL 4
	BAF2 111	HV	TEM	WILCOX+HOLT 1969.SEE 3
	SAPPHIRE 00.1	CVD	OM,XD	YIM+STOFKO 1972.SEE ZNS 2
	MICA 00.1	SP	RED	YURASOVA ET AL 1968.SEE 3

Z= 48 OVERGROWTH		SUBSTRATE	PREP	OBSERV	REFERENCE
CADMIUM SELENIDE	2	ZNTE 110	CVD	OM,XD	ARNOLD ET AL 1970. SEE ZNTE 2
		ZNTE 110	VG	OM,RED, XD	GASHIN,PA+SIMASHKEVICH,AV 1973 PHYS.STAT.SOLIDI A19, 379
		CDSE	HV	RED,TED	KALINKIN,IP+SERGEEVA,LA+ALESKOVSKII,VB+ STRAKHOV,LP 1962 SOVIET PHYS.-SOLID STATE 3,1922
		GE 111	CVD	OM,RED	RATCHEVA-STAMBOLIEVA ET AL 1972.SEE CDS 2
		GE 111	VG	OM,XD	RATCHEVA-STAMBOLIEVA,TM+DRAGIEVA,ID+DJOGLEV,DH+ TCHISTYAKOV,YU.D+KRASULIN,GA 1974 PHYS.STAT.SOLIDI A21,703
		ZNSE 001,111,CDS	HV	RED	SERGEEVA ET AL 1965.SEE ZNSE 2
		PBSE 001	HV	TEM	YAGI ET AL 1970.SEE PD 1
	3	NACL 001,110,111	HV	RED,TED	DHERE,NG+GOSWAMI,A 1969 INDIANJPUREAPPLPHYS.7,398
		NACL 111,KCL 111, KBR 111	HV	OM,TEM	KALINKIN ET AL 1962.SEE 2
		NACL 001,111	HV	TEM	KALINKIN,IP+SERGEEVA,LA+ALESKOVSKII,VB+ STRAKHOV,LP 1963 SOVIET PHYS.-CRYST.8,360
		NACL 001,110	HV	OM,TEM	KALINKIN,IP+SERGEEVA,LA+ALESKOVSKII,VB+ STRAKHOV,LP 1963 SOVIET PHYS.-SOLID STATE 5,86
		NACL 001	HV	RED,TED	SEMILETOV 1956.SEE CDS 3
		NACL 001,111	HV	RED,TEM	YASUDA,Y 1968 JAPAN.J.APPL.PHYS.7,1171
	4	YMNO3 00.1	UHV	RED,LED	ABERDAM,D+BOUCHET,G+DUCROS,P+DAVAL,J+GRUNBERG,G 1969 SURFACE SCI.14,121
		CAF2 111	HV	RED,REM	ALTMAN,LV+VORONTSOVA,EN+RUBAN,YUV+TIKHOMIROV,GP 1968 SOVIET PHYS.-CRYST.12,601
		MICA 00.1	HV	TED	BRUNNSCHWEILER,A 1966 NATURE (LONDON) 209,493
		MICA 00.1	HV	RED,TED	DHERE+GOSWAMI 1969.SEE 3
		BAF2 111	HV	RED	LUDEKE,R+PAUL,W 1967 PHYS.STATUS SOLIDI 23,413
		BAF2 111	HV	RED	LUDEKE,R+PAUL,W 1967 2-6 SEMICOND.COMPOUNDS 1967 INT.CONF.(DG.THOMAS,ED.),123.BENJAMIN,NEW YORK
		SAPPHIRE 00.1	CVD	OM,RED, XD	MANASEVIT+SIMPSON 1971.SEE ZNS 4
		MICA 00.1	HV	RED	MURAVJEVA ET AL 1970.SEE ZNS 4
		MICA 00.1,SAPPH.	HV	RED	MURAVJEVA ET AL 1972.SEE CDS 4
		CAF2 111	CVD	OM,RED	RATCHEVA-STAMBOLIEVA ET AL 1972.SEE CDS 2
		SAPPHIRE 00.1	HV	OM,RED	RATCHEVA-STAMBOLIEVA,TM+TCHISTYAKOV,YU.D+ KRASULIN,GA+VANYUKOV,AV+DJOGLEV,DH 1973 PHYS.STAT.SOLIDI 16,315
		SAPPHIRE 00.1, MICA 00.1	VG	OM,RED	RATCHEVA-STAMBOLIEVA ET AL 1974. SEE 2
		SAPPHIRE 00.1	VG	OM	RATCHEVA-STAMBOLIEVA,TM+KRASULIN,GA+ TCHISTYAKOV,YU.D+DRAGIEVA,ID+DJOGLEV,DH 1974 PHYS.STAT.SOLIDI A22,593
		MICA 00.1	HV	RED,TED	SEMILETOV 1956.SEE CDS 3
		MICA 00.1	HV	RED	SERGEEVA ET AL 1971.SEE CDS 4
		MICA 00.1	HV	OM,RED	SERGEEVA ET AL 1972.SEE ZNSE 4
CADMIUM TELLURIDE	2	GE 001,110,111	HV	TEM	ABDALLA,MI+HOLT,DB 1973 PHYS.STAT.SOLIDI A17,267
		GAAS	CVD	RED	AITKHOZHIN,SA+TEMIROV,YU.SH 1970 KRISTALLOGRAFIYA 15,1057
		GAAS	CVD	OM,RED	ALFEROV ET AL 1965.SEE GAP 2B
		CDS 00.1	CVD	OM	PAORICI,C+PELOSI,C+ZUCALLI,G 1972 PHYS.STATUS SOLIDI A13,95
		CDTE 001,110,111	VG	OM	SARAIE,J+AKIYAMA,M+TANAKA,T 1972 JAPAN.J.APPL.PHYS.11,1758
		CDS 00.1	CVD	OM,XD	WEINSTEIN ET AL 1965.SEE CDS 2
	3	NACL 001,110,111	HV	RED,TED	DHERE+GOSWAMI 1969.SEE CDSE 3
		NACL 001,KCL 001, KBR 001	HV	TED	NOVICK,FT+RUMSH,MA+ZIMKINA,TM 1963 SOVIET PHYS.-CRYST.8,295
		NACL 111	---	TED	RUMSH,MA+NOVIK,FT+ZIMKINA,TM 1963 SOVIET PHYS.-CRYST.7,711
		NACL 001	HV	RED,TED	SEMILETOV 1956.SEE CDS 3
		NACL 001	HV	TEM	SELLA,C+SURYANARAYANAN,R 1968*4ERCEM 1,397
	4	MICA,SAPPH.,CAF2	CVD	RED	AITKHOZHIN+TEMIROV 1970.SEE 2
		MICA 00.1	HV	RED,TED	DHERE+GOSWAMI 1969.SEE CDSE 3
		BAF2 111	HV	RED	LUDEKE+PAUL 1967.SEE CDSE 4
		SAPPH. 00.1, SPINEL 111,BEO10.1	CVD	OM,RED, XD	MANASEVIT+SIMPSON 1971.SEE ZNS 4
		MICA 00.1	HV	RED	MURAVJEVA ET AL 1970.SEE ZNS 4
		MICA 00.1,SAPPH.	HV	RED	MURAVJEVA ET AL 1972.SEE CDS 4
		MICA 00.1	HV	TEM	PAPARODITIS,C+BERNARD,J+SELLA,C+DARMAGNA,D 1966*ISTFCL, 733
		MICA 00.1	---	TED	RUMSH ET AL 1963.SEE 3
		MICA 00.1	HV	RED,TED	SEMILETOV 1956.SEE CDS 3
		MICA 00.1	HV	TEM,XD	SHIOJIRI,M+SUITO,E 1964 JAPAN.J.APPL.PHYS.3,314
		MICA 00.1	HV	TEM	SUITO,E+SHIOJIRI,M 1963 J.ELECTRONMICROSC.12,134/ SEE ALSO HOLT,DB 1966 J.MATER.SCI.1,280
		MICA 00.1	HV	OM,RED	YEZHOVSKY,YU.K+KALINKIN,IP 1973 THIN SOLID FILMS 18,127

Z= 48-49	OVERGROWTH		SUBSTRATE	PREP	OBSERV	REFERENCE
	CADMIUM SULFIDE-SELENIDE	2	GE 001,110,111	CVD	OM,RED, EP,XD	KAMADJIEV,PR+MLADJOV,LK+GOSPODINOV,MM 1972 THIN SOLID FILMS 12,135
			CDS	CVD	OM,EP	RUFER,H+BILLE,J 1972 PHYS.STATUS SOLIDI A14,147
	CADMIUM-CHROMIUM SULPHIDE	4	SAPPHIRE 10.2, SPINEL 001,110,111	CVD	---	PINCH,HL+EKSTROM,L 1970 RCA REV.31,629
	CADMIUM SULPHIDE-INDIUM PHOSPHIDE	2	GAAS 111 + CDS	CVD	OM,XD	YIM ET AL 1970.SEE ZNS-GAP 2
	CADMIUM-MERCURY TELLURIDE	2	CDTE	VG+ CG	OM	COHEN-SOLAL,G+MARFAING,Y+BAILLY,F+RODOT,M 1965 C.R.ACAD.SCI.261,931
			CDTE 111	VG+ CG	OM,RED, TED,XD	COHEN-SOLAL,G+MARFAING,Y+BAILLY,F 1966 REV.PHYS.APPL.1,11
			CDTE 111	SP		COHEN-SOLAL ET AL 1970. SEE HGTE 2
			CDTE	VG	OM	TUFTE,ON+STELZER,EL 1969 J.APPL.PHYS.40,4559
			CDTE 111	SP		ZOZIME,A+SELLA,C+COHEN-SOLAL,G 1972 THIN SOLID FILMS 13,373
		3	NACL 001	SP	RED,TEM	ZOZIME ET AL 1972.SEE 2
	CADMIUM CHLORIDE	1	W 001	UHV	LED,AES	MORRISON+LANDER 1969.SEE KBR 1
	CADMIUM IODIDE	4	CDI2	SG	OM,XD	JAIN,RK+TRIGUNAYAT,GC 1968 J.CRYST.GROWTH 2,267
			MICA 00.1	HV	TED	KIRCHNER,F 1932 ANN.PHYS.(LEIPZIG) 76,576
49	INDIUM	2	SI 111	UHV	LED	LANDER+MORRISON 1964,1965.SEE AL 2
			MOS2 00.1	HV	TEM	POCZA,JF+BARNA,A+BARNA,PB 1969JVACSCITECHN.6,472
		3	NACL 001	SP	RED	KHAN,IH 1968 SURFACE SCI.9,306
	INDIUM PHOSPHIDE	2	INP 001,111	LPE	OM,ER	ASTLES,MG+SMITH,FGH+WILLIAMS,EW 1973 J.ELECTROCHEM.SOC.120, 1750
			INP 111	LPE	OM	BROWN,KE 1973 J.CRYST.GROWTH 20,161
			INP 001	CVD	OM	CLARKE,RC+JOYCE,BD+WILGOSS,WHE 1970 SOLID-STATE COMMUN.8,1125
			INAS 001,111, GAP 111,GAAS 111	CVD	OM,SEM, XD,XM	HALLAIS,J+SHEMALI,C+FABRE,E 1972 J.CRYST.GROWTH 17,173
			GE 001,110,111	CVD	RED	HOLLOWAY,H 1968 J.PHYS.CHEM.SOLIDS 29,1977
			INP	CVD	XD	JOYCE ET AL 1971.SEE GA-INP 2
			GAAS 111	CVD	OM,RED	MANASEVIT+SIMPSON 1973.SEE GA-INAS 2
			GAAS	CVD	XD	OLDHAM,WG+MILNES,AG 1963 SOLID-ST.ELECTRON.6,121
			GE 001,110,111	CVD	OM,RED	PERRIN,J+CAPELLA,L 1968 C.R.ACAD.SCI.267,218
			GE 001,110,111	HV	OM,RED, XD	RICHARDS ET AL 1963,1964.SEE ALSB 2
			GAAS 001	CVD	OM,RED, XD	SEKI,H+KINOSHITA,M 1968 JAPAN.J.APPL.PHYS.7,1142
			INP	LPE	---	SHAY,JL+BACHMANN,KJ+BUEHLER,E 1974 APPL.PHYS.LETT.24,192
			INAS 001	CVD	OM,XD	TIETJEN,JJ+MARUSKA,HP+CLOUGH,RB 1969 J.ELECTROCHEM.SOC.116,492
		4	SAPPHIRE 00.1	CVD	OM,RED	MANASEVIT+SIMPSON 1973.SEE GA-INAS 2
	INDIUM ARSENIDE	2	GAAS 001,111,311	CVD	RED	BAUER,GE 1967 Z.NATURFORSCH.22A,284
			INAS 001	CVD	OM	CRONIN,GR+BORRELLO,SR 1967 J.ELECTROCHEM.SOC.114,1078
			GAAS 001	CVD	OM,XD	CRONIN,GR+CONRAD,RW+BORRELLO,SR 1966 J.ELECTROCHEM.SOC.113,1336
			GAAS 001,110,111	HV	OM,RED, XD	GODINHO,N+BRUNNSCHWEILER,A 1970 SOLID-ST.ELECTRON.13,47
			GAAS 111	CVD	OM,RED, XD,EP	KASANO+IIDA 1967.SEE GAAS 2A
			GAAS 001,111	CVD	OM,RED, XD	MANASEVIT+SIMPSON 1973.SEE GA-INAS 2
			GAAS 001	CVD	XD	MCCARTHY,JP 1967 SOLID-STATE COMMUN.5,5
			GAAS 001,110	CVD	OM,EP, XD	MCCARTHY,JP 1967 SOLID-ST.ELECTRON.10,649
			GAAS 111	CVD	OM,REC	NENTWICH,G 1966 Z.NATURFORSCH.21A,816
			GE 001,110,111	HV	OM,RED, XD	RICHARDS ET AL 1963,1964.SEE ALSB 2
			ZNTE 111	LPE	RED,EP	TAMURA ET AL 1972. SEE ZNTE 2
			INAS 001	CVD	OM,XD	TIETJEN ET AL 1969.SEE INP 2
			GAAS 110	HV	OM,RED	VLASOV,VA+SEMILETOV,SA 1969SOVIETPHYSCRYST.13,580
		4	SAPPHIRE 00.1	CVD	OM,RED, XD	MANASEVIT+SIMPSON 1973.SEE GA-INAS 2
	INDIUM ANTIMONIDE	2	INSB 001,111	HV	XD	HOLLOWAY,H+RICHARDS,JC+BOBB,LC+PERRY,JR.J+ ZIMMERMAN,E 1966 J.APPL.PHYS.37,4694
			INSB	LPE	OM	KUMAGAWA,M+WITT,AF+LICHTENSTEIGER,M+GATOS,HC 1973 J.ELECTROCHEM.SOC.120,583
			CDTE	UHV	RED	LEFLOCH,G 1968 THIN SOLID FILMS 2,383

Z= 49-50 OVERGROWTH		SUBSTRATE	PREP	OBSERV	REFERENCE
INDIUM ANTIMONIDE(CONT)	2	CDTE 111	LPE	OM,ER, XD	MCNTEGU,B+MAYET,L+CASTET,L 1971 THIN SOLID FILMS 8,183
		GE 001,110,111	HV	OM,RED, XD	RICHARDS ET AL 1963,1964.SEE ALSB 2
		GAAS 110	HV	OM,RED	VLASOV+SEMILETOV 1969.SEE INAS 2
	3	NACL 001,110,111	HV	RED,TED	BARUA,KC+GOSWAMI,A 1967 INDIANJPUREAPPLPHYS.5,480
		NACL 001	SP	TED,XD	FRANCOMBE,MH+FLOOD,JJ+TURNER,GLE 1962*5ICEM 1,DD-8
		NACL 001	SP	RED	KHAN 1968.SEE IN 3
		NACL 001	SP	TED	YURASOVA ET AL 1966.SEE CDS 3
	4	MICA 00.1	HV	RED,TED	BARUA+GOSWAMI 1967.SEE 3
		MICA 00.1	HV	TEM	JUHASZ,C+ANDERSON,JC 1964 PHYS.LETT.12,163
		MICA 00.1	HV, UHV	TEM	JUHASZ,C+ANDERSON,JC 1967*3IVC 2,233
INDIUM PHOSPHIDE-ARSENIDE	2	GAAS 001,110,111	CVD	EP,XD	ALLEN,HA 1970 J.ELECTROCHEM.SOC.117,1417
		GAAS 001,110,111, GAP 111	CVD	OM,ER, XD	ALLEN,HA+MEHAL,EW 1970 J.ELECTROCHEMSOC.117,1081
		INAS	CVD	EP	BUCKMELTER,JR+KENNEDY,JK 1973 J.ELECTROCHEM.SOC.120,133
		GAP 111,GAAS 111, INAS 001,111	CVD	OM,SEM, XD,XM	HALLAIS ET AL 1972.SEE INP 2
		INAS 001	CVD	OM,XD	TIETJEN ET AL 1969.SEE INP 2
INDIUM ARSENIDE-ANTIMONIDE	2	INAS111,INSB 111, INAS-SB 111	LPE	OM,EP, XD	STRINGFELLOW,GB+GREENE,PE 1971 J.ELECTROCHEM.SOC.118,805
INDIUM TELLURIDE	3-4	NACL 001,110,111, MICA 00.1	HV	RED,TED	BARUA,KC+GOSWAMI,A 1970 INDIANJPUREAPPLPHYS.8,258
INDIUM OXIDES	4	MICA 00.1	CVD	RED	RYABOVA+SAVITSKAYA 1968.SEE ZNO 4
50 TIN	1	NB 110	UHV	LED	JACKSON+HOOKER 1969.SEE NB-SN 1
		MO 001	UHV	REM,LED, AES	JACKSON,AG+HOOKER,MP 1971 SURFACE SCI.27,197
		AG 111	HV	RED	NEWMAN 1957.SEE NI 1
	2	ZNS 110,PBS 001	HV	RED	CURZON,AE 1962 CRYOGENICS 2,334
		SI 111	UHV	LED	ESTRUP,PJ+MORRISON,J 1964 SURFACE SCI.2,465
		SNTE 001	---	---	JESSER,WA+VAN DER MERWE,JH 1971 PHIL.MAG.24,295
		SNTE 001	HV, UHV	TEM	TAKAYANAGI,K+YAGI,K+KOBAYASHI,K+HONJO,G 1972 J.PHYS.SOC.JAPAN 32,580
		SNTE 001	UHV	TEM	VINCENT,R 1968*4ERCEM 1,389
		SNTE 001	UHV	TEM	VINCENT,R 1969 PHIL.MAG.19,1127
		SNTE 001	UHV	TEM	YAGI,K+TAKAYANAGI,K+HONJO,G 1972 J.PHYS.SOC.JAPAN 32,1445
	3	NACL 001	HV, UHV	TEM	SHIMAOKA,G+KOMORIYA,G 1969*27EMSA, 122
		NACL 001	UHV	RED,TEM	SHIMAOKA,G+KOMORIYA,G 1970 J.VAC.SCI.TECHN.7,178
		NACL 001	UHV	RED,TEM	SHIMAOKA,G+KOMORIYA,G+CHANG,SC 1970*7ICEM 2,443
		NACL 001	UHV	RED,TEM	VOOK,RW 1961 J.APPL.PHYS.32,1557 + 2474/ 1962 IBID.33,2498
		NACL 001	UHV	TEM	VOOK,RW 1962*5ICEM 1,B-7
		NACL 001	UHV	TEM	VOOK,RW 1964 ACTA MET.12,197/ 1965*3ERCEM A,231
TIN-ANTIMONY ALLOYS	3	NACL 001,110,111	HV	TEM	PATEL,AR+MYSOREWALA,DV 1972 THINSOLIDFILMS 9,181
TIN SULPHIDE	3	NACL 001	HV	RED,TED	BAOACHHAPE,SB+GOSWAMI,A 1964 INDIAN J.PURE APPL.PHYS.2,250
		NACL001+AMORPH.C, SIO,SIO2,BATIO3	HV	RED,TED	CHOPRA 1969.SEE AG 3
		NACL 001	HV	TED	MARIANO,AN+CHOPRA,KL 1967 APPL.PHYS.LETT.10,282
	4	MICA00.1+AMORPH.C, SIO,SIO2,BATIO3	HV	RED,TED	CHOPRA 1969.SEE AG 3
TIN SELENIDE	3	NACL 001	HV	TED	MARIANO+CHOPRA 1967.SEE SNS 3
	4	TRIGLYCINE SULPH.	HV	OM	VLASOV ET AL 1970.SEE AGCL 4
TIN TELLURIDE	2	PBSE 001	HV	TEM	YAGI ET AL 1971.SEE PD 1
	3	KCL 001	HV	XD	CARDONA,M+GREENAWAY,DC 1964 PHYS.REV.133,A1685
		NACL 001	HV	RED,TED	GOSWAMI,A+JOG,RH 1969 INDIANJ.PUREAPPLPHYS.7,273
		KCL 001	HV	---	GREEN ET AL 1973.SEE AU 2
		NACL 001	HV	---	OTA,Y+ZEMEL,JN 1969 J.VAC.SCI.TECHN.6,558
		KCL 001	UHV	TEM	STOEMENOS+VINCENT 1972.SEE GETE 2
		KCL 001	UHV	---	VINCENT 1969.SEE SN 2

Z=50-55 OVERGROWTH		SUBSTRATE	PREP	OBSERV	REFERENCE
TIN TELLURIDE	4	MICA 00.1	HV	RED,TED	GOSWAMI+JOG 1969.SEE 3
TIN OXIDES	4	TIO2 110	CVD	OM,RED, XD	CASLAVSKA,V+ROY,R 1969 J.APPL.PHYS.40,3414
51 ANTIMONY	3	KCL 001	HV	TEM	BAHADUR,K+CHAUDHARY,KL 1968*4ERCEM 1,405/ 1969 APPL.PHYS.LETT.15,277
		KCL 001	HV	TEM	SCHUZ,W 1970 PHYS.STATUS SOLIDI A1,K81
ANTIMONIUM SULPHIDE	3	NACL 001	HV	RED,TED	BADACHHAPE,SB+GOSWAMI,A 1967 INDIAN J.PURE APPL.PHYS.5,477
ANTIMONIUM TELLURIDE	3-4	NACL 110,111, MICA 00.1	HV	RED,TED	GADGIL,LH+GOSWAMI,A 1969 J.VAC.SCI.TECHN.6,591
52 TELLURIUM	1	CU 001	UHV	LED	ANDERSON,DE+ANDERSON,S 1970 SURFACE SCI.23,311
		CU 111	UHV	LED,ES	ANDERSON,S+MARKLUND,I+MARTINSON,J 1969 SURFACE SCI.12,269
	2	SE 10.0	HV	OM,TEM	BAMMES ET AL 1972.SEE SE 3
	3	KBR 001	HV	TEM	CAPERS,MJ+WHITE,M 1971 THIN SOLID FILMS 8,317
		LICL 001,LIBR001, NABR 001,NAI 001, KCL 001,KBR 001, KI 001,RBCL 001, RBBR 001,RBI 001	HV	TEM	DICK,E+SCHUZ,W 1972 Z.NATURFORSCH.27A,789
		KBR 001	HV	TEM	NABITOVICH ET AL 1971.SEE SE 3
		KCL 001,KBR 001	HV	TEM	SCHUZ,W+DICK,E 1970 Z.NATURFORSCH.25A,456
	4	MICA 00.1	HV	TEM	WEIDMANN,EJ+ANDERSON,JC 1971 THINSOLIDFILMS 7,265
53 IODINE	2	SI 001,111	SG	RED	HENDERSON,RC+POLITO,WJ 1969 SURFACE SCI.14,473
54 XENON	1	CU 001	UHV	LED	CHESTERS+PRITCHARD 1971.SEE CO 1
		CU 001,110,111, AG 110,111,211	UHV	LED	CHESTERS,MA+HUSSAIN,M+PRITCHARD,J 1973 SURFACE SCI. 35,161
		IR 001	UHV	LED	IGNATIEV,A+JONES,AV+RHODIN,TN 1972 SURFACE SCI.30,573
		PD 001	UHV, VG	LED	PALMBERG,PW 1971 SURFACE SCI.25,598
	2	GRAPHITE 00.1	VG	TEM	BALL+VENABLES 1970.SEE AR 2
		GRAPHITE 00.1	HV	TEM	KRAMER+VENABLES 1972.SEE KR 2
		GRAPHITE 00.1	VG	LED	LANDER 1967.SEE GERMANIUM IODIDES 2
		GRAPHITE 00.1	VG	LED	LANDER+MORRISON 1967.SEE GERMANIUM IODIDES 2
		GRAPHITE 00.1	VG	LED	MORRISON,J+LANDER,JJ 1966 SURFACE SCI.5,163
		GRAPHITE 00.1	HV	TEM	VENABLES,JA+BALL,DJ 1968 J.CRYST.GROWTH 3,4,180
		GRAPHITE 00.1	UHV	TEM	VENABLES,JA+ENGLISH,CA 1971 THINSOLIDFILMS 7,369
55 CESIUM	1	NI 111	UHV	LED	CALLCOTT,TA+MACRAE,AU 1969 PHYS.REV.178,986
		W 001,110	UHV	---	CARROLL+MAY 1972.SEE NA 1
		W 110	UHV	LED	FEDORUS,AG+NAUMOVETS,AG 1970 SURFACE SCI.21,426
		NI 110	UHV	LED	GERLACH+RHODIN 1969(A,B).SEE NA 1
		W 001,110	UHV	LED,AES	MACRAE,AU+MULLER,K+LANDER,JJ+MORRISON,J 1969 SURFACE SCI.15,483
		W 001	UHV	LED	MACRAE,AU+MULLER,K+LANDER,JJ+MORRISON,J+ FHILLIPS,JC 1969 PHYS.REV.LETT.22,1048
		W 001	UHV	LED,AES	THOMAS+HAAS 1973.SEE K 1
		RE 00.1	VG	OM	WEBSTER,HF 1967 J.APPL.PHYS.38,3700
	2	GRAPHITE 00.1	VG	LED	LANDER 1967.SEE GERMANIUM IODIDES 2
		GRAPHITE 00.1	VG	LED	LANDER+MORRISON 1967.SEE GERMANIUM IODIDES 2
CESIUM FLUORIDE	2	PBS 001	SG	OM	ROYER 1928.SEE LICL 2
	3	NACL 001,KCL 001	SG	OM	ROYER 1928.SEE LICL 2
	4	CACO3 100	HV	RED	SCHULZ 1952(B).SEE LIF 4
CESIUM CHLORIDE	1	FE	SG	OM	JOHNSON 1951.SEE NACL 1
		W 001,110	UHV	LED,AES	MORRISON+LANDER 1969.SEE KBR 1
		AG 001,111	HV, SG	RED	SCHULZ 1952(C).SEE LIF 1
		AG 001	SG	OM	SLOAT+MENZIES 1931.SEE LICL 3
	3	LIF 001,NACL 001, KCL 001, KI 001	HV	RED	LUDEMANN,H 1957 Z.NATURFORSCH.12A,226
		LIF 001,NACL 001, KBR 001	HV	RED	SCHULZ,LG 1951(C) ACTA CRYST.4,487

Z= 55-74 OVERGROWTH		SUBSTRATE	PREP	OBSERV	REFERENCE
CESIUM CHLORIDE	4	MICA 00.1, CACO3 100	HV, SG	RED	SCHULZ 1951(C).SEE 3
		NANO3 100, CACO3 100	HV, SG	RED	SCHULZ 1952(B).SEE LIF 4
CESIUM BROMIDE	1	AG 001,111	HV, SG	RED	SCHULZ 1952(C).SEE LIF 1
	3	KCL 001	HV	RED	LUDEMANN 1957.SEE CSCL 3
		LIF 001	HV	RED	SCHULZ,LG 1950 J.APPL.PHYS.21,942
		LIF 001,NACL 001, KBR 001	HV	RED	SCHULZ 1951(C).SEE CSCL 3
	4	MICA 00.1	SG	OM	LISGARTEN 1954.SEE NH4CL 4
		MICA 00.1	HV	RED	SCHULZ 1951(C).SEE CSCL 3
		NANO3 100, CANO3 100	HV, SG	RED	SCHULZ 1952(B).SEE LIF 4
CESIUM IODIDE	3	LIF 001,NACL 001, KBR 001	HV	RED	SCHULZ 1951(C).SEE CSCL 3
	4	MICA 00.1	SG	OM	LISGARTEN 1954.SEE NH4CL 4
		MICA 00.1	HV	RED	SCHULZ 1951(C).SEE CSCL 3
56 BARIUM	1	MO 110	UHV	LED	BONDARENKO,BV+MAKHOV,VI 1971 SOVIET PHYS.-SOLID STATE 12,1522
BARIUM FLUORIDE	3-4	MICA00.1,LIF 001, NACL 001,110,111	HV	RED,TEM, REM	BAUER 1956.SEE LIF 3
BARIUM TITANATE	1	AU 001	HV	RED	MULLER,EK+NICHOLSON,BJ+TURNER,G.L≠E. 1962 BRIT.J.APPL.PHYS.13,486/ 1963.SEE CATIO3 1
		PT 111	MG	OM,XD	TADA,O+SHINTANI,Y+YOSHIDA,Y 1969 JAPPLPHYS.40,498
	3	LIF 001,NAF 001	HV	RED	MULLER ET AL 1962.SEE 1/ 1963. SEE CATIO3 1
		LIF 001	HV	RED	NAKAMURA,T+MIDORIKAWA,M 1966 J.PHYS.SOC.JAPAN.21,1453
BARIUM STANNATE, BARIUM CERATE	3	LIF 001,NAF 001	HV	RED	MULLER ET AL 1963.SEE CATIO3 1
BARIUM OXIDE	4	MGO 001	VG	OM	SPROULL,RL+DASH,WC+TYLER,WW+MOORE,AR 1951 REV.SCI.INSTR.22,410
62 SAMARIUM SELENIDE, SAMARIUM TELLURIDE	4	CAF2 111	HV	TED,XD	PAPARODITIS,C+SURYANARAYANAN,R 1972 J.CRYST.GROWTH 13/14,389
63 EUROPIUM SELENIDE	3-4	NACL 001,CAF2 111	HV	TED,XD	PAPARODITIS+SURYANARAYANAN 1972.SEE SMSE 4
EUROPIUM TELLURIDE	4	CAF2 111	HV	TED,XD	PAPARODITIS+SURYANARAYANAN 1972.SEE SMSE 4
EUROPIUM-LEAD TELLURIDE	3	NACL 001	HV	TED,XD	PAPARODITIS+SURYANARAYANAN 1972.SEE SMSE 4
EUROPIUM OXIDES	4	CAF2 111	SP	XD	LEE,K+SUITS,JC 1970 J.APPL.PHYS.41,954
64 GADOLINIUM	1	MO 110	UHV	LED	BONDARENKO,BV+MAKHOV,VI 1971 SOVIET PHYS.-SOLID STATE 12,1525
70 YTTERBIUM SELENIDE, YTTERBIUM TELLURIDE	4	CAF2 111	HV	TED,XD	PAPARODITIS+SURYANARAYANAN 1972.SEE SMSE 4
71 LUTETIUM NITRIDE	4	SAPPHIRE 10.2	CVD	XD	DISMUKES ET AL 1970.SEE SCN 4
73 TANTALUM	3	NACL 001	SP	RED,TED, XD	CHOPRA ET AL 1967.SEE MO 3
	4	MGO 001	HV	TED	BERTING,FM+LAWLESS,KR 1969*27EMSA, 114
		MGO 001	HV	TEM	DENBIGH,PN+MARCUS,RB 1966 J.APPL.PHYS.37,4325
		MGO 001	HV	TEM	MARCUS,RB 1966*6ICEM 1,503/ 1966JAPPLPHYS.37,3121
		MGO 001	UHV	TEM	MARCUS,RB+QUIGLEY,S 1968 THIN SOLID FILMS 2,467
		SAPPH. 10.2,00.1	UHV	OM,RED, REM	O≠NEAL,JE 1972 THIN FILMS 2,119
		SAPPH. 10.2,00.1	UHV	RED,REM	O≠NEAL ET AL 1972.SEE VANADIUM 4
TANTALUM CARBIDES	3-4	NACL 001, MGO 001, CAF2 111	HV	TEM,XD	NAIKI,T+NINOMIYA,M+IHARA,M 1972 JAPAN.J.APPL.PHYS.11,1106
74 TUNGSTEN	1	MO	CVD	OM,SEM, XD,ED	GILLARDEAU ET AL 1971.SEE MO 1
	3	NACL 001	SP	RED,TED, XD	CHOPRA ET AL 1967.SEE MO 3

Z= 74-79 OVERGROWTH		SUBSTRATE	PREP	OBSERV	REFERENCE
TUNGSTEN	4	SAPPHIRE 11.3	CVD	XD	BARNETT,GD+MILLER,A+PULLIAM,GR+WARREN,RG 1963*AIMEMAEM 19,263
		SAPPH. 70 OFF C, 00.1	CVD	RED,REM, SEM,XD	MAYADAS,AF+CUOMO,JJ+ROSENBERG,R 1969 J.ELECTROCHEM.SOC.116,1742
		SAPPH. 10.1,10.2, 11.3,00.1	CVD	XD,TEM	MILLER,A+MANASEVIT,HM+FORBES,DH+CADOFF,IB 1966 J.APPL.PHYS.37,2921
TUNGSTEN CARBIDES	1	W 110	VG	LED	BAUER 1967.SEE AG 1
		W 001,110,111	CG	LED	BOUDART,M+OLLIS,DF 1969*4IMATC, 63-1
		W 112	CG	LED	CHEN,JM+PAPAGEORGOPOULOS,CA 1970 SURFACE SCI.20,195
TUNGSTEN OXIDES	3	LIF 001,NAF 001	HV	RED	MULLER ET AL 1963.SEE CATIO3 3
77 IRIDIUM	1	IR	UHV	FIM	GRAHAM ET AL 1972.SEE RH 1
		MO 110	UHV	FIM	WHITMELL,DS 1968 SURFACE SCI.11,37
78 PLATINUM	1	AU 111	HV	TEM	BASSET ET AL 1958.SEE NI 1
		AG	HV	TEM	BRAME+EVANS 1958.SEE RH 1
		W	HV	FIM	DURAI RAGHAVAN,NV+BAYUZICK,RJ 1972 J.VAC.SCI.TECHN.9,784
		IR	HV, UHV	FIM	GRAHAM,WR+HUTCHINSON,F+NADAKAVUKAREN,JJ+REED,DA+ SCHWENTERLY,SW 1969 J.APPL.PHYS.40,3931
		RH,IR	UHV	FIM	GRAHAM ET AL 1972.SEE RH 1
		AU 111	ED	TEM	JESSER,WA+MATTHEWS,JW+KUHLMANN-WILSDORF,D 1966 APPL.PHYS.LETT.9,176/ SEE ALSO LAWLESS 1965 IN NI 1
		W 001	UHV	RED,XD	JOWETT,CW+DOBSON,PJ+HOPKINS,BJ 1969 SURFACE SCI.17,474
		AU 001	UHV	TEM	MATTHEWS,JW 1966 PHIL.MAG.13,1207
		AU 001,ETC.	UHV	TEM	MATTHEWS,JW+JESSER,WA 1967 ACTA MET.15,595
		W,IR	HV	FIM	MURR,LE+INAL,OT+SINGH,HP 1971*29EMSA, 214
		MO,W,IR	HV, UHV	FIM	MURR,LE+INAL,OT+SINGH,HP 1972THINSOLIDFILMS 9,241
		W 110	UHV	FIM	NAUMOVETS,AG+FEDORUS,AG 1968 SOVIET PHYS.-SOLID STATE 10,627
		AU 111	HV	TEM	PASHLEY ET AL 1957.SEE CO 1
		AG	HV	TEM,XD	REICHELT,K+SCHOBER,T+VIEHWEG,J 1973 J.CRYST.GROWTH 18,312
		IR	ED	FIM	RENDULIC,KD+MULLER,EW 1967 J.APPL.PHYS.38,550
		AU 001	HV	TEM	SCHOBERT 1969.SEE PD 1
		W	UHV	FIM	SUGATA,E+ISHII,S+MASUI,K 1971 SURFACE SCI.24,612
	2	MOS2 00.1	HV	RED	KAINUMA 1951.SEE NI 2
		SI 111	SP	XD	KAWAMURA,T+SHINODA,D+MUTA,H 1967 APPL.PHYS.LETT.11,101
	3	NACL 001	UHV	TEM	MATTHEWS 1966.SEE FE 3
		NACL 001	HV	TEM	MURR ET AL 1971,1972.SEE 1
		NACL 001	HV	TEM	SUMNER,GG 1965 PHIL.MAG.12,767/ 1966 SURFACE SCI.4,313
		KCL 001	HV	RED	THIRSK 1950.SEE AG 3
	4	MICA 00.1	UHV	TED	ALLPRESS ET AL 1966.SEE NI 4
		MICA 00.1		OM	VOTAVA,E 1953 NATURWISSENSCHAFTEN 40,290
PLATINUM-SILICON ALLOYS	2	SI 001,111	CG	TEM	ANDERSON,RM 1972*30EMSA, 518
		SI 001,111	CG	RED,TEM, XD	ANDERSON,RM+DASH,S 1969*27EMSA, 120
		SI 111	SP	XD	KAWAMURA ET AL 1967.SEE PT 2
		SI 001,111	SP+ CG	RED,TEM, SEM,XD	SINHA,AK+MARCUS,RB+SHENG,TT+HASZKO,SE 1972 J.APPL.PHYS.43,3637
		SI 111	HV, SP	RED,TED, XD	WALKER,GA+WNUK,RC+WOODS,JE 1970 J.VAC.SCI.TECHN.7,543
PLATINUM-SILVER ALLOYS	1	AG 111	HV	RED	GLOSSOP+PASHLEY (UNPUBLISHED).SEE NI 1
79 GOLD	1	AG 001,111	HV	TEM	ALLPRESS+SANDERS 1964.SEE AG 3
		AG 001,111	HV	TEM	BASSETT,GA+PASHLEY,DW 1958-59 J.INSTMETALS 87,449
		AG 001,111	HV	TEM	BASSET ET AL 1959.SEE CU 18
		CU,W	UHV		BAUER+POPPA 1972. SEE AG 1
		AG 001	HV	TED	BRUCK 1936.SEE AL 3
		AG	HV	TEM	BRAME+EVANS 1958.SEE RH 1
		ZN 00.1	HV	RED	COUDERC ET AL 1959.SEE CU 18
		AG 001	HV	TEM,XD, XM,IGH	DATZ,S+NOAK,CD+NOGGLE,TS+APPLETON,BR+LUTZ,HO 1969 PHYS.REV.179,315
		AG 001	UHV	RED,TEM, REM,LED	DENNIS,PNJ+DOBSON,PJ 1972 SURFACE SCI.33,187
		AG 111	HV, ED	TEM	DICKSON+PASHLEY 1962.SEE AG 4
		AG 111	HV, ED	TEM	DICKSON,EW+JACOBS,MH+PASHLEY,DW 1965 PHIL.MAG.11,575
		AG 111,AG 111 + C, C REPLICA AG 111	HV	TED	DUHLER,I+KITTL,P 1969(A) J.MATER.SCI.4,89

Z= 79 OVERGROWTH		SUBSTRATE	PREP	OBSERV	REFERENCE
GOLD (CONT.)	1	C REPLICA AG 111	HV	TEM	DUMLER,I+KITTL,P 1969(B) ANNSOC.CI.ARGENT,188,177
		AG 111 + C, SIO	HV	TED	DUMLER,I+MARRAPODI,MR 1972 THINSOLIDFILMS 12,279
		CU 111,FE 110	ED	RED	FINCH ET AL 1947.SEE CO 1
		AG 001,111	HV	TEM,PCH, ICH	GIBSON,WM+RASMUSSEN,JB+AMBROSIUS-OLESEN,P+ ANDREEN,CJ 1968 CAN.J.PHYS.46,551
		CU 111	HV	RED	HAASE 1956.SEE FE 1
		AG 111	HV, ED	TEM	JACOBS,MH+PASHLEY,DW 1962*5ICEM 1,DD-4
		AG 111	UHV	TEM	JAEGER,H+MERCER,PD+SHERWOOD,RG 1969 SURFACE SCI.13,349
		NI 111	HV	RED	KUBO+MIYAKE 1948.SEE NI 1
		CU 111	HV	RED	LAFOURCADE,L+LARROQUE,P+NGUYEN QUAT TI 1959 C.R.ACAD.SCI.249,390
		CU 001,111	ED	RED,TEM	LAWLESS 1965.SEE NI 1
		CU3AU	ED	TEM	MARCINKOWSKI,MJ+BAKER,AJ+FISHER,RM 1972 PHYS.STATUS SOLIDI A12,431
		PD 001,AG 001	UHV	TEM	MATTHEWS 1966.SEE PT 1
		AG	UHV	TEM	MATTHEWS,JW 1967 PHYSICS OF THIN FILMS (G.HASS+RE.THUN,EDS.) 4,137.ACADEMIC PRESS.
		AG 111	HV	TEM	MISSIROLI 1972.SEE NACL 4
		W	UHV	FEM,FIM	MONTAGU-POLLOCK,HM+RHODIN,TN+SOUTHON,MJ 1968 SURFACE SCI.12,1
		AG 111	HV	RED	NEWMANN 1957.SEE NI 1
		CU 110,111	HV	RED	NGUYEN QUAT TI ET AL 1961.SEE CO 1
		AG 001,AU 001	UHV	LED	PALMBERG+RHODIN 1967.SEE AG 3
		CU 001,PD 001	UHV	LED,AES	PALMBERG+RHODIN 1968(A).SEE CU 4/ 1968(C).SEE CU-AU 1
		AG 111	HV, UHV	TEM	PASCARD ET AL 1972(B).SEE NI-FE 1
		AG 111	HV	TEM	PASHLEY,DW 1959(B) PHIL.MAG.4,324/ 1960 PROC.ROY.SOC.,SER.A 255,218
		AG 001,111	HV	TEM	PASHLEY,DW+STOWELL,MJ 1963 PHIL.MAG.8,1605
		AG 111	HV	TEM	PENG,JOC 1972*30EMSA, 634
		AG	HV	TEM	POPPA,H 1962*5ICEM 1,GG-14
		PD 111	HV	TEM	POPPA 1964.SEE PD 1
		MO 001	UHV	RED,REM	PRICE,CH 1973*31EMSA, 38
		AG 111	HV	XD,PCH	REICHELT+LUTZ 1971.SEE AG 4
		AG 001	HV, UHV	TEM	SCHOBER 1969.SEE PD 1
		AU 001	UHV	REM,XD	SCHWOEBEL,RL 1964 SURFACE SCI.2,356
		AU 110,111	UHV	TEM	SCHWOEBEL,RL 1966 J.APPL.PHYS.37,2515
		NI001,PD001,AG001	HV	TED	SHIRAI ET AL 1961.SEE NI 1
		AG 111	HV	XD	SMITH,JR.JN+SALTSBURG,H 1964 J.CHEM.PHYS.40,3585
		CU 001,111	ED	XD,RED, TEM	THOMPSON+LAWLESS 1969.SEE NI 1
		PD 001	HV	TEM	YAGI ET AL 1970.SEE PD 1
		PD 001,111	HV, UHV	TEM	YAGI ET AL 1971.SEE PD 1
		AG 111	HV	TEM	ZIGNANI ET AL 1968.SEE AG 4
	2	SI,PBS,PBSE,PBTE	HV, UHV		BAUER+POPPA 1972. SEE AG 1
		GRAPHITE 00.1	UHV	TEM	DARBY,TP+WAYMAN,CM 1970 PHYS.STATUS SOLIDI A1,729
		PBS 001	UHV	RED,LED, TEM	GREEN ET AL 1971.SEE AG 2
		PBS 001, PBSE 001 PBTE 001,SNTE 001	HV	RED,TEM	GREEN,AK+DANCY,J+BAUER,J 1973 J.VAC.SCI.TECHN.10,494
		PBS 001	UHV	RED,LED, TEM	HANAWA,T+TAKEDA,K 1969 APPL.PHYS.LETT.15,360
		PBS 001	UHV	TEM	HEINEMANN,K+POPPA,H 1972 APPL.PHYS.LETT.20, 122
		GRAPHITE 00.1	SG	TEM	HOCART,R+OBERLIN,A 1954 C.R.ACAD.SCI.239,1228
		MOS2 00.1	HV	TEM	HONJO+YAGI 1969.SEE AG 2
		MOS2 00.1	HV	TEM	JACOBS,MH+STOWELL,MJ 1965 PHIL.MAG.11,591
		MOS2 00.1	HV	TEM	JACOBS,MH+PASHLEY,DW+STOWELL,MJ 1966 PHIL.MAG.13,129
		MOS2 00.1	UHV	TEM	JESSER,WA+KUHLMANN-WILSDORF,D 1967 J.APPL.PHYS.38,5128
		MOS2 00.1	UHV	TEM	JESSER,WA+KUHLMANN-WILSDORF,D 1968 ACTA MET.16,1325
		MOS2 00.1	HV	RED	KAINUMA 1951.SEE NI 2
		MOS2 + C,SIOX	HV	TED	KHIDR,MS+IGNACZ,P+POCZA,JF 1968*2CTF, 45
		ZNS 110	HV	RED	KUBO+MIYAKE 1948.SEE NI 1
		SI 111	UHV	TEM	MASSON,A+KERN,R 1968 J.CRYST.GROWTH 2,227
		MOS2 00.1	HV	TEM	MATTHEWS 1972(A).SEE CU 1
		MOS2 00.1	HV	TEM	PASHLEY,DW+STOWELL,MJ 1966 J.VAC.SCI.TECHN.3,156/ 1966*6ICEM 1,487
		MOS2 00.1	HV	TEM	PASHLEY ET AL 1964.SEE AG 2
		PBSE 001	UHV	TEM	POPPA ET AL 1974. SEE AG 2
		MOS2 00.1	HV	TEM	STOWELL,MJ 1969*CMPSS, 461
		MOS2 00.1	HV	TEM	STOWELL,MJ+LAW,TJ 1966 PHYS.STATUS SOLIDI 16,117
		MOS2 00.1	UHV	TEM	STOWELL,MJ+LAW,TJ 1968 PHYS.STATUS SOLIDI 25,139
		MOS2 00.1	UHV	TEM	VALDRE,U+ROBINSON,EA+PASHLEY,DW+STOWELL,MJ+LAW,TJ 1970 J.PHYSICS.E=SCI.INSTRUM.3, 501

Z= 79 OVERGROWTH		SUBSTRATE	PREP	OBSERV	REFERENCE
GOLD	3	NACL 001,KCL 001, KBR 001,KI 001	UHV	TEM	ADAM,RW 1966 Z.NATURFORSCH.21A,497/ 1968 IBID.23A,1526
		NACL 001,111	HV	TEM	ADAMSKY 1966.SEE AG 3
		NACL 001	HV	TEM	ADAMSKY+LE BLANC 1963.SEE NI 3
		NACL 001,111	HV	TEM	ADAMSKY+LE BLANC 1965.SEE AG 3
		NACL 001	HV	TEM	BASSET+PASHLEY 1958-59.SEE 1
		NACL 001	HV	TEM	BASSET ET AL 1959.SEE CU 18
		LIF 001,NACL 001, KCL 001	UHV		BAUER+POPPA 1972.SEE AG 1
		NACL 001,KCL 001	UHV	RED,TEM	BAUER,E+GREEN,AK+KUNZ,KM 1966 APPLPHYSLETT.8,248
		NACL 001,KCL 001	HV, UHV	TEM	BETHGE 1969.SEE AG 3
		NACL 001	HV	TEM	BIRJEGA,MI+GLODEANU,F+POPESCU-POGRION,NG+ TEODORESCU,IA+TOPA,V 1972 THINSOLID FILMS 10,307
		NACL 001	HV, UHV	RED,TED, XD	BRINE+YOUNG 1963.SEE CU 3
		NACL 001	HV	TED	BRUCK 1936.SEE AL 3
		NACL 001	SP	TEM	CAMPBELL+STIRLAND 1964.SEE AG 3
		NACL 001	HV	TEM	CATLIN,A+WALKER,WP+LAWLESS,KR 1960 ACTAMET.8,734
		NACL 001	HV	TEM,PCH	CHADDERTON,LT+ANDERSON,MG 1968 THIN FILMS 1,229
		NACL 001	HV	TEM	CHOPRA 1965.SEE CU 3
		NACL 001 + SIO, SIO2,BATIO3	HV	RED,TED	CHOPRA 1966.SEE AG 3
		NACL 001,KCL 001, KBR 001	HV	TED	CONJEAUD,P 1959 C.R.ACAD.SCI.248,566
		NACL 001,KCL 001, KBR 001	HV	TEM	CONJEAUD,P+SELLA,C 1959 C.R.ACAD.SCI.248,1680
		NACL 001	HV, UHV	TEM	CONNEL,RA 1967 J.APPL.PHYS.38,2397
		NACL 001 + C	HV	TEM	DISTLER ET AL 1971.SEE AG 3
		NACL	UHV	TEM	DONOHOE,AJ+ROBINS,JL 1972 J.CRYST.GROWTH 17,70
		C REPLICA NACL001	HV	TED	DUMLER+KITTL 1969(A).SEE 1
		NACL 001	HV	TEM	ESTEBAN,MF+ROJO,JM 1973 THIN SOLID FILMS 15, S7
		NACL 001	SP	TEM	FRANCOMBE,MH+SCHLACTER,M 1964. SEE FRANCOMBE,MH 1966*NATOUTF, 29
		NACL 001	HV	XD	FUKAMACHI,M+NAGAKURA,S+OKETANI,S 1967 ACTA MET.15,1402
		NACL 001	HV	TEM	GERASIMOV,YM+DISTLER,GI 1968 NATURWISSENSCHAFTEN 55,132
		LIF 001,NACL 001, KBR 001	HV, UHV	TED	GILLET,M+GILLET,E 1966*6ICEM 1,633
		NACL 001	HV	TED	GILLET,E+GILLET,M 1966 C.R.ACADSCI.,SER.B 262,359
		NACL 001	HV, UHV	TEM	GILLET,E+GILLET,M 1969 THIN SOLID FILMS 4,171
		NACL 001	UHV	TEM	GILLET,E+GILLET,M 1973 THIN SOLID FILMS 15,249
		NACL 001,LIF 001, KCL 001,KI 001	HV	RED	GOTTSCHE 1956.SEE AL 3
		NACL 001, CA-DOPED NACL 001	HV, UHV	TEM	GREEN,A+BAUER,E+DANCY,J 1969*CMPSS, 479
		LIF 001	UHV	RED,TEM	GREEN ET AL 1970.SEE AL 3
		NACL 001	UHV	TEM	GRUNBAUM,E+MATTHEWS,JW 1965PHYS.STAT.SOLIDI 9,731
		LIF 001,110,111, 112	HV	XD	HALL+THOMPSON 1961.SEE CU 3
		NACL 001,KCL 001, KI 001	HV	RED	HARSDORFF 1964.SEE CU 3
		NACL 001,KCL 001, KI 001	HV, UHV	RED,TED	HARSDORFF 1968(A).SEE CU 3
		NACL 001	HV, UHV	---	HARSDORFF 1968(B).SEE AG 2
		NACL 001,KCL 001, KBR 001,KI 001, NACL 001 + C	UHV	RED,TEM	HARSDORFF,M+ADAM,RW+SCHMEISSER,H 1970 KRISTALL UND TECHNIK 5,279
		NACL 001+C, SIO2	HV	TEM	HAYEK+SCHWABE 1970.SEE AG 3
		NACL 001	UHV	TEM,XD	HENNING,CAO 1966 SOLID-ST.COMMUN.4,439
		NACL 001	UHV	OM,TEM, XD	HENNING,CAO 1968 SURFACE SCI.9,277 + 296
		KCL 001	HV, UHV	TEM	HENNING,CAO 1969 SURFACE SCI.12,308
		NACL 001 + C, KCL 001 + C	UHV	TEM	HENNING,CAO 1970 NATURE 227,1129
		NACL 001	UHV	TEM	HENNING,CAO+VERMAAK,JS 1969 APPL.PHYS.LETT.15,3
		NACL 001,KCL 001	UHV	---	HENNING,CAO+VERMAAK,JS 1970 PHIL.MAG.22,281
		NACL 001,KCL 001	UHV	TEM	HENNING,CAO+LOMBAARD,JC+BOTHA,JC 1969 APPL.PHYS.LETT.14,109
		LIF 001	HV, UHV	TED	HONJO,G 1964-65 UNPUBLISHED WORK. SEE SATO ET AL 1969 IN AG 4
		NACL	HV	TEM	HONJO,G 1964*ICTFBB, 189
		NACL 001	UHV	TEM	INO,S 1966 J.PHYS.SOC.JAP.21,346
		NACL 001	--	---	INO,S 1969 JPHYSSOCJAP.26,1559/ 1969 IBID.27,941
		NACL 001,KCL 001	UHV	TEM	INO+OGAWA 1966.SEE AL 3/ 1967 JPHYSSOCJAP.22,1365
		NACL 001	HV	TEM	INO ET AL 1962,1964.SEE CU 3
		NACL 001	UHV	TEM	INUZUKA,T+UEDA,R 1968(A) APPL.PHYS.LETT.13,3
		NACL 001,KCL 001	UHV	TEM	INUZUKA,T+UEDA,R 1968(B) J.PHYS.SOC.JAP.25,1299
		NACL 001	HV	TEM	JAUNET+SELLA 1965.SEE AL 3

Z= 79 OVERGROWTH		SUBSTRATE	PREP	OBSERV	REFERENCE
GOLD (CONT.)	3	LIF 001,NAF 001, NACL 001,NABR 001, NAI 001,KF 001, KCL 001,KBR 001, KI 001,RBCL 001, RBBR 001,RBI 001	UHV	TEM	JESSER+MATTHEWS 1969.SEE CU 3
		KCL 001,KBR 001, KI 001	HV	TED	KATO 1968.SEE AL 3
		NACL 001	HV	RED	KEHOE 1957.SEE CU 3
		KCL 001	UHV	---	KERN,R+MASSON,A+METOIS,JJ 1971 SURFACE SCI.27,483
		NACL 111	HV	RED,TEM	KOCH ET AL 1972.SEE CU 3
		NACL 001	HV	TEM	KOMODA,T 1968 JAPAN.J.APPL.PHYS.7,27
		NACL 001,KCL 001, KI 001	UHV	RED,TEM	KUNZ ET AL 1966.SEE AL 3/ SEE ALSO BAUER E+ GREEN,AK+KUNZ,KM+POPPA,H 1966*ISTFCL, 135
		NACL 001	HV, UHV	TEM	LEWIS,B+JORDAN,MR 1970 THIN SOLID FILMS 6,1
		NACL 001	HV, UHV	TEM	LOMBAARD,JC+KOTZE,IA+HENNING,CAO 1974 SOLID-ST.COMMUN.14,217
		NACL 001	HV	TEM	MARUYAMA,S+KIHO,H 1968 J.PHYS.SOC.JAP.25,1392
		KCL 001	UHV	TEM	MASSON,A+METOIS,JJ+KERN,R 1968 C.R.ACAD.SCI., SER.B 267,64/ 1968 J.CRYST.GROWTH 3/4,196/ 1970 C.R.ACAD.SCI.,SER.B 271,235 + 298/ 1971 SURFACE SCI.27,463/ 1971 ADV.IN EPITAXY AND ENDOTAXY(H.G.SCHNEIDER,ED.) 2,103. VEB DEUTSCHER VERLAG FUR GRUNDSTOFFINDUSTRIE,LEIPZIG
		NACL 001	HV	TEM	MATTHEWS 1959.SEE CU 3
		NACL 001	UHV	TEM	MATTHEWS,JW 1965 PHIL.MAG.12,1143
		NACL 001	UHV	TEM	MATTHEWS,JW 1965 APPL.PHYS.LETT.7,131
		NACL 001	UHV	TEM	MATTHEWS 1966.SEE FE 3
		NACL 001	UHV	RED,TEM	MATTHEWS,JW+GRUNBAUM,E 1964 APPLPHYS.LETT.5,106/ 1965 PHIL.MAG.11,1233
		KCL 001	UHV	TEM	METOIS,JJ 1973 SURFACE SCI.36,269
		KCL 001	UHV	TEM	METOIS ET AL 1972(A).SEE AL 3
		NACL 001	HV	TEM	MIHAMA,K 1969 J.VAC.SCI.TECHN.6,480
		NACL 001	HV	RED,TEM, REM	MIHAMA,K+AOE,H 1966*6ICEM 1,523
		NACL 001	HV	TEM	MIHAMA,K+TANAKA,M 1968 J.CRYST.GROWTH 2,51
		NACL 001	HV	TEM	MIHAMA,K+YASUDA,Y 1966 J.PHYS.SOC.JAPAN 21,1166
		NACL 001	HV	TEM	MIHAMA,K+MIYAHARA,H+AOE,H 1967 J.PHYS.SOC.JAPAN 23,785
		NACL 111	HV	TEM	MISSIROLI 1972.SEE NACL 4
		NACL 001	HV	XD	MIWA+ANNAKA 1954.SEE AL 3
		NACL 001	UHV	TEM	MURAYAMA,Y 1972 THIN SOLID FILMS 12,287
		NACL 001	UHV	TEM	MURR 1964.SEE NI 3
		NACL 001	HV, UHV	TEM	MURR+INMAN 1966.SEE AL 3
		NACL 001	HV	TEM	NAGASAWA+OGAWA 1960. SEE AG 3
		NACL 001	HV	PCH	NELSON,RS 1967 PHIL.MAG.15,845
		NACL 001,KCL 001	UHV	TEM	OGAWA,S+INO,S 1969 J.VAC.SCI.TECHN.6,527
		NACL 001,KCL 001	UHV	TEM	OGAWA,S+INO,S 1972 J.CRYST.GROWTH 13/14,48
		NACL 001	HV	RED,TEM	OGAWA ET AL 1955.SEE NI 3
		KCL 001	UHV	TEM	OGAWA ET AL 1966.SEE AL 3
		NACL 001 + MNCL2, PBCL2,KCL	HV	TEM	PALATNIK,LS+KOSEVITSCH,VM+SOKOL,AA 1968*4ERCEM 1,401
		KCL 001	UHV	TEM,LED	PALMBERG ET AL 1967(A).SEE AG 3
		KCL 001	UHV	TEM,LED, AES	PALMBERG ET AL 1968.SEE CU 3
		NACL 001 + C	HV	TEM	PAUNOV,M+HARSDORFF,M 1972 ZNATURFORSCH.27A,1381
		NACL 001	UHV	TEM	QUINTANA,MC+SACEDON,JL 1972 THINSOLIDFILMS 14,149
		NACL 001	UHV	TEM	QUINTANA,MC+SACEDON,JL 1973 THINSOLIDFILMS 17,311
		NACL 001	UHV	XD	QUINTANA,MC+SACEDON,JL+MENDIOLA,J 1973 THIN SOLID FILMS 17,319
		KCL 001	UHV	TEM	RHODIN ET AL 1969. SEE AG 3
		NACL 001,NABR001, NAF 001,KCL 001, KBR 001	UHV	TEM	ROOS+VERMAAK 1972.SEE AG 3
		NACL 001	UHV	TEM	SATO,H+SHINOZAKI,S 1970 J.APPL.PHYS.41,3165
		NACL 001	HV	TEM	SCHLOTTERER 1965.SEE NI 3
		NACL 001	HV	TEM	SELLA+TRILLAT 1964.SEE AL 3
		NACL 001	HV	TEM	SHARMA,SK 1970 THIN SOLID FILMS 6,17
		NACL 001	HV	TED	SHIRAI 1943(A).SEE NI 3
		NACL 001	HV	TEM	STIRLAND,DJ 1966 APPL.PHYS.LETT.8,326
		NACL 001	HV	TEM	STIRLAND,DJ 1967/68 THIN SOLID FILMS 1,447
		KF 001	HV	TEM	STIRLAND,DJ 1969 APPL.PHYS.LETT.15,88
		NACL 001	HV, UHV	TEM	TOTH+CICOTTE 1968.SEE AG 3
		NACL	HV	RED,TED	TRILLAT,JJ+TERAO,N+TERTIAN,L+GERVAIS,H 1955 C.R.ACAD.SCI.240,1557
		NACL 001 + C	HV	---	TROFIMOV,VI+GORODETSKII,AE 1972 SOVIET PHYS-SOLID STATE 14,760
		NACL 001	UHV	TEM	UEDA,R+INUZUKA,T 1968 J.CRYST.GROWTH 3,4,191
		NACL 001	UHV	TEM	UEDA,R+INUZUKA,T 1971 J.CRYST.GROWTH 9,79
		NACL 001	UHV	TEM	VERMAAK,JS+HENNING,CAO 1970 PHIL.MAG.22,269
		NACL 001,KCL 001	UHV	---	VERMAAK+HENNING 1970.SEE AG 3

Z= 79-80	OVERGROWTH		SUBSTRATE	PREP	OBSERV	REFERENCE
	GOLD	4	MICA 00.1	HV	TEM	ADAMSKY 1966.SEE AG 3
			MICA 00.1	HV	TEM	ADAMSKY+LEBLANC 1965.SEE AG 3
			MICA 00.1	UHV	TEM	ALLPRESS+SANDERS 1967.SEE NI 4
			MICA 00.1	UHV	TEM	ALLPRESS ET AL 1966.SEE NI 4
			MGO 001	UHV		BAUER+POPPA 1972.SEE AG 1
			MGO 001	HV, UHV	RED,TED, XD	BRINE+YOUNG 1963.SEE CU 3
			MICA 00.1	HV	TEM	CHOPRA 1965.SEE CU 3
			MICA 00.1 + C, SIO,SIO2,BATIO3	HV	RED,TED	CHOPRA 1969.SEE AG 3
			MICA 00.1	SP	RED	CHOPRA,KL+BOBB,LC 1964*ICTFBB, 373
			MICA 00.1	SP	RED	CHOPRA,KL+BOBB,LC+FRANCOMBE,MH 1963 J.APPL.PHYS.34,1699
			MICA 00.1	HV	XD	FUKAMACHI ET AL 1967.SEE 3
			MGO 001	UHV	RED,TEM	GREEN ET AL 1970.SEE AL 3
			MICA 00.1	HV	OM,XD	HALL+THOMPSON 1961.SEE CU 3
			MICA 00.1	UHV	OM,TEM	HINES,RL 1964 J.PHYSIQUE 25,134
			MGO 001	HV, UHV	TED	HONJO,G 1964-65. SEE 3
			MGO 001,111	HV	TEM	HONJO+YAGI 1969.SEE AG 2
			MICA 00.1	UHV	TEM	JAEGER ET AL 1969.SEE 1
			MICA 00.1	HV	RED	KIRCHNER+RUDIGER 1937.SEE AG 3
			MICA 00.1	HV	TEM	MATTHEWS 1972(A).SEE CU 1
			MGO 001	UHV	TEM	MOORHEAD,RD+POPPA,H 1972*30EMSA ,516
			MICA 00.1	HV	PCH	NELSON 1967.SEE 3
			MGO 001	UHV	LED	PALMBERG+RHODIN 1967.SEE AG 3
			MGO 001	UHV	TEM,LED, AES	PALMBERG ET AL 1967(B),1968.SEE AG 4
			MICA 00.1	UHV	TEM	POPPA,H+HEINEMANN,K+GRANT ELLIOT,A 1971 J.VAC.SCI.TECHN.8,471
			MICA 00.1, MGO 001,111	UHV	TEM,LED, AES	POPPA,H+MOORHEAD,RD+HEINEMANN,K 1972 NUCL.INSTR.METH.102,521
			MICA 00.1	HV	XD,PCH	REICHELT+LUTZ 1971.SEE AG 4
			MGO 001	UHV	TEM,LED	RHODIN ET AL 1969. SEE AG 3
			MICA00.1,CAF2 111, CACO3 100	HV	RED	RUDIGER 1937.SEE PD 4
			MGO 001	HV	TEM	SATO,H+SHINOZAKI,S 1970 SURFACE SCI.22,229
			MGO 001	HV	TEM	SATO+SHINOZAKI 1971.SEE AG 4
			MGO 001	HV	TEM	SATO ET AL 1969.SEE AG 4
			MGO 001	HV	RED	TULL 1951.SEE AG 4
			ZNO 00.1	UHV	TEM	WASSERMAN,EF+POLACEK,KA 1970 APPLPHYSLETT.16,259/ 1971 SURFACE SCI.28,77
	GOLD-NICKEL ALLOYS	3	NACL 001	SP	TED	FRANCOMBE,MH+KHAN,IH+FLOOD,JJ+SCHLACTER,M 1964*ICTFBB, 91
			NACL 001	HV	TEM	FUCANO,Y 1961 J.PHYS.SOC.JAPAN 16,1195
			NACL 001	SP	RED,TED	KHAN,IH+FRANCOMBE,MH 1963 NATURE 199,800/ 1965 J.APPL.PHYS.36,1699
		4	MICA 00.1	SP	TED	FRANCOMBE ET AL 1964.SEE 3
			MICA 00.1	SP	RED	KHAN+FRANCOMBE 1963,1965.SEE 3
	GOLD-ZINC ALLOYS	3	NACL 001	HV	TED,XD	IWASAKI,H+HIRABAYASHI,M+FUJIWARA,K+WATANABE,D+ OGAWA,S 1960 J.PHYS.SOC.JAPAN.15,1771
	GOLD-MANGANESE ALLOYS	3	NACL 001,110,111	HV	TED	SATO,H+TOTH,RS+SHIRANE,G+COX,DE 1966 J.PHYS.CHEM.SOLIDS 27,413
			NACL 001	HV	TED	WATANABE,D 1958 J.PHYS.SOC.JAPAN.13,535/ 1960 IBID.15,1030
80	MERCURY SULPHIDE	2	CDS 11.0	VG	TEM,REM	CAVENEY 1968.SEE ZNS 2
		3	NACL 001	HV	RED,TED	AGGARWAL,PS+GOSWAMI,A 1963 INDIAN J.PURE APPL.PHYS.1,122
	MERCURY SELENIDE	3	NACL 001,110,111	HV	RED,TED	BARUA,KC+GOSWAMI,A 1970 JAPAN.J.APPL.PHYS.9,705
		4	MICA 00.1	HV	RED,TED	BARUA+GOSWAMI 1970.SEE 3
			MICA 00.1	HV	RED	DEOKAR,VD+GOSWAMI,A 1970 INDIANJPUREAPPLPHYS.8,93
	MERCURY TELLURIDE	2	CDTE 001,110,111, 211	VG	RED,EP	ANTCLIFFE,GA+KRAUS,H 1969 JPHYSCHEMSOLIDS 30,243
			CDTE 111	VG	OM,RED, XD	COHEN-SOLAL,G+MARFAING,Y 1965 C.R.ACAD.SCI.260,4190
			GETE,PBTE	VG+ CG	---	COHEN-SOLAL ET AL 1966.SEE CD-HGTE 2
			CDTE 111	SP		COHEN-SOLAL,G+DEL VALLE,J+SELLA,J 1970 LE VIDE 147,409
			CDTE 111	VG+ CG	RED,EP, XD	MARFAING,Y+COHEN-SOLAL,G+BAILLY,F 1967*1ICCG, 549
			CDTE 111	VG+ CG	RED,TED, REM	SELLA,C+COHEN-SOLAL,G+BAILLY,F 1967 C.R.ACAD.SCI.264,179
		3	NACL 001,110,111	HV	RED,TED	BARUA+GOSWAMI 1970.SEE HGSE 3

Z=80-82	OVERGROWTH		SUBSTRATE	PREP	OBSERV	REFERENCE
	MERCURY TELLURIDE	4	MICA 00.1	HV	RED,TED	BARUA+GOSWAMI 1970.SEE HGSE 3
81	THALLIUM	1	AG 111	HV	RED	NEWMANN 1957.SEE NI 1
	THALLIUM SELENIDE	3	NACL 001	HV	RED,TED	BARUA,KC+GOSWAMI,A 1969 SURFACE SCI.14,415
	THALLIUM TELLURIDE	3	NACL 001	HV	RED,TED	BARUA+GOSWAMI 1969.SEE TL2SE 3
	THALLIUM-BISMUTH SELENIDE	3	NACL 001	HV	TED	MAN,LI+SEMILETOV,SA 1963 SOVIET PHYS-CRYST.7,686
	THALLIUM CHLORIDE	1	AG 001,AU 001	HV	RED	KHAN,IH 1960 PROC.PHYS.SOC.LONDON 76,507
		2	ZNS 110,PBS 001	HV	RED	KHAN 1960.SEE 1
		3	LIF 001,NAF 001, NACL 001,KCL 001, KBR 001,KI 001, RBBR 001,RBI 001	HV	RED	KHAN 1960.SEE 1
			NACL 001,KCL 001, KBR 001	HV	RED	LUDEMANN 1957.SEE CSCL 3
			NACL 001	HV	RED	PASHLEY 1952.SEE AGCL 3
			NACL 001,KBR 001	HV	RED	SCHULZ 1951(C).SEE CSCL 3
			NACL 001,KCL 001, KBR 001,KI 001	HV	RED	UNGELENK,J 1962 NATURWISSENSCHAFTEN 49,252/ 1963 PHYS.KONDENS.MATERIE.1,152
		4	MICA 00.1	HV	RED	KHAN 1960.SEE 1
			MICA 00.1	HV	RED	PASHLEY 1952.SEE AGCL 3
			MICA 00.1	HV	RED	SCHULZ 1951(C).SEE CSCL 3
	THALLIUM BROMIDE	1	AG 001,AU 001	HV	RED	KHAN 1960.SEE TLCL 1
		2	ZNS 110,PBS 001	HV	RED	KHAN 1960.SEE TLCL 1
		3	LIF 001,NAF 001, NACL 001,KCL 001, KBR 001,KI 001, RBBR 001,RBI 001	HV	RED	KHAN 1960.SEE TLCL 1
			LIF 001,NACL 001, KBR 001	HV	RED	SCHULZ 1951(C).SEE CSCL 3
			NACL 001,KCL 001, KBR 001,KI 001	HV	RED	UNGELENK 1962,1963.SEE TLCL 3
	THALLIUM IODIDE	3	NACL 001,KCL 001, KBR 001,KI 001	HV	RED	KHAN 1960.SEE TLCL 1
			LIF 001,NACL 001, KBR 001	HV	RED	SCHULZ 1951(C).SEE CSCL 3
			NACL 001, KCL 001 KBR 001,KI 001	HV	RED	UNGELENK 1962,1963. SEE TLCL 3
82	LEAD	1	NI 111	UHV	LED	AL KHOURY NEMEH,E+CINTI,RC+NGUYEN,TTA 1972 SURFACE SCI.30,697
			AU 001	UHV	LED,AES	BIBERIAN,JP+RHEAD,GE 1973 J.PHYS.F=METAL PHYS.3, 675
			PB 001,110,111	ED	OM,XD	BICELLI+POLI 1966.SEE CU 1A
			W 001	UHV	LED	GORODETSKII,DA+YAS≠KO,AA 1972 SOVIET PHYS.SOLID STATE 14,636
			AG 111	HV	RED	GRUNBAUM,E 1958 PROC.PHYS.SOC.72,459
			CU 001,110,111	UHV	OM,LED, AES	HENRION,J+RHEAD,GE 1972 SURFACE SCI.29, 20
			W	UHV	FEM	MELMED,AJ 1965 J.CHEM.PHYS.42,3332
			AG 111	HV	RED	NEWMAN 1957.SEE NI 1
			NI 001,110,111	UHV	LED	PERDEREAU,J+SZYMERSKA,I 1972 SURFACE SCI.32,247
			CU 001,111	LPE	OM,XD	SANG,H+MILLER,WA 1970 J.CRYST.GROWTH 6,303
			AG 001	HV	TED	SHIRAI ET AL 1961.SEE NI 1
			AU 111	UHV	TEM	SNYMAN,HC+BOSWELL,FW 1974 SURFACE SCI.41,21
		2	MOS2 00.1	HV	TEM	COOPERSMITH,B+CURZON,AE+KIMOTO,K+LISGARTEN,ND 1966*ISTFCL, 83
			SI 111	UHV	LED	ESTRUP+MORRISON 1964.SEE SN 2
			MOS2 00.1	UHV	TEM	PASHLEY,DW+STOWELL,MJ+ROBINSON,EA+LAW,TJ+VALDRE,U 1968*4ERCEM 1,387
			MOS2 00.1	UHV, MG	TEM	STOWELL,MJ+LAW,TJ+SMART,J 1970 PROC.ROY.SOC.,SER.A 318,231
			PBTE 001	UHV	LED	STRONGIN ET AL 1973.SEE AL 2
		4	MICA 00.1	UHV	OM,RED	WYATT,PW+YELON,A 1972 J.APPL.PHYS.43,1989
	LEAD SULPHIDE	2	GE 001,111	SG	XD	DAVIS,JL+NORR,MK 1966 J.APPL.PHYS.37,1670
			PBSE 001	HV	TEM	MATTHEWS,JW 1961 PHIL.MAG.6,1347/ 1963IBID.8,711/ 1971 IBID.23,1405/ 1972(B).SEE NI 1
			PBS 001	HV	---	SCHOOLAR,RB 1970 APPL.PHYS.LETT.16,446
			CDS 001,110	SG	OM,RED	WATANABE,S+MITA,Y 1969 J.ELECTROCHEM.SOC.116,989

Z= 82 OVERGROWTH		SUBSTRATE	PREP	OBSERV	REFERENCE
LEAD SULPHIDE	3	NACL 001+C,SIOX	HV	TEM	BARNA ET AL 1969.SEE ZNS 3
		NACL 001	HV	XD	CARDONA+GREENAWAY 1964.SEE SNTE 3
		NACL 001 + C,SIO, SIO2,BATIO3	HV	RED,TED	CHOPRA 1969.SEE AG 3
		NACL 001 + C	HV	TEM	DISTLER,GI+KOBZAREVA,SA+GERASIMOV,YM 1968 J.CRYST.GROWTH 2,45
		NACL 001,110,111, 443	HV	RED,TED	ELLEMAN,AJ+WILLMAN,H 1948 PROC.PHYS.SOC.61,164
		KCL 001	HV	---	GREEN ET AL 1973.SEE AU 2
		NACL 001	HV	TEM	MATTHEWS,JW+ISEBECK,K 1963 PHIL.MAG.8,469
		NACL 001,KCL 001, KBR 001	HV	TEM	NABITOVICH ET AL 1971.SEE SE 3
		NACL 001	HV	TEM	NORDEN,H 1965*3ERCEM, 397
		KCL 001	VG	REM,SEM, EP,XD	PAIC,M+PAIC,V 1972 J.MATER.SCI.7,1260
		KCL 001	VG	XD	PAIC,M+PAIC,V+DUH,K+ZEMEL,JN 1972 THIN SOLID FILMS 12,419
		NACL 001	HV	TEM	PALATNIK,LS+SOROKIN,VK+ZOZULA,LP 1969 SOVIET PHYS.-SOLID STATE 11,1027/ 1970 IBID.12,181
		NACL 001	HV	XD	SCHOOLAR,RD+ZEMEL,JN 1964 J.APPL.PHYS.35,1848
		NACL 001	HV	OM,RED, TED	SEMILETOV,SA 1964 SOVIET PHYS.-CRYST.9,65
		NACL 001	HV	RED	SEMILETOV,SA+VORONINA,IP+KORTUKOVA,EI 1966 SOVIET PHYS.-CRYST.10,429
		NACL 001	HV	RED,TEM	WILSON,AD+NEWMAN,RC+BULLOUGH,R 1963 PHIL.MAG.8,2035
	4	MICA 00.1+C,SIO, SIO2,BATIO3	HV	RED,TED	CHOPRA 1969.SEE AG 3
		MICA 00.1	SG	TEM,REM	DISTLER,GI+DARYUSINA,SA 1962SOVIETPHYSCRYST.7,87
		MICA 00.1	HV	---	EGERTON,RF+JUHASZ,C 1967 BRITJAPPL.PHYS.18,1009
		MGO 001	HV	TEM	HONJO+YAGI 1969.SEE AG 2
		MICA 00.1	HV	TEM	SATO,H 1972 THIN SOLID FILMS 11,343
		MGO 001	HV	TEM	YAGI,K+KOBAYASHI,K+HONJO,G 1969 J.APPL.PHYS.40,3857
LEAD SELENIDE	2	PBS 001	HV	TEM	HONJO+YAGI 1969.SEE AG 2
		PBS 001	HV	TEM	MATTHEWS 1961,1963. SEE PBS 2
		PBSE 001	UHV	TEM	POPPA ET AL 1972.SEE AU 4
		PBS 001	HV	TEM	WILSON ET AL 1963.SEE PBS 2
		PBS 001	HV	TEM	YAGI ET AL 1970.SEE PD 1
		PBS 001,PBTE 001	HV	TEM	YAGI ET AL 1971.SEE PD 1
	3	NACL 001,KCL 001	HV	XD	CARDONA+GREENAWAY 1964.SEE SNTE 3
		NACL 001	HV	RED	CHOPRA,KL+DUFF,RH+MALLIO,WJ 1970*7ICEM 2,425
		NACL 001,KCL 001, KBR 001	HV	TED,XD	GOBRECHT,H+BOETERS,KE+FLEISCHER,HJ 1965 Z.PHYSIK 187,232
		KCL 001	HV	---	GREEN ET AL 1973.SEE AU 2
		NACL 001	HV	TEM	MATTHEWS 1971.SEE PBS 2
		NACL 001,KCL 001, KBR 001	HV	TEM	NABITOVICH ET AL 1971.SEE SE 3
		NACL 001,KCL 001	HV	---	PALATNIK,LS+SOROKIN,VK 1966 SOVIET PHYS.-SOLID STATE 8,869
		NACL 001	HV	TEM	PALATNIK ET AL 1969.SEE PBS 3
		NACL 001	HV	TEM	PALATNIK,LS+ZOZULYA,LP+KOSEVICH,VM 1970 SOVIET PHYS.-SOLID STATE 11,1460
		NACL 001,KCL 001	HV	---	SCHOOLAR,RB+LOWNEY,JR 1971 JVAC.SCI.TECHN.8,224
		NACL 001	HV	OM,RED, TED	SEMILETOV 1964.SEE PBS 3
		NACL 001,LIF 001	HV	OM,RED	SEMILETOV,SA+VORONINA,IP 1964 SOVIET PHYS.DOKLADY 8,960
		NACL 001	HV	RED	SEMILETOV ET AL 1966.SEE PBS 3
		NACL 001	HV	OM,RED	VORONINA,IP+SEMILETOV,SA 1964 SOVIET PHYS.-SOLID STATE 6,1204
		NACL 001	HV	RED,TEM	WILSON ET AL 1963.SEE PBS 3
	4	MICA 00.1	HV	RED	CHOPRA ET AL 1970.SEE 3
		MICA 00.1	HV	---	EGERTON+JUHASZ 1967.SEE PBS 4
		BAF2 111,SRF2 111	UHV	REM,XD	HOHNKE,DK+KAISER,SW 1974 J.APPL.PHYS.45,892
		MGO 001	HV	TEM	HONJO+YAGI 1969.SEE AG 2
		MICA 00.1	HV	TED,REM, XD	POH,KJ+ANDERSON,JC 1969 THIN SOLID FILMS 3,139
		MICA 00.1	HV	TEM	SATO 1972.SEE PBS 4
		MGO 001	HV	TEM	YAGI ET AL 1969.SEE PBS 4
LEAD TELLURIDE	2	PB-SNTE	HV	--	WALPOLE,JN+CALAWA,AR+RALSTON,RW+HARMAN,TC+ MC VITTIE,JP 1973 APPL.PHYS.LETT.23,620
		PBTE 001,111	LPE	OM,XB	WAGNER,JW+THOMPSON,AG 1970 J.ELECTROCHEM.SOC.117,936
		PBS 001	HV	TEM	YAGI ET AL 1970.SEE PD 1
		PBS 001,PBSE 001	HV	TEM	YAGI ET AL 1971.SEE PD 1

Z= 82-83 OVERGROWTH		SUBSTRATE	PREP	OBSERV	REFERENCE
LEAD TELLURIDE	3	KBR 001	HV	XD	CARDONA+GREENAWAY 1964.SEE SNTE 3
		NACL 001	HV	RED	CHOPRA ET AL 1970.SEE PBSE 3
		NACL 001	SP	TED,XD	FRANCOMBE ET AL 1962.SEE INSB 3
		KCL 001	HV	---	GREEN ET AL 1973.SEE AU 2
		NACL 001	HV	TEM	JAYADEVAIAH,TS+KIRBY,RE 1970 THINSOLIDFILMS 6,343
		KCL 001	HV	TEM	KOSEVICH,VM+PALATNIK,LS+SOROKIN,VK 1969 SOVIET PHYS.-SOLID STATE 10,1936
		NACL 001	HV	TEM	LEWIS,B+STIRLAND,DJ 1968 J.CRYST.GROWTH 3,4,200
		NACL 001	HV	TEM,REM	MYERS,JH+MORRISS,RH+DECK,RJ 1971JAPPLPHYS.42,5578
		NACL 001,KCL 001, KBR 001	HV	TEM	NABITOVICH ET AL 1971.SEE SE 3
		NACL 001,KCL 001	HV	XD	NUCCIOTTI,A+DE STEFANO,P+MASCHERETTI,P+ SAMOGGIA,G 1972 PHYS.STATUS SOLIDI A12,193
		NACL 001,KCL 001	HV	---	PALATNIK+SOROKIN 1966.SEE PBSE 3
		NACL 001	HV	OM,RED, XD	PALATNIK,LS+SOROKIN,VK+LEBEDEVA,MV 1965 SOVIET PHYS.-SOLID STATE 7,1374
		NACL 001	HV	TEM	PALATNIK,LS+KOSEVICH,VM+ZOZULYA,LP+SOROKIN,VK 1969 SOVIET PHYS.-DOKLADY 13,884
		NACL 001	HV	TEM	PALATNIK ET AL 1969.SEE PBS 3
		NACL 001,KCL 001	HV	TEM	PALATNIK ET AL 1970.SEE PBS 3
		NACL 001	HV	TEM	SELLA+SURYANARAYANAN 1968.SEE CDTE 3
		NACL 001	HV	OM,RED, TED	SEMILETOV 1964.SEE PBS 3
		NACL 001,KCL 001	HV	OM,RED	SEMILETOV,SA+VORONINA,IP 1965 SOVIET PHYS.-CRYST.9,405
		NACL 001	HV	RED	SEMILETOV ET AL 1966.SEE PBS 3
		NACL 001	HV	TEM	SUMNER,GG+REYNOLDS,LL 1969 J.VAC.SCI.TECHN.6,493
		NACL 001,KCL 001	HV	OM,RED	VORONINA,IP+SEMILETOV,SA 1964 SOVIET PHYS.-SOLID STATE 6,1494
	4	MICA 00.1	HV	RED	CHOPRA ET AL 1970.SEE PBSE 3
		MICA 00.1	HV	TEM,REM, XD	EGERTON,RF 1969 PHIL.MAG.20,547
		MICA 00.1	HV	---	EGERTON+JUHASZ 1967.SEE PBS 4
		MICA 00.1	HV	TEM,REM	GREEN 1971.SEE KCL 4
		BAF2 111	HV	REM,XD	HOLLOWAY,H+LOGOTHETIS,EM 1971 J.APPLPHYS.42,4522
		CAF2 111	HV	TEM	JORDAN,MR+STIRLAND,DJ 1971 THINSOLID FILMS 8,221
		MICA 00.1	HV	RED,XD	MAKINO,Y 1964 J.PHYS.SOC.JAPAN.19,580
		MICA 00.1	HV	TEM	SATO 1972. SEE PBS 4
		TRIGLYCINE SULPH.	HV	OM	VLASOV ET AL 1970.SEE AGCL 4
LEAD SELENIDE-TELLURIDE	3	NACL 001	HV	XD	BIS,RF+ZEMEL,JN 1966 J.APPL.PHYS.37,228
LEAD-TIN SULPHIDE	3	NACL 001	HV	TED	MARIANO+CHOPRA 1967.SEE SNS 3
LEAD-TIN SELENIDE	3	KCL 001	HV	EP,XD	STRAUSS,AJ 1967 PHYS.REV.157,608
	4	BAF2 111,SRF2 111	UHV	REM,XD	HOHNKE+KAISER 1974. SEE PBSE 4
LEAD-TIN TELLURIDE	2	GE 111	SP	XD	CORSI,C+ALFIERI,I+PETROCCO,G 1974 APPL.PHYS.LETT.24,484
		PB-SNTE 001,110, 111	LPE	OM	LONGO ET AL 1972. SEE GAAS 2A
		PBTE 001,110,111	VG	OM,RED, SEM,XD	RCLLS,W+LEE,R+EDDINGTON,RJ 1970 SOLID-ST.ELECTRON.13,75
		PB-SNTE 001,111	LPE	OM,XD	THOMPSON,AG+WAGNER,JW 1971 PHYS.STAT.SOL.A5,439
		PBTE 001	LPE	OM	TOMASETTA,LR+FONSTAD,CG 1974APPL.PHYS.LETT.24,567
	3	NACL 001	HV	OM,XD	BIS,RF+DIXON,JR+LOWNEY,JR 1972 JVACSCITECHN,9,226
		KCL 001	HV	OM,XD	BYLANDER,EG 1966 MATER.SCI.ENG.1,190
		NACL 001	SP	XD	CORSI ET AL 1974.SEE 2
		NACL 001,KCL 001	HV	XD	FARINRE,TO+ZEMEL,JN 1970 J.VAC.SCI.TECHN.7,121
		NACL 001	HV	XD	NUCCIOTTI ET AL 1972.SEE PBTE 3
	4	BAF2	SP	XD	CORSI ET AL 1974.SEE 2
		BAF2 111	HV	RED,XD	HOLLOWAY,H+LOGOTHETIS,EM+WILKES,E 1970 J.APPL.PHYS.41,3543
		BAF2 111	HV	REM,XD	HOLLOWAY+LOGOTHETIS 1971.SEE PBTE 4
		CAF2 111,BAF2 111	HV	XD	TAO,TF+WANG,CC 1972 J.APPL.PHYS.43,1313
83 BISMUTH	1	CU 001	UHV	LED,AES	DELAMARE,F+RHEAD,GE 1973 SURFACE SCI.35,172
		CU 111	UHV	LED,AES	DELAMARE,F+RHEAD,GE 1973 SURFACE SCI.35,185
	3	NACL 001,KCL 001	HV	TEM,REM	HERRMANN,W+REIMER,L 1965 Z.NATURFORSCH 20A,1050
		NACL 001	HV	TED	ZAV≠YALOVA,AA+IMAMOV,RM 1969 SOVIET PHYS.CRYST.14,305
	4	MICA 00.1	HV	TEM	DUGGAL,VP+RUP,R 1969 J.APPL.PHYS.40,492
		MICA 00.1	HV	RED	DUGGAL,VP+RUP,R+TRIPATHI,P 1966APPLPHYSLETT.9,293
		MICA 00.1	HV	TEM,REM	HERRMANN ET AL 1965.SEE 3
		MICA 00.1	HV	TEM	SAUVAGE,M+WILLAIME,C 1971 PHYS.STAT.SOL.A5,K147
		MICA 00.1	HV	TEM	UNGER+STOLZ 1971.SEE MNBI 4

Z= 83-90 OVERGROWTH		SUBSTRATE	PREP	OBSERV	REFERENCE
BISMUTH SULPHIDE	3	NACL 001	HV	RED,TED	BADACHHAPE+GOSWAMI 1967.SEE ANTIMONIUM SULPH. 3
BISMUTH SELENIDE	3	NACL 001,110,111	HV	RED,TEM	DHERE,NG+GOSWAMI,A 1972 J.VAC.SCI.TECHN.9,523
	4	MICA 00.1	HV	RED	DHERE+GOSWAMI 1972.SEE 3
BISMUTH TELLURIDE	3	NACL 111	SP	TED	FRANCOMBE,MH 1964 PHIL.MAG.10,989
		NACL 001	SP	TED,XD	FRANCOMBE ET AL 1962.SEE INSB 3
		NACL 001	SP	RED,TED	FRANCOMBE ET AL 1964.SEE AU-NI 3
BISMUTH SILICATE,FERRITE, ZINCATE,GALLATE	4	BI12GEO20	LPE	OM	BALLMAN,AA+BROWN,H+TIEN,PK+MARTIN,RJ 1973 J.CRYST.GROWTH 20,251
BISMUTH TITANATES	1	PT 001,110	SP	RED,XD	TAKEI,WJ+FORMIGONI,NP+FRANCOMBE,MH 1969 APPL.PHYS.LETT.15,256/ 1970 J.VAC.SCI.TECHN.7,442
	4	BI12GEO20	LPE	OM	BALLMAN ET AL 1973.SEE BISMUTH SILICATE 4
		MGO 001,110	SP	RED,XD	TAKEI ET AL 1960,1970. SEE 1
		MGO 110	SP	OM,XD	WU,SY+TAKEI,WJ+FRANCOMBE,MH+CUMMINS,SE 1972 FERROELECTRICS 3,217
BISMUTH OXIDES	3	NACL 001	HV	TED	ZAV≠YALOVA+IMAMOV 1969.SEE BI 3
90 THORIUM		W 001	UHV	LED	ANDERSON,J+ESTRUP,PJ+DANFORTH,WE 1965 APPL.PHYS.LETT.7,122
		W 001	UHV	LED	ESTRUP,PJ+ANDERSON,J 1967 SURFACE SCI.7,255/ 1967 IBID.8,101
		W 001	UHV	LED	ESTRUP,PJ+ANDERSON,J+DANFORTH,WE 1966 SURFACE SCI.4,286
		W 001	UHV	OM,LED, AES	POLLARD,JH 1971 SURFACE SCI.20,269
		TA 001	UHV	OM,REM, LED	POLLARD,JH+DANFORTH,WE 1968 J.APPL.PHYS.39,4019/ 1969*4IMATC, 71-1
ORGANIC MATERIALS ANTHRAQUINONE		NACL 001,NACL + C, PVC REPL.NACL 001			DARLING,DF+FIELD,BO 1973 SURFACE SCI.34,420
		NACL 001, PVC REPL.NACL 001	HV	OM	DISTLER,GI+KOBZAREVA,SA 1969 NATURWISSENSCHAFTEN 56, 325
		TRIGLYCINE SULPH. NACL,MICA	HV	OM	KOBZAREVA,SA+DISTLER,GI 1971 J.CRYST.GROWTH 10,269
		TRIGLYCINE SULPHATE 010	HV	OM	KOBZAREVA,SA+DISTLER,GI+KONSTANTINOVA,VP 1970 SOVIET PHYS.-CRYST.15,431
FLAVANTHRONE,IDANTHRONE, PYRANTHRONE,VIOLANTHRONE		MICA 00.1	HV	TEM	SUITO,E+UYEDA,N+ASHIDA,M 1962 NATURE 194,273
		MICA 00.1	HV	TEM	UYEDA,N+ASHIDA,M+SUITO,E 1962*5ICEM 1,GG-15
METALLO-PHTALOCYANINES		MICA 00.1	HV	TEM	SUITO ET AL 1962.SEE FLAVANTHRONE,ETC.
		MICA 00.1	HV	TEM	UYEDA ET AL 1962.SEE FLAVANTHRONE,ETC./ 1965 J.APPL.PHYS.36,1453
		KCL 001,KBR 001, MICA 00.1	HV	TEM	UYEDA,N+ASHIDA,M+SUITO,E 1966*6ICEM 1,485
POLYFLUORETHYLENES		KCL 001	MG	TEM	TANI,Y+ISHIHARA,N 1962*5ICEM 1,GG-17
TETRACYANOQUINOMETHANES		NACL 001,112	VG	OM,TEM	CHAUDHARI,P+SCOTT,BA+LAIBOWITZ,RB+TOMKIEWICZ,Y+ TORRANCE,JB 1974 APPL.PHYS.LETT.24,439
		KCL 001	HV	TEM	UYEDA,N+KOBAYASHI,T+SUITO,E 1970*7ICEM 2,433
BENZENE DERIVATIVES		PT 111	VG	LED	GLAND,JL+SOMORJAI,GA 1973 SURFACE SCI.38,157
		GRAPHITE 00.1	VG	LED	LANDER 1967.SEE GERMANIUM IODIDES 2
		GRAPHITE 00.1	VG	LED	LANDER+MORRISON 1966,1967.SEE GERMANIUM IODIDES 2
		AG 001,CU 001, NACL 001	SG	OM	NEUHAUS,A+NOLL,W 1944 NATURWISSENSCHAFTEN 32,76
NAPHTALENE		PT 111	VG	LED	GLAND+SOMORJAI 1973. SEE BENZENE DERIVATIVES
PYRIDINE		PT 111	VG	LED	GLAND+SOMORJAI 1973. SEE BENZENE DERIVATIVES

INDEX

T

W

MATERIALS SCIENCE AND TECHNOLOGY

A. S. Nowick and B. S. Berry, Anelastic Relaxation in Crystalline Solids, 1972

E. A. Nesbitt and J. H. Wernick, Rare Earth Permanent Magnets, 1973

W. E. Wallace, Rare Earth Intermetallics, 1973

J. C. Phillips, Bonds and Bands in Semiconductors, 1973

J. H. Richardson and R. V. Peterson (editors), Systematic Materials Analysis, Volumes I, II, and III, 1974; IV, 1978

A. J. Freeman and J. B. Darby, Jr. (editors), The Actinides: Electronic Structure and Related Properties, Volumes I and II, 1974

A. S. Nowick and J. J. Burton (editors), Diffusion in Solids: Recent Developments, 1975

J. W. Matthews (editor), Epitaxial Growth, Parts A and B, 1975

J. M. Blakely (editor), Surface Physics of Materials, Volumes I and II, 1975

G. A. Chadwick and D. A. Smith (editors), Grain Boundary Structure and Properties, 1975

John W. Hastie, High Temperature Vapors: Science and Technology, 1975

John K. Tien and George S. Ansell (editors), Alloy and Microstructural Design, 1976

M. T. Sprackling, The Plastic Deformation of Simple Ionic Crystals, 1976

James J. Burton and Robert L. Garten (editors), Advanced Materials in Catalysis, 1977

Gerald Burns, Introduction to Group Theory with Applications, 1977

L. H. Schwartz and J. B. Cohen, Diffraction from Materials, 1977

Zenji Nishiyama, Martensitic Transformation, 1978

Paul Hagenmuller and W. van Gool (editors), Solid Electrolytes: General Principles, Characterization, Materials, Applications, 1978

G. G. Libowitz and M. S. Whittingham, Materials Science in Energy Technology, 1978

Otto Buck, John K. Tien, and Harris L. Marcus (editors), Electron and Positron Spectroscopies in Materials Science and Engineering, 1979

Lawrence L. Kazmerski (editor), Polycrystalline and Amorphous Thin Films and Devices, 1980

Manfred von Heimendahl, ELECTRON MICROSCOPY OF MATERIALS: AN INTRODUCTION, 1980

O. Toft Sørensen (editor), NONSTOICHIOMETRIC OXIDES, 1981

M. Stanley Whittingham and Allan J. Jacobson (editors), INTERCALATION CHEMISTRY, 1982

A. Ciferri, W. R. Krigbaum, and Robert B. Meyer (editors), POLYMER LIQUID CRYSTALS, 1982